Introduction to D

Introduction to
Differential Equations

Rabindra Kumar Patnaik
Pro-Vice-Chancellor
The ICFAI University at Tripura
Tripura

PHI Learning Private Limited
New Delhi-110001
2009

Rs. 225.00

INTRODUCTION TO DIFFERENTIAL EQUATIONS
Rabindra Kumar Patnaik

© 2009 by PHI Learning Private Limited, New Delhi. All rights reserved. No part of this book may be reproduced in any form, by mimeograph or any other means, without permission in writing from the publisher.

ISBN-978-81-203-3603-2

The export rights of this book are vested solely with the publisher.

Published by Asoke K. Ghosh, PHI Learning Private Limited, M-97, Connaught Circus, New Delhi-110001 and Printed by Jay Print Pack Private Limited, New Delhi-110015.

Contents

Preface ix

Chapter 1 **Nature of Differential Equations** 1–14

    1.1 Introduction *1*
    1.2 Basic Concepts *3*
    1.3 Solution of Differential Equation *5*
    1.4 Some Examples from Fields of Application *7*
    1.5 Summary *11*
        Exercises 1.5 12

Chapter 2 **First Order Differential Equations** 15–37

    2.1 Introduction *15*
    2.2 Variables Separable *15*
        Exercises 2.2 18
    2.3 Homogeneous Equations *19*
        Exercises 2.3 20
    2.4 Exact Equations *21*
        Exercises 2.4 29
    2.5 Linear Equations *30*
        Exercises 2.5 32
    2.6 Equations Reducible to Known Forms *33*
        Exercises 2.6 35

Chapter 3 **Linear Differential Equations of Higher Order** 38–73

    3.1 Introduction *38*
    3.2 Second Order Linear Differential Equations *38*
    3.3 Method of Solving Homogeneous Equations with Constant Coefficients *42*
        Exercises 3.3 44
    3.4 Method of Solving Homogeneous Equations with Variable Coefficients *45*
        Exercises 3.4 50
    3.5 Method of Solving Nonhomogeneous Equations *51*
        Exercises 3.5 57

3.6 Reduction of Order 58
    Exercises 3.6 59
3.7 Series Solution Method 60
    Exercises 3.7 72

Chapter 4 Properties of Solutions 74–85

4.1 Introduction 74
4.2 Zeros of a Solution 74
4.3 Sturm Comparison Theorem 81
    Exercises 4.3 84

Chapter 5 Special Functions and Equations 86–142

5.1 Introduction 86
5.2 Gamma and Beta Functions 86
5.3 Gamma Function 87
5.4 Domain of Gamma Function 87
5.5 Beta Function 88
5.6 Relationship between Gamma and Beta Functions 89
5.7 Digamma Function 90
    Exercises 5.7 93
5.8 Hypergeometric Functions 95
5.9 Properties of Hypergeometric Function 97
    Exercises 5.9 101
5.10 Orthogonal Polynomials 102
5.11 Orthogonal Functions 102
5.12 Simple Sets of Polynomials 102
5.13 Properties of Simple Set of Polynomials 103
     Exercises 5.13 106
5.14 Bessel Functions 107
5.15 Bessel Equation 107
5.16 Properties of Bessel Functions 110
     Exercises 5.16 115
5.17 Legendre Polynomials 117
5.18 Legendre's Equation 118
5.19 Polynomial Solutions 118
5.20 Properties of Legendre Polynomials 119
5.21 Legendre Series 123
     Exercises 5.21 124
5.22 Chebyshev Polynomials 126
5.23 Properties of Chebyshev Polynomials 128
     Exercises 5.23 130
5.24 Hermite Polynomials 131
5.25 Properties of Hermite Polynomials 132
     Exercises 5.25 137
5.26 Laguerre Polynomials 138
5.27 Properties of Laguerre Polynomials 139
     Exercises 5.27 142

Chapter 6   Systems of First Order Linear Equations   143–159

- 6.1 Introduction *143*
- 6.2 Linear Systems *144*
- 6.3 Relationship *146*
- 6.4 Application to Electrical Circuits *147*
- 6.5 Solutions *149*
- 6.6 Method of Obtaining Solutions *150*
- 6.7 Homogeneous Linear Systems *151*
- 6.8 Nonhomogeneous Linear Systems *155*
  Exercises 6.8 *158*

Chapter 7   Total Differential Equations   160–180

- 7.1 Introduction *160*
- 7.2 Condition for Integrability of $P\,dx + Q\,dy + R\,dz = 0$ *160*
- 7.3 Geometrical Interpretation of $P\,dx + Q\,dy + R\,dz = 0$ *164*
- 7.4 Methods of Solving Total Differential Equations *164*
- 7.5 Solution by Inspection *164*
  Exercises 7.5 *166*
- 7.6 One Variable Assumed to be Constant *167*
  Exercises 7.6 *169*
- 7.7 Method of Auxiliary Equations *169*
  Exercises 7.7 *171*
- 7.8 Homogeneous Equations *171*
  Exercises 7.8 *173*
- 7.9 Non-integrable Equations *173*
  Exercises 7.9 *174*
- 7.10 Simultaneous Equations *175*
- 7.11 Geometrical Interpretation *175*
- 7.12 Solution of Simultaneous Equations *176*
  Exercises 7.12 *179*

Chapter 8   The Laplace Transform   181–209

- 8.1 Introduction *181*
- 8.2 Definition and Existence of Laplace Transform *181*
  Exercises 8.2 *190*
- 8.3 Properties of Laplace Transform *191*
  Exercises 8.3 *202*
- 8.4 Convolution *204*
  Exercises 8.4 *208*

Chapter 9   Sturm–Liouville Boundary Value Problems and Fourier Series   210–233

- 9.1 Introduction *210*
- 9.2 Sturm–Liouville Problem *210*
  Exercises 9.2 *214*

9.3 Fourier Series *215*
Exercises 9.3 *220*
9.4 Fourier Sine and Cosine Series *221*
Exercises 9.4 *224*
9.5 Convergence of Fourier Series *225*
Exercises 9.5 *232*

Chapter 10 Partial Differential Equations   **234–254**

10.1 Introduction *234*
10.2 Basic Concepts *234*
10.3 One-Dimensional Wave Equation *236*
Exercises 10.3 *241*
10.4 One-Dimensional Heat Equation *241*
Exercises 10.4 *249*
10.5 Laplace Equation *249*
Exercises 10.5 *253*

Suggested Readings   **255**

Answers to Exercises   **257–279**

Index   **281–282**

Preface

The aim of this book is to present the elements of differential equations in a form suitable for the use of undergraduate students of mathematics, with emphasis on developing problem-solving skills but nevertheless supported by adequate theory without going into the intricacies of some involved proofs. In its present form, the book has been developed from a course offered by the author to the engineering students of BITS Pilani. It will therefore be useful to engineering students of all disciplines as part of their course in engineering mathematics.

Efforts have been made to provide students with a good grounding in the fundamentals of the subject. Chapter 1 introduces the students to some basic concepts of ordinary and partial differential equations, linear and nonlinear differential equations, initial and boundary value problems, etc.

Chapter 2 deals with ordinary differential equations of the first order and first degree. The method of solution by separation of variables, and the solutions of equations that are homogeneous, linear, exact, non-exact, or reducible to one of these forms, are discussed in detail with illustrative examples.

Different methods of solving various linear differential equations of second order—whether of constant or variable coefficients, normal or non-normal form, or homogeneous or nonhomogeneous form—are discussed at length in Chapter 3. Qualitative properties of solutions of linear differential equations are discussed in Chapter 4 by direct analysis of the equation itself.

No book on differential equations is complete without a treatment of special functions and special equations. Chapter 5 is devoted to the detailed study of special functions like the gamma function, beta function, hypergeometric function and Bessel function, and of special equations like the Legendre equation, Chebyshev equation, Hermite equation and Laguerre equation. To understand better the orthogonal polynomials like Legendre, Chebyshev, Hermite and Laguerre, a discussion on orthogonal polynomials is introduced. Here the general properties of orthogonal polynomials are treated in detail.

Simultaneous differential equations in two variables with constant coefficients are discussed in Chapter 6. Methods of solving systems of first order linear equations, both homogeneous and nonhomogeneous, form an important part of this chapter. Total differential equations of the form $Pdx + Qdy + Rdz = 0$ and solutions of simultaneous total differential equations are discussed in Chapter 7.

Chapter 8 is devoted to a detailed study of Laplace transforms. Existence theorem, convolution theorem and applications of Laplace transform to differential equations form an important part of this chapter.

Sturm–Liouville boundary value problems and Fourier series, the most important tools for the solutions of partial differential equations, are discussed in Chapter 9. These two concepts are treated adequately for comprehending the solutions of partial differential equations in the next chapter.

Chapter 10, the final chapter, is devoted to a brief introductory discussion on partial differential equations. The basic concepts of partial differential equations are discussed in this chapter. Also discussed is a technique known as the 'method of separation of variables' to study the solutions of three fundamental equations of mathematical physics, namely the wave equation, heat equation and Laplace equation, under certain boundary and initial conditions.

A large number of illustrative examples are woven into the text to help the reader in securing a better understanding of the conceptual ideas involved. Numerous exercises are given at the end of many sections for the reader to practice and to test his comprehension and ability. Answers to exercises are given at the end of the book.

There remains the pleasant duty of acknowledging the help received in preparing this book. I have consulted various books for developing the content of my book. I would also like to acknowledge the initiation, interest and inspiration I received from all my colleagues at BITS, Pilani, especially Dr M.V. Tamhankar, without whose help this book probably would not have seen the light of day. I am thankful to all of them for their valuable suggestions.

I was very fortunate to have had many excellent students, who with their questions and comments contributed much to the clarity of exposition of this book. My special thanks to all of them.

I owe a lot to my wife Deepa, our son Anup and daughter Anuli, who continuously lent me their support throughout the long period of preparation of this book. I gratefully acknowledge their contribution in the making of this book. I am also thankful to the staff of PHI Learning for their kind and skilful assistance in developing this edition from my handwritten manuscript.

I take full responsibility for all errors that may have crept in this book. I will appreciate the effort if such errors are brought to my notice. Constructive criticism, comments and suggestions from the readers for improving the book shall also be welcomed.

Rabindra Kumar Patnaik

CHAPTER 1

Nature of Differential Equations

1.1 INTRODUCTION

What is a differential equation? A differential equation is an equation involving one dependent variable and its derivatives with respect to one or more independent variables. For example

$$\frac{d^3y}{dx^3} - 3\frac{d^2y}{dx^2} + 2\frac{dy}{dx} - y = \sin 2x \tag{1.1}$$

$$\frac{dy}{dx} = \frac{\sqrt{x}}{\sqrt{(1+y^2)}} \tag{1.2}$$

$$dy = \cos x \, dx \tag{1.3}$$

$$x\frac{\partial u}{\partial x} + y\frac{\partial u}{\partial y} + u = 0 \tag{1.4}$$

$$\frac{\partial^2 u}{\partial t^2} = a^2 \frac{\partial^2 u}{\partial x^2} \tag{1.5}$$

$$\frac{\left[1 + (dy/dx)^2\right]^{3/2}}{\dfrac{d^2y}{dx^2}} = k \tag{1.6}$$

$$\left(\frac{d^2y}{dx^2}\right)^3 - xy\frac{dy}{dx} + y = x \ln x \tag{1.7}$$

$$h^2\left(\frac{\partial^2 u}{\partial x^2} + \frac{\partial^2 u}{\partial y^2}\right) = \frac{\partial u}{\partial t} \tag{1.8}$$

$$\left(x^2 + y^2\right)\frac{dy}{dx} - 2xy = 0 \tag{1.9}$$

1

$$\frac{\partial^2 u}{\partial x^2} + \frac{\partial^2 u}{\partial y^2} + \frac{\partial^2 u}{\partial z^2} = 0 \qquad (1.10)$$

are all differential equations.

Why differential equations? Mathematical models are developed in science and engineering to aid in the understanding of physical phenomena. These models often yield an equation that relates an unknown function and some of its derivatives, which are nothing but differential equations. Many of the general laws of nature find their most natural expression in the language of differential equations.

It was Isaac Newton (1642–1727), the English mathematician, and Gottfried Wilhelm Leibniz (1646–1716), the German mathematician, who gave birth to the subject of differential equations in 1675. Leibniz forged a powerful tool of the integral sign and Newton classified differential equations of the first order. A large number of great mathematicians of the past three centuries such as Fermat, Bernoulli, Euler, Lagrange, Laplace, Gauss, Abel, Hamilton, Liouville, Reimann, Poincare and others developed this subject and brought it to the present form, which is the source of most of the ideas and theories that constitute higher analysis. Since the time of Isaac Newton, differential equations have been of fundamental importance in the application of mathematics to physical sciences. Lately, differential equations have gained increasing importance in the biological sciences and social sciences too.

The following example of a model developed in calculus may illuminate the above remarks. According to Newton's second law of motion, the acceleration of a body of mass m is proportional to the total force F acting on it, with $1/m$ as the constant of proportionality, so that we have

$$a = \frac{F}{m} \quad \text{or} \quad ma = F \qquad (1.11)$$

If that body of mass m falls freely under the influence of gravity alone, then the only force acting on it is mg, where g is the acceleration due to gravity, which can be considered constant on the surface of the earth. If y is the distance from some fixed point down to the body, then its acceleration is $\dfrac{d^2 y}{dt^2}$, and Eq. (1.11) becomes

$$m\frac{d^2 y}{dt^2} = mg \quad \text{or} \quad \frac{d^2 y}{dt^2} = g \qquad (1.12)$$

If we change the situation by assuming that air exerts a resisting force proportional to the velocity, then the total force acting on the body is $mg - k\dfrac{dy}{dt}$, and Eq. (1.11) becomes

$$m\frac{d^2 y}{dt^2} = mg - k\frac{dy}{dt} \qquad (1.13)$$

where k is a constant of proportionality.

Equations (1.12) and (1.13) are the differential equations that express the essential attribute of the physical processes under consideration.

Our main objective is to find a solution or solutions, if exist, for a differential equation. To achieve our objective, we need some common term to start with.

1.2 BASIC CONCEPTS

A differential equation involving ordinary derivatives with respect to a single independent variable is called an *Ordinary Differential Equation*. For example, Eqs. (1.1), (1.2), (1.3), (1.6), (1.7) and (1.9) are ordinary differential equations.

A differential equation involving partial derivatives with respect to more than one independent variable is called a *Partial Differential Equation*. Equations (1.4), (1.5), (1.8) and (1.10) are examples of partial differential equation.

The *order* of a differential equation is the order of the highest order derivative present in the equation. Equations (1.2), (1.3), (1.4), (1.9) are of first order; Eqs. (1.5), (1.6), (1.7), (1.8), (1.10) are of second order; and Eq. (1.1) is of third order.

The *degree* of a differential equation is the highest exponent of highest order derivative which occurs in it, after the equation is converted to the form free from radicals and fractions. For example, Eq. (1.6) when freed from radicals and fractions takes the form

$$\left[1+\left(\frac{dy}{dx}\right)^2\right]^3 = k^2\left(\frac{d^2y}{dx^2}\right)^2 \qquad (1.14)$$

Thus, Eq. (1.6) or (1.14) is of second degree. Equations (1.1) to (1.5), and (1.8) to (1.10) are of first degree, whereas Eq. (1.7) is of third degree.

In order to study differential equations, they are classified in terms of order, degree and type—ordinary or partial. For example, Eq. (1.6) is an ordinary second order and second degree differential equation, whereas Eq. (1.5) is a partial second order and first degree differential equation. It will be useful to further classify differential equations as either linear or nonlinear.

When, in an ordinary or partial differential equation, the dependent variable and its derivatives occur to the first degree only, and not as higher powers or products, we call the equation *linear*. Naturally, the coefficients of a linear equation are either constants or functions of the independent variable or variables.

If an ordinary differential equation is not linear, then we call it *nonlinear*. If a partial differential equation is not linear, then it can be either quasi-linear, semi-linear or non-linear. Thus, an ordinary linear differential equation can be written in the form

$$a_n(x)\frac{d^n y}{dx^n} + a_{n-1}(x)\frac{d^{n-1} y}{dx^{n-1}} + \cdots + a_0(x)y = g(x) \qquad (1.15)$$

where $a_n(x)$, $a_{n-1}(x)$, ..., $a_0(x)$, called *coefficients*, and $g(x)$, called the *right hand side function*, are all continuous real-valued functions of x defined on an interval

I and do not depend on *y*. *y* is an *n*th differentiable function of *x* defined on the same interval *I*. Equations (1.1), (1.3), (1.12) and (1.13), are linear, whereas Eqs. (1.2), (1.6), (1.7) and (1.9) are non-linear.

If, in Eq. (1.15), $a_n(x) \neq 0$ in *I*, then the equation is called a *normal equation*; otherwise it is called a *non-normal* equation. For example, the equation

$$(\sin x)\frac{d^2y}{dx^2} - 3\frac{d^2y}{dx^2} + e^x \frac{dy}{dx} + (\cos x)y = \tan x \qquad (1.16)$$

is a normal differential equation if we take the interval of definition $I = (0, \pi)$. On the other hand, it is a nonnormal differential equation if we take $I = [0\ \pi]$ or $I = [0, \pi)$.

Equation (1.15), in which $g(x)$ is not zero for all $x \in I$, is called a *non-homogeneous* equation; otherwise it is called a *homogeneous* equation. Equation (1.1) defined on the interval $I = \left[-\frac{3\pi}{2}, +\frac{3\pi}{2}\right]$ is a non-homogeneous equation, whereas the equation

$$\frac{d^3y}{dx^3} - 3\frac{d^2y}{dx^2} + 3\frac{dy}{dx} - y = 0 \qquad (1.17)$$

defined in the same interval is a homogeneous equation.

If the values of the unknown function or dependent variable *y* and its $(n-1)$ dervatives for Eq. (1.15) are specified at a particular point $x = x_0$, then the system consisting of the differential equation and the values $y(x_0), y'(x_0), y''(x_0), \ldots, y^{n-1}(x_0)$ is called an *Initial Value Problem*. For example, the equation

$$\frac{d^2y}{dx^2} + 2\frac{dy}{dx} + 2y = 0 \qquad (1.18)$$

together with the values $y(0) = 1, y'(0) = -1$, is an initial value problem.

The specific values $y(x_0) = y_0, y'(x_0) = y_1, \ldots, y^{(n-1)}(x_0) = y_{n-1}$, where $y_0, y_1, \ldots, y_{n-1}$ are given constants, are called *Initial Conditions*.

In practice, we often come across differential equations together with specified values of the dependent variable *y* and its derivatives at two or more given points. Such specified values are called *Boundary Conditions*, and the differential equation along with these boundary conditions is called a *Boundary Value Problem*. For example, the equation

$$\frac{d^4y}{dx^4} + ky = q \qquad (1.19)$$

together with $y(0) = y'(0) = 0$

$y(L) = y''(L) = 0$

where *k*, *q* and *L* are given constants, is a boundary value problem.

With these basic definitions, we are now in a position to start our main problem, that is, the problem of finding the solution of a differential equation. But then the natural question is: what exactly is the meaning of a solution of a differential equation? In the next section, we shall find an answer to this question and also to questions like: (i) Under what condition(s) does the solution of a given ordinary differential equation exist? (ii) If the solution exists, is it a unique solution?

1.3 SOLUTION OF DIFFERENTIAL EQUATION

An ordinary differential equation of nth order has the general form

$$F\left(x, y, \frac{dy}{dx}, \frac{d^2y}{dx^2}, \ldots, \frac{d^ny}{dx^n}\right) = 0$$

or
$$F\left(x, y, y', y'', \ldots, y^{(n)}\right) = 0 \qquad (1.20)$$

Usually it is a simple task to verify that a given function $y = y(x)$ satisfies an equation like Eq. (1.20). All that is required is to compute the derivatives of $y(x)$ and to show that $y(x)$ together with its derivatives, when substituted in Eq. (1.20), reduce it to an identity in x. If such a function $y(x)$ exists, we call it a *solution* of Eq. (1.20). It is usually assumed that $y(x)$ is an nth differentiable, real-valued function of x defined on an interval I.

Example 1.1 We see that $y = cx^5$ is a solution of the differential equation $\frac{dy}{dx} - \frac{5}{x}y = 0$, for any constant c.

Example 1.2 In the same way we see that $y = e^{2x}$, $y = \sin x$, $y = \cos x$ are solutions of the differential equation $\sin x \frac{d^3y}{dx^3} - 2\frac{d^2y}{dx^2} + \frac{dy}{dx} - 2y = 0$ and more generally that $y = c_1 e^{2x} + c_2 \sin x + c_3 \cos x$ is also a solution for every choice of the constants c_1, c_2 and c_3.

Example 1.3 By differentiating $(y^2 + 2x^2)^3 = cy^2$ and rearranging the result, we can see that it is a solution of the equation $\frac{dy}{dx} = \frac{3xy}{(x^2 - y^2)}$, for every value of the constant c.

Example 1.4 The equation $\left(\frac{dy}{dx}\right)^2 + y^2 + 1 = 0$ has no real solution.

Example 1.5 The equation $\cos\left(\frac{dy}{dx}\right) = 3$ has no real solution.

The above examples suggest many important and interesting facts about the solutions of a differential equation. These are as follows:

1. The solutions of a differential equation may or may not exist. We have to find conditions under which the solutions exist.
2. A differential equation may have more than one solution; it may even have infinitely many solutions. It is natural to ask: Under what conditions does a differential equation have a unique solution?
3. Example 1.1 shows that the solution of a first order differential equation contains one arbitrary constant, whereas Example 1.2 shows that the solution of a third order differential equation contains three arbitrary constants. Is it accidental? No! This is because a differential equation of order n will, in general, possess a solution involving n arbitrary constants. This solution is called the *general solution* of the differential equation. There is a more precise mathematical theorem to this effect. The statement and proof of this theorem is beyond the scope of this book. However, this theorem for an ordinary nth order linear differential equation will be discussed in Chapter 3. Further, any solution which is obtained from the general solution by giving particular values to the arbitrary constants is called a *particular solution*. For example, $y = c_1 e^{2x} + c_2 \sin x + c_3 \cos x$ is the general solution of the differential equation in Example 1.2, whereas e^{2x}, $\sin x + \cos x$, $\sin x$ and $\cos x$ are a few particular solutions.
4. A differential equation can have solutions which are not explicit functions of the independent variable but are implicitly defined (as in Example 1.3), and sometimes it may be difficult or impossible to express the dependent variable explicitly in terms of the independent variable.

Thus, many natural questions arise regarding a solution of a given differential equation: given a differential equation, whether it has a solution, is the solution unique, what are the conditions under which it will have a solution of a particular nature, and so on. We shall not even dare to attempt to answer any of these questions in this book: on the contrary, we shall be concerned merely with an exposition of the methods of solving some particular classes of differential equations, and their solutions will be expressed by the ordinary algebraic, trigonometric, hyperbolic, exponential and logarithmic functions. Thus, for us a differential equation always has a solution.

In order to discuss and develop different methods for obtaining solutions to a given differential equation, we classify them first in terms of ordinary and partial differential equations and then in terms of order and degree. We shall first start with ordinary differential equations. The major part of this book deals with ordinary differential equations. Later, we shall discuss partial differential equations. In the ordinary differential equation, we first take up first order differential equations. When we discuss second or higher order ordinary differential equations, we concentrate only on linear differential equations. Before we actually discuss the methods of obtaining solutions to ordinary differential equations, we give in the following section some examples to indicate how differential equations are important and useful.

1.4 SOME EXAMPLES FROM FIELDS OF APPLICATION

Example 1.6 A certain population of bacteria is known to grow at a rate proportional to the amount present in a culture that provides enough food and space. Initially there are 250 bacteria, and after 7 hours, 800 bacteria are observed in the culture. We have to derive an expression to determine the number of bacteria present in the culture at any time t.

Let $N(t)$ denote the number of bacteria present in the culture at any time t. The growth rate is $\dfrac{dN}{dt}$, which is proportional to N. Thus

$$\frac{dN}{dt} = kN$$

where k is a constant of proportionality. It is also given that at $t = 0$, $N = 250$, and at $t = 7$, $N = 800$.

Therefore, our problem is to solve the boundary value problem

$$\frac{dN}{dt} = kN$$

together with the boundary conditions $N(0) = 250$, $N(7) = 800$.

Example 1.7 A metal bar at a temperature of 100°C is placed in a room at a constant temperature of 0°C. If after 20 minutes the temperature of the bar is 50°C, then we have to find an expression for the tempoerature of the bar at any time.

To solve this problem, we have to use Newton's law of cooling, which states that the rate of change of the temperature T of the body is proportional to the temperature difference between the body and its surrounding medium. If T_m is the temperature of the surrounding medium, then Newton's law of cooling implies

$$\frac{dT}{dt} = -k(T - T_m) \tag{1.21}$$

where k is a positive constant of proportionality, and the negative sign on the right hand side indicates the sign of cooling.

Thus in our problem, Eq. (1.21) becomes

$$\frac{dT}{dt} = -k(T - 0)$$

$$\Rightarrow \frac{dT}{dt} + kT = 0$$

and we have to solve this equation together with the boundary conditions when $t = 0$, $T = 100$, and when $t = 20$, $T = 50$, to get an expression for the temperature of the bar at any time.

8 Introduction to Differential Equations

Example 1.8 Derive the differential equation governing the following system by using Hooke's law together with Newton's second law of motion:

A spring with a mass m attached to its lower end is suspended vertically from a mounting and allowed to come to rest in an equilibrium position. The system is then set in motion by releasing the mass with an initial velocity v_0 at a distance x_0 below its equilibrium position and simultaneously applying to the mass an external force $F(t)$ in the downward direction. Refer to Fig. 1.1.

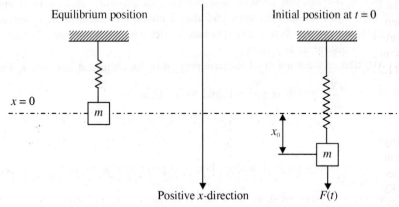

Fig. 1.1 Example 1.8.

In this system, we choose the downward direction as the positive direction, and the origin as the centre of gravity of the mass in the equilibrium position. We assume that air resistance is present and is proportional to the velocity of the mass.

At any time t, there are three forces acting on the system, viz. (1) the force $F(t)$, measured in the positive direction, (2) a restoring force F_s due to Hooke's law, which states that the restoring force of a spring is proportional to the displacement of the spring from its normal length; thus $F_s = -kx$, where $k > 0$ is the constant of proportionality called *spring constant*, and the negative sign indicates F_s is acting upwards when x is positive and downwards when x is negative, (3) a force F_a due to air resistance given by $F_a = -a\dfrac{dx}{dt}$, where $a > 0$ is a constant of proportionality, and the negative sign indicates that F_a always acts in the direction opposite to that of velocity, and consequently tends to retard, or damp, the motion of the mass.

Now we apply Newton's second law, which states that the rate of change of momentum of a body is equal to the force acting on that body.

Here, momentum is given by mass into its velocity, i.e. $m\dfrac{dx}{dt}$.

Thus we have
$$\frac{d}{dt}\left(m\frac{dx}{dt}\right) = F_s + F_a + F(t)$$

$$\Rightarrow m\frac{d^2x}{dt^2} = -kx - a\frac{dx}{dt} + F(t)$$

$$\Rightarrow m\frac{d^2x}{dt^2} + a\frac{dx}{dt} + kx = F(t) \tag{1.22}$$

with the initial conditions $x(0) = x_0$, $\left.\dfrac{dx}{dt}\right|_0 = v_0$, since the system starts at $t = 0$ with an initial velocity v_0 and from an initial position x_0.

In Eq. (1.22) the force g due to gravity does not appear explicitly as we have chosen the centre of gravity of the mass as our origin, though it is present. However, if we choose the terminal end-point of the unstretched spring before the mass m is attached, as the origin, indicated by the dotted line in Fig. 1.1, then Eq. (1.22) will become

$$m\frac{d^2x}{dt^2} + a\frac{dx}{dt} + kx = g + F(t) \tag{1.23}$$

Example 1.9 Using Kirchoff's loop law, derive a differential equation for the current I in a simple circuit consisting of a resistor, a capacitor, an inductor, and an electromotive force (normally a battery or a generator) connected in series.

Kirchoff's loop law states that the algebraic sum of the voltage drops in a simple closed electric circuit is zero.

The circuit is shown in Fig. 1.2, where R is the resistance in ohms, C is the capacitance in farads, L is the inductance in henries, $E(t)$ is the electromotive force (emf) in volts, and I is the current in amperes.

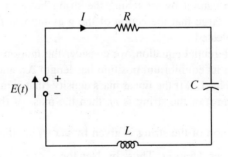

Fig. 1.2 Example 1.9.

We know that the voltage drops across a resistor, a capacitor, and an inductor are respectively RI, $\dfrac{1}{C}q$ and $L\dfrac{dI}{dt}$, where q is the charge on the capacitor and $I = \dfrac{dq}{dt}$, and the voltage drop across an emf is $-E(t)$. Thus, by Kirchoff's loop law, we have

$$RI + \frac{1}{C}q + L\frac{dI}{dt} - E(t) = 0$$

which becomes, by differentiating with respect to t

$$R\frac{dI}{dt} + \frac{1}{C}\frac{dq}{dt} + L\frac{d^2I}{dt^2} - \frac{dE(t)}{dt} = 0$$

$$\Rightarrow L\frac{d^2I}{dt^2} + R\frac{dI}{dt} + \frac{1}{C}I = \frac{dE(t)}{dt} \qquad (1.24)$$

Example 1.10 In this example, we derive the differential equation governing transverse vibrations of an elastic string, which is stretched to length l and then fixed at the end-points. For covenience, we take the fixed end-points to be $x = 0$ and $x = 1$, and the stretched string along the x-axis. The string is drawn aside into a certain curve $y = y(x)$ in the xy-plane and released at a certain instant, say $t = 0$, and then allowed to vibrate. Now, the problem is to determine the differential equation governing the vibrations of the string, i.e. the deflection $u(x, t)$ at any point x and at any time $t > 0$.

In order to obtain the equation of motion, we usually have to make some simplifying assumptions.

(i) The string performs a small transverse motion in a vertical plane, that is, each point of the string has a constant x-coordinate, so that the deflection and the slope at every point of the string remain small in absolute value, and the y-coordinate depends only on x and the time t.

(ii) The mass of the string per unit length is constant (that is, it is a homogeneous string), the string is perfectly elastic, and it does not offer any resistance to bending.

(iii) The tension caused by stretching the string before fixing it at the end points is so large that the action of the gravitational force on the string can be neglected.

To obtain the differential equation, we consider the motion of a small portion of the string, which in its equilibrium position has length Δx, and the forces acting on it are as shown in Fig. 1.3. If the linear mass density (that is mass per unit length of the undeflected string) of the string is m, then the mass of the small portion is $m\Delta x$. Also, the deflection of the string is given by $u(x, t)$, so that $\frac{\partial u}{\partial t}, \frac{\partial^2 u}{\partial t^2}$ are the string's velocity and acceleration. Thus, by Newton's second law of motion the transverse force F acting on the small portion of the string is given by

$$F = m\Delta x \frac{\partial^2 u}{\partial t^2}$$

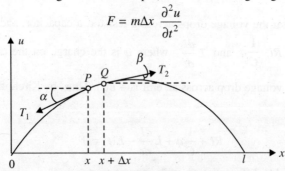

Fig. 1.3 Example 1.10.

Nature of Differential Equations

Since the string does not offer any resistance to bending (as it is flexible), the tension at a point is directed along the tangent to the curve of the string. Let T_1 and T_2 be the tensions at the points P and Q and the small portion. As there is no motion in the horizontal direction, the horizontal components of the tension must be constant. Thus we have from Fig. 1.3

$$T_1 \cos \alpha = T_2 \cos \beta = T \quad \text{which is a constant} \tag{1.25}$$

In the vertical direction there are two forces, namely the vertical components $-T_1 \sin \alpha$ and $T_2 \sin \beta$ of the tensions T_1 and T_2 respectively. Here the minus sign is due to the fact that the component at P is downwards. Since the motion is only due to tension, we have

$$T_2 \sin \beta \, T_1 \sin \alpha = m \Delta x \frac{\partial^2 u}{\partial t^2}$$

$$\Rightarrow \frac{T_2 \sin \beta}{T_2 \cos \beta} - \frac{T_1 \sin \alpha}{T_1 \cos \alpha} = \frac{m \Delta x}{T} \frac{\partial^2 u}{\partial t^2} \quad \text{using (1.25)}$$

$$\Rightarrow \tan \beta - \tan \alpha = \frac{m \Delta x}{T} \frac{\partial^2 u}{\partial t^2}$$

$$\Rightarrow \left(\frac{\partial u}{\partial x}\right)_{x+\Delta x} - \left(\frac{\partial u}{\partial x}\right)_x = \frac{m \Delta x}{T} \frac{\partial^2 u}{\partial t^2}$$

since $\tan \alpha$ and $\tan \beta$ are respectively the slopes of the curve of the string at x and $x + \Delta x$. Dividing by Δx, we have

$$\frac{1}{\Delta x}\left[\left(\frac{\partial u}{\partial x}\right)_{x+\Delta x} - \left(\frac{\partial u}{\partial x}\right)_x\right] = \frac{m}{T} \frac{\partial^2 u}{\partial t^2}$$

Now making $\Delta x \to 0$, we get the required equation of motion of the vibrating elastic string as

$$\frac{\partial^2 u}{\partial x^2} = \frac{m}{T} \frac{\partial^2 u}{\partial t^2}$$

$$\Rightarrow a^2 \frac{\partial^2 u}{\partial x^2} = \frac{\partial^2 u}{\partial t^2} \tag{1.26}$$

where $a^2 = \frac{T}{m}$ is a positive constant.

This is the well-known one-dimensional wave equation.

1.5 SUMMARY

Before we go to the next chapter to discuss first order differential equations, let us recall what we have covered in this chapter.

Introduction to Differential Equations

We first discussed the definition of and necessity for differential equations; basic definitions like ordinary differential equations and partial differential equations; order and degree of a differential equation; linear and non-linear differential equations; normal and non-normal differential equations; homogeneous and non-homogeneous differential equations; and initial value problems and boundary value problems. Further, we discussed solutions of a differential equation; the solution of an *n*th order differential equation, which contains *n* arbitrary constants, called general solutions, and a particular solution. Many physical situations like population growth, cooling and heating, spring motion, electrical circuit and vibrating string motion were discussed.

Exercises 1.5

Determine in Exercises 1 to 10 whether the given differential equation is ordinary or partial, linear or non-linear, and homogeneous or non-homogeneous, and also the order and degree of the equation.

1. $y - 5x(y')^4 = e^x + 1$

2. $y''' + 2(y'')^2 + y' = \cos x$

3. $\dfrac{\partial^2 z}{\partial x^2} + \dfrac{\partial^2 z}{\partial y^2} = x^2 + y$

4. $5x\dfrac{d^2 y}{dx^2} + 3x^2 \dfrac{dy}{dx} - (\sin x)y = 0$

5. $\dfrac{\partial^2 y}{\partial t^2} - 4\dfrac{\partial^2 y}{\partial x^2} = 0$

6. $5\dfrac{dy}{dt} - 3y\left(\dfrac{dy}{dt}\right)^7 = t$

7. $\left(\dfrac{\partial z}{\partial x}\right)^2 + \left(\dfrac{\partial z}{\partial y}\right)^2 + 2x\dfrac{\partial z}{\partial x} + 2y\dfrac{\partial z}{\partial y} + 1 = 0$

8. $t\dfrac{d^2 y}{dt^2} + t^2 \dfrac{dy}{dx} - (\sin t)\sqrt{y} = t^2 - t + 1$

9. $z^2\left[1 + \left(\dfrac{\partial z}{\partial x}\right)^2 + \left(\dfrac{\partial z}{\partial y}\right)^2\right] = 0$

10. $y\dfrac{d^2 x}{dy^2} = y^2 + 1$

Nature of Differential Equations

11. Radium decomposes at a rate proportional to the amount present. Derive a differential equation for the amount of radium present at time t.

12. A red hot steel rod is suspended in air which remains at a constant temperature of 24°C. Find a differential equation for the temperature of the rod as a function of time.

13. A particle of mass m moves along a straight line (the x-axis) while subjected to (i) a force proportional to its displacement x from a fixed point 0 in its path and directed toward 0, and (ii) a resisting force proportional to its velocity. Write a differential equation for the motion of the particle.

14. Find the differential equation of the family of circles of fixed radius r with centres on the x-axis.

15. Form the differential equation representing all tangents to the parabola $y^2 = 2x$.

16. One hundred grams of cane sugar in water is being coverted into dextrose at a rate which is proportional to the unconverted amount. Find a differential equation expressing the rate of conversion after t minutes.

17. A torpedo is fired from a ship and travels in a straight path just below the water's surface. Derive the differential equation governing the motion of the torpedo, if the water retards the torpedo with a force proportional to its speed.

18. A 128 lb weight is attached to a spring having a spring constant of 64 lb/ft. The weight is put in motion with no initial velocity by displacing it by 6 inches above the equilibrium position and by simultaneously applying to the weight an external force $F(t) = 8 \sin 4t$. Derive a differential equation governing the subsequent vibrations of the spring if there is no air resistance.

19. A simple series RCL circuit has $R = 180 \, \Omega$, $C = 1/280$ F, $L = 20$ H, and applied voltage $E(t) = 10 \sin t$. Derive a differential equation for the change on the capacitor at any time t. Use Fig. 1.2.

20. A steel ball weighing 128 lb is suspended from a spring, by which the spring stretches 2 ft from its natural length. The ball is then put in motion with no initial velocity by displaying it by 6 inches above the equilibrium position. Derive a differential equation governing the subsequent vibrations of the spring if there is no air resistance. Use Fig. 1.1.

21. If $y(x)$ is the solution of a differential equation, determine which of the conditions are initial conditions and which are boundary conditions on $y(x)$.
    (a) $y(0) = 0$, $y(1) = 0$
    (b) $y(0) = 0$, $y'(0) = 1$
    (c) $y(0) = 0$, $y'(0) = 0$, $y''(\pi) = 0$
    (d) $y(\pi/8) = 0$, $y'(\pi/8) = 2$
    (e) $y(0) = 1$, $y'(\pi/2) = 2$

Find the solutions of the differential equations given in Exercises 22 to 30.

22. $\dfrac{dy}{dx} = y^2 + 1$

23. $\dfrac{dy}{dt} = -\dfrac{y}{t}$

24. $y\,dy + (y^2 + 1)\,dx = 0$

25. $\dfrac{dy}{dx} = \sec y \tan x$

26. $\dfrac{dy}{dx} = y^2 \quad y(0) = 4$

27. $y\,dx + x\,dy = 0$

28. $y\,dx - x\,dy = 0$

29. $2xy\,dx + (1 + x^2)\,dy = 0 \quad y(0) = 1$

30. $\dfrac{dy}{dx} - 3y = 6 \quad y(0) = 0$

CHAPTER 2

First Order Differential Equations

2.1 INTRODUCTION

In this chapter, we shall discuss different methods of solving ordinary first order differential equations. The general form of such a differential equation is

$$p_n(x,y)\left(\frac{dy}{dx}\right)^n + p_{n-1}(x,y)\left(\frac{dy}{dx}\right)^{n-1} + \cdots + p_1(x,y)\frac{dy}{dx} + p_0(x,y) = 0 \quad (2.1)$$

where $p_0(x, y)$, ..., $p_n(x, y)$ are functions of x and y.

No systematic procedure exists for obtaining the solution of such an equation with arbitrary forms of the functions $p_0(x, y)$, ..., $p_n(x, y)$ even for $n = 1$. However, there are certain standard types of first order equations for which methods of obtaining solutions are available. In this chapter, we shall first discuss different methods of solving a few of the first order and first degree differential equations, and then the differential equations of the first order but not of the first degree, that have many applications.

The general form of a first order and first degree differential equation is

$$p_1(x,y)\frac{dy}{dx} + p_0(x,y) = 0 \quad \text{or} \quad p_1(x,y) + p_0(x,y)\frac{dx}{dy} = 0 \quad (2.2)$$

Equation (2.2) suggests that in a first order and first degree equation, there is no distinction between dependent and independent variables. Either one can be considered the dependent variable and the other one the independent variable. We start with the simplest form.

2.2 VARIABLES SEPARABLE

If it is possible to express a first order and first degree equation in the form

$$f(y)\,dy + g(x)\,dx = 0$$

then the equation can be solved by integration. Then the general solution is

$$\int f(y)\,dy + \int g(x)\,dx = c \qquad (2.3)$$

Since the equation is solved by separating the variables, the method is called *variables separable*.

Example 2.1 Solve the equation $y - x\dfrac{dy}{dx} = y^2 + \dfrac{dy}{dx}$.

The given equation can be written as

$$y - y^2 = (x+1)\dfrac{dy}{dx}$$

$$\Rightarrow \dfrac{dy}{y - y^2} = \dfrac{dx}{x+1}$$

$$\Rightarrow \dfrac{dy}{y(1-y)} = \dfrac{dx}{x+1}$$

$$\Rightarrow \left(\dfrac{1}{y} + \dfrac{1}{1-y}\right)dy = \dfrac{dx}{x+1}$$

Integrating both the sides, we get

$$\ln y - \ln(1-y) = \ln(x+1) + c$$

$$\Rightarrow \ln\left(\dfrac{y}{1-y}\right) = \ln k(x+1) \quad \text{taking } c = \ln k$$

$$\Rightarrow y = k(1-y)(x+1) \text{ is the required solution.}$$

Example 2.2 Solve $\dfrac{dy}{dx} = \sin(x+y) + \cos(x+y)$.

Putting $x + y = v$, the equation reduces to

$$\dfrac{dv}{dx} - 1 = \sin v + \cos v$$

$$\Rightarrow \dfrac{dv}{1 + \sin v + \cos v} = dx$$

$$\Rightarrow \int \dfrac{dv}{1 + \sin v + \cos v} = x + c \quad \text{integrating both the sides}$$

$$\Rightarrow \int \dfrac{dv}{2\cos^2 \dfrac{y}{2} + 2\sin\dfrac{v}{2}.\cos\dfrac{v}{2}} = x + c$$

First Order Differential Equations

$$\Rightarrow \int \frac{dv}{2\cos^2 \frac{v}{2}\left(1+\tan\frac{v}{2}\right)} = x + c$$

$$\Rightarrow \ln\left(1+\tan\frac{v}{2}\right) = x + c$$

$$\Rightarrow \ln\left(1+\tan\frac{x+y}{2}\right) = x + c \quad \text{is the required solution.}$$

Example 2.3 Solve the equation $y' = f(ax + by + c)$ by using the substitution $z = ax + by + c$, where a, b and c are constants.

$$z = ax + by + c \Rightarrow \frac{dz}{dx} = a + b\frac{dy}{dx}$$

$$\Rightarrow \frac{dy}{dx} = \frac{1}{b}\left(\frac{dz}{dx} - a\right)$$

Now, $\qquad y' = f(ax + by + c)$

$$\Rightarrow \frac{1}{b}\left(\frac{dz}{dx} - a\right) = f(z)$$

$$\Rightarrow \frac{dz}{bf(z)+a} = dx$$

$$\Rightarrow x + c = \int \frac{dz}{bf(z)+a}$$

which can be integrated in specific cases to obtain the solution.

Example 2.4 Solve the equation $\dfrac{dy}{dx} = \dfrac{x+y+4}{x+y-6}$.

Put $\qquad x + y = v$

$$\Rightarrow 1 + \frac{dy}{dx} = \frac{dv}{dx}$$

Now, $\qquad \dfrac{dy}{dx} = \dfrac{x+y+4}{x+y-6}$

$$\Rightarrow \frac{dv}{dx} - 1 = \frac{v+4}{v-6}$$

$$\Rightarrow \frac{dv}{dx} = \frac{2v-2}{v-6}$$

Introduction to Differential Equations

$$\Rightarrow \frac{v-6}{v-1} dv = 2dx$$

$$\Rightarrow \int dv - 5\int \frac{dv}{v-1} = 2\int dx$$

$$\Rightarrow v - 5 \ln(v-1) = 2x + c$$

$$\Rightarrow y - x = 5 \ln(x + y - 1) + c$$

Exercises 2.2

Determine whether the variables in the equations given in Exercises 1, 2 and 3 can be separated or not. Justify your answer.

1. $\sin y' = x$

2. $y^2 + \dfrac{dy}{dx} = \dfrac{y+1}{xy}$

3. $e^{y'} = x$

Solve the equations in Exercises 4 through 14.

4. $(1+x)y\, dx + (1-y)x\, dy = 0$

5. $\dfrac{dy}{dx} = \dfrac{1+y^2}{1+x^2}$

6. $\sec^2 x \tan y\, dx + \sec^2 y \tan x\, dy = 0$

7. $\dfrac{dy}{dx} = \dfrac{x(2\ln x + 1)}{\sin y + y \cos y}$

8. $\dfrac{dy}{dx} = e^{x-y} + x^2 e^{-y}$

9. $\dfrac{x\, dx + y\, dy}{x\, dy - y\, dx} = \left(\dfrac{a^2 - x^2 - y^2}{x^2 + y^2}\right)^{1/2}$

10. $y \sin x e^{\cos x}\, dx + y^{-1} dy = 0$

11. $\dfrac{dy}{dx} = (x+y)^2$

12. $\dfrac{dy}{dx} = \sin^2(x - y + 1)$

13. $\left(\dfrac{x+y-a}{x+y-b}\right)\dfrac{dy}{dx} = \dfrac{x+y+a}{x+y+b}$

14. $(x-y)^2 \dfrac{dy}{dx} = a^2$

15. If $\dfrac{dy}{dx} = F\left(\dfrac{ax+by+c}{a'x+b'y+c'}\right)$ be such that $ab' = a'b$, then find a substitution that reduces the equation to one in which the variables are separable.

2.3 HOMOGENEOUS EQUATIONS

If a first order and first degree differential equation can be written in the form

$$\dfrac{dy}{dx} = F\left(\dfrac{y}{x}\right)$$

then that equation is called a *homogeneous equation*. Such an equation can be solved by changing the dependent variable from y to v through the substitution $y = vx$. Then the differential equation will be reduced to the variables separable form.

Example 2.5 Solve $x \cos\dfrac{y}{x}(y\, dx + x\, dy) = y \sin\dfrac{y}{x}(x\, dy - y\, dx)$.

The above equation can be written in the form

$$\dfrac{dy}{dx} = \dfrac{\dfrac{y}{x}\cos\dfrac{y}{x} + \dfrac{y^2}{x^2}\sin\dfrac{y}{x}}{\dfrac{y}{x}\sin\dfrac{y}{x} - \cos\dfrac{y}{x}}$$

Putting $y = vx$, $\dfrac{dy}{dx} = x\dfrac{dv}{dx} + v$, we get

$$x\dfrac{dv}{dx} + v = \dfrac{v\cos v + v^2 \sin v}{v \sin v - \cos v}$$

$$\Rightarrow x\dfrac{dv}{dx} = \dfrac{2v \cos v}{v \sin v - \cos v}$$

$$\Rightarrow \dfrac{v \sin v - \cos v}{2v \cos v} dv = \dfrac{dx}{x}$$

$$\Rightarrow \left(\dfrac{1}{2}\tan v - \dfrac{1}{2v}\right) dv = \dfrac{dx}{x}$$

$$\Rightarrow \ln \sec v - \ln v = 2\ln x + \ln k$$

$$\Rightarrow \ln \frac{\sec v}{v} = \ln k \, x^2 \Rightarrow \sec v = k v x^2$$

$$\Rightarrow \sec \frac{y}{x} = k \, xy \quad k \text{ is the constant of integration.}$$

Example 2.6 If $ab' \neq a'b$, show that constants h and k can be chosen in such a way that the substitution $x = z - h$, $y = w - k$ reduces

$$\frac{dy}{dx} = F\left(\frac{ax + by + c}{a'x + b'y + c'}\right)$$

to a homogeneous equation.

We have $\quad x = z - h \quad y = w - k \quad \dfrac{dy}{dx} = \dfrac{dw}{dz}$

Then the equation reduces to

$$\frac{dw}{dz} = F\left(\frac{az + bw - ah - bk + c}{a'z + b'w - a'h - b'k + c}\right)$$

we choose h, k such that $ah + bk - c = 0$ and $a'h + b'k - c = 0$, so that we have

$$\frac{dw}{dz} = F\left(\frac{az + bw}{a'z + b'w}\right),$$ which is a homogeneous equation and can be solved by putting $w = vz$.

Exercises 2.3

Solve the following differential equations:

1. $\dfrac{dy}{dx} + \dfrac{x^2 + 3y^2}{3x^2 + y^2} = 0$

2. $x^2 \, ydx - (x^3 + y^3) dy = 0$

3. $y^2 + x^2 \dfrac{dy}{dx} = xy \dfrac{dy}{dx}$

4. $x \dfrac{dy}{dx} + \dfrac{y^2}{x} = y$

5. $(x^2 + y^2) \dfrac{dy}{dx} = xy$

6. $(x^2 - y^2)\dfrac{dy}{dx} = xy$

7. $x\dfrac{dy}{dx} - y = x\sqrt{x^2 + y^2}$

8. $(x^3 - 3xy^2)dx = (y^3 - 3x^2y)dy$

9. $(x - y)dx = (x + y + 1)dx$

10. $\dfrac{dy}{dx} = \dfrac{x + 2y - 3}{2x + y - 3}$

11. $\dfrac{dy}{dx} = \dfrac{y - x + 1}{y + x + 5}$

12. $\dfrac{2x + 9y - 20}{2y + 6x - 10} = \dfrac{dy}{dx}$

13. $(2x + 3y - 5)dy + (3x + 2y - 5)dx = 0$
14. $(x - y - 2)dx + (x - 2y - 3)dy = 0$
15. $(2x + y)dx + (x - 2y)dy = 0$

2.4 EXACT EQUATIONS

A differential equation

$$M(x, y)\,dx + N(x, y)\,dy = 0 \qquad (2.4)$$

is said to be exact if there exists a function $f(x, y)$ such that

$$df(x, y) = M(x, y)\,dx + N(x, y)\,dy \qquad (2.5)$$

where $M(x, y)$, $N(x, y)$ and $f(x, y)$ are continuous functions and have continuous first partial derivatives on some rectangle of the (x, y) plane. Further, $f(x, y)$ has continuous second partial derivatives in the same region.

For example, $x\,dy + y\,dx = 0$ and $\sin x \cos y\,dy + \cos x \sin y\,dx = 0$ are exact differential equations, because $d(xy) = x\,dy + y\,dx$ and $d(\sin x \sin y) = \sin x \cos y\,dy + \cos x \sin y\,dx$, whereas $\sin x \cos y\,dx - \sin y \cos x\,dy = 0$ is not an exact differential equation, since there does not exist a function $f(x, y)$ such that $df(x, y) = \sin x \cos y\,dx - \sin y \cos x\,dy$.

Now, if Eq. (2.4) is exact, then there exists a function $f(x, y)$ satisfying the condition specified in Eq. (2.5).

Also, we know

$$df(x, y) = \dfrac{\partial f}{\partial x}dx + \dfrac{\partial f}{\partial y}dy \qquad (2.6)$$

Comparing the relations in Eqs. (2.5) and (2.6), we obtain a necessary condition for Eq. (2.4) to be exact, which is

$$\frac{\partial f}{\partial x} = M \quad \text{and} \quad \frac{\partial f}{\partial y} = N$$

i.e.
$$\frac{\partial M}{\partial y} = \frac{\partial}{\partial y}\left(\frac{\partial f}{\partial x}\right) = \frac{\partial^2 f}{\partial x\, \partial y} = \frac{\partial}{\partial x}\left(\frac{\partial f}{\partial y}\right) = \frac{\partial N}{\partial x} \qquad (2.7)$$

The condition specified in Eq. (2.7) is also sufficient for Eq. (2.4) to be exact, i.e., if Eq. (2.7) holds, then Eq. (2.4) is exact. For this, let

$$\int M(x,y)\,dx = u$$

$$\Rightarrow \frac{\partial u}{\partial x} = M$$

$$\Rightarrow \frac{\partial}{\partial y}\left(\frac{\partial u}{\partial x}\right) = \frac{\partial M}{\partial y} = \frac{\partial N}{\partial x} \qquad \text{by Eq. (2.7)}$$

$$\Rightarrow \frac{\partial N}{\partial x} = \frac{\partial}{\partial y}\left(\frac{\partial u}{\partial x}\right) = \frac{\partial}{\partial x}\left(\frac{\partial u}{\partial x}\right)$$

$$\Rightarrow N = \frac{\partial u}{\partial y} + g'(y) \qquad \text{by integration}$$

where $g'(y)$, the constant of integration, is a function of y alone, since it should vanish under differentiation with respect to x.
Thus

$$M\,dx + N\,dy = \frac{\partial u}{\partial x}\,dx + \frac{\partial u}{\partial y}\,dy + g'(y)\,dy = du + dg = d(u+g) \qquad (2.8)$$

which is an exact differential. Thus we have proved the following theorem:

Theorem 2.1 The differential equation $M(x,y)\,dx + N(x,y)\,dy = 0$ is an exact differential iff $\dfrac{\partial M}{\partial y} = \dfrac{\partial N}{\partial x}$.

If the differential equation shown in Eq. (2.4) is exact, then we know that a function $f(x,y)$ exists such that Eq. (2.5) holds. So we get the solution by integrating Eq. (2.5).

i.e.
$$f(x,y) = \int M\,dx + g(y) = c \qquad (2.9)$$

You can carry out the integrations in Eq. (2.9) as follows: First integrate $M(x, y)$ with respect to x, treating y as constant. Then integrate those terms in $N(x, y)$ which do not contain x. Add these two expressions and equate to a constant to obtain the solution. The reason for this procedure is obvious if we examine Eq. (2.8).

Example 2.7 Solve the equation

$$(x^2 - 4xy - 2y^2)dx + (y^2 - 4xy - 2x^2)dy = 0$$

Here

$$M(x, y) = x^2 - 4xy - 2y^2$$

$$\Rightarrow \frac{\partial M}{\partial y} = -4x - 4y$$

and

$$N(x, y) = y^2 - 4xy - 2x^2$$

$$\Rightarrow \frac{\partial N}{\partial x} = -4y - 4x$$

thus $\dfrac{\partial M}{\partial y} = \dfrac{\partial N}{\partial x}$, and hence the equation is exact.

Then

$$f(x, y) = \int M dx + g(y) = \int (x^2 - 4xy - 2y^2)\, dx + g(y)$$

$$= \frac{x^3}{3} - 2x^2 y - 2y^2 x + g(y)$$

$$\Rightarrow \frac{\partial f}{\partial y} = -2x^2 - 4yx + \frac{dg}{dy}$$

$$\Rightarrow N(x, y) = -2x^2 - 4yx + \frac{dg}{dy}$$

$$\Rightarrow y^2 - 4xy - 2x^2 = -2x^2 - 4yx + \frac{dg}{dy}$$

$$\Rightarrow y^2 = \frac{dg}{dy} \Rightarrow g(y) = \frac{y^3}{3}$$

$$\Rightarrow \int y^2 dy = \frac{y^3}{3}$$

Thus, to get $g(y)$ we have to integrate with respect to y those terms in $N(x, y)$ which are free from x.

Now the solution is

$$\frac{x^3}{3} - 2x^2y - 2xy^2 + \frac{y^3}{3} = c$$

Example 2.8 Solve $(x + \sin y)\, dx + (x \cos y - 2y)\, dy = 0$.

Here

$$M(x, y) = x + \sin y$$

$$\Rightarrow \frac{\partial M}{\partial y} = \cos y$$

and $\qquad N(x, y) = x \cos y - 2y \Rightarrow \dfrac{\partial N}{\partial x} = \cos y$

Thus the equation is exact and we proceed as follows to get the solution.
We are looking for a function $f(x, y)$ which satisfies

$$\frac{\partial f}{\partial x} = M \quad \text{and} \quad \frac{\partial f}{\partial y} = N$$

i.e.

$$\frac{\partial f}{\partial x} = x + \sin y$$

Integrating this with respect to x, we find

$$f(x, y) = \frac{x^2}{2} + x \sin y + g(y)$$

where $g(y)$ is constant of integration

This gives

$$\frac{\partial f}{\partial y} = x \cos y + g'(y)$$

$$\Rightarrow x \cos y - 2y = x \cos y + g'(y)$$
$$\Rightarrow g'(y) = -2y$$
$$\Rightarrow g(y) = -y^2$$

Hence, $f(x, y) = \dfrac{x^2}{2} + \sin y - y^2 = c$ is the required solution.

Example 2.8 justifies the method applied in Example 2.7 to solve an exact equation.

Next the natural question is: "What if the equation is not exact?". For example, $x\, dy - y\, dx = 0$ is not an exact equation, since $\dfrac{\partial (-y)}{\partial y} = -1 \neq 1 = \dfrac{\partial x}{\partial x}$.

However, if we multiply this equation by $\dfrac{1}{x^2}$ or $\dfrac{1}{y^2}$ or $\dfrac{1}{xy}$, then the resulting equation is exact.

First Order Differential Equations

Can every nonexact equation be made exact by multiplying it by an appropriate function $I(x, y)$? If Eq. (2.4) is not exact and $I(x, y)$ is a function such that

$$I(x, y) M(x, y) dx + I(x, y) N(x, y) dy = 0 \qquad (2.10)$$

is exact, then the function $I(x, y)$ is called an *Integrating Factor*. What is the relationship between the solutions of Eqs. (2.4) and (2.10)? All these questions are answered by the following theorem.

Theorem 2.2 If Eq. (2.4) is not exact, then there exists a function $I(x, y)$ such that Eq. (2.10) is exact and Eqs. (2.4) and (2.10) have the same solution.

We shall not prove this theorem, but explain the ideas contained in the theorem by an example.

Example 2.9 Solve $x\, dy - y\, dx = 0$.

Since this is not an exact equation, let us multiply this equation by

$$I(x, y) = \frac{1}{x^2 + y^2}.$$

Then

$$\frac{x}{x^2 + y^2} dy - \frac{y}{x^2 + y^2} dx = 0$$

which is exact, for

$$\frac{x\, dy - y\, dx}{x^2 + y^2} = 0$$

$$\Rightarrow d\left(\tan^{-1} \frac{y}{x}\right) = 0$$

$$\Rightarrow \tan^{-1}\left(\frac{y}{x}\right) = c$$

is the solution of the given equation.

The above theorem, however, does not help us in finding an integrating factor, though the above example suggests that the number of integrating factors for an equation is infinite.

We give below some rules for finding an integrating factor for a nonexact differential equation given by Eq. (2.4). The proofs of these rules are left for the reader as exercises.

Rule 1: Sometimes an integrating factor can be found by inspection. In the above example, integrating factors were obtained by inspection.

Example 2.10 Solve $(1 + xy)y\, dx + (1 - xy)x\, dy = 0$.

As such this equation is not exact. However, if we rewrite this as

Introduction to Differential Equations

$$(y\,dx + x\,dy) + \left(xy^2\,dx - x^2y\,dy\right) = 0$$

$$\Rightarrow d(xy) + xy^2\,dx - x^2y\,dy = 0$$

$$\Rightarrow \frac{d(xy)}{x^2y^2} + \frac{dx}{x} - \frac{dy}{y} = 0 \qquad \text{dividing by } x^2y^2$$

$$\Rightarrow -\frac{1}{xy} + \ln x - \ln y = c \qquad \text{by integrating}$$

$$\Rightarrow -\frac{1}{xy} + \ln \frac{x}{y} = \ln k$$

$$\Rightarrow \ln \frac{x}{y} - \ln k = \frac{1}{xy} = \ln e^{1/xy}$$

$$\Rightarrow \frac{x}{ky} = e^{1/xy}$$

$$\Rightarrow x = ky\,e^{1/xy}$$

is the required solution.

Rule 2: If Eq. (2.4) is homogeneous and nonexact such that $Mx + N \neq 0$, then $\dfrac{1}{Mx + Ny}$ is an integrating factor.

Example 2.11 Solve $(x^2y - 2xy^2)\,dx - (x^3 - 3x^2y)\,dy = 0$.

This can be rewritten as

$$(x^2y - 2xy^2)\,dx + (3x^2y - x^3)\,dy = 0$$

This equation is nonexact and homogeneous, and

$$(x^2y - 2xy^2)\,x + (3x^2y - x^3)\,y \neq 0$$

$$\Rightarrow x^2y^2 \neq 0$$

So let us multiply the given equation by $\dfrac{1}{x^2y^2}$ and we get

$$\left(\frac{1}{y} - \frac{2}{x}\right) dx + \left(\frac{3}{y} - \frac{x}{y^2}\right) dy = 0, \text{ which is exact. Its solution by the usual method is given by}$$

$$\frac{x}{y} + \ln \frac{y^3}{x^2} = c$$

Rule 3: If Eq. (2.4) is not exact but of the form

$$f_1(xy)y\,dx + f_2(xy)x\,dy = 0$$

and $Mx - Ny \neq 0$, then $\dfrac{1}{Mx - Ny}$ is an integrating factor.

What happens if $Mx - Ny = 0$?

Example 2.12   Solve $(xy^2 - y) \, dx - x \, dy = 0$.

This is not an exact differential equation (check!). Rewriting this, we have $(xy - 1)y \, dx - (1) \, x \, dy = 0$.

We take $I(x, y) = \dfrac{1}{(xy - 1)xy + xy} = \dfrac{1}{x^2 y^2}$.

Then we have

$$\dfrac{(xy - 1)y}{x^2 y^2} dx - \dfrac{x}{x^2 y^2} dy = 0$$

$$\Rightarrow \dfrac{xy - 1}{x^2 y} dx - \dfrac{1}{xy^2} dy = 0$$

which is exact. Check!

Its solution by usual method is $y = -\dfrac{1}{(x \ln |cx|)}$

Rule 4:   If Eq. (2.4) is not exact and

$$\dfrac{\dfrac{\partial M}{\partial y} - \dfrac{\partial N}{\partial x}}{N} = h(x)$$

is a function of x only, then an integrating factor is $e^{\int h(x) dx}$.

Example 2.13   Solve $(x^2 + y^2) \, dx - 2xy \, dy = 0$.

This is not an exact differenitial equation. However,

$$\dfrac{\dfrac{\partial M}{\partial y} - \dfrac{\partial N}{\partial x}}{N} = \dfrac{2y + 2y}{-2xy} = -\dfrac{2}{x}$$

and integrating factor is $e^{-\int (2/x) dx} = 1/x^2$.

Multiplying this on both the sides of the equation, we get

$$\dfrac{x^2 + y^2}{x^2} dx - \dfrac{2y}{x} dy = 0$$

which is exact.

Its solution by usual method is $x^2 - y^2 = cx$.

Rule 5: If Eq. (2.4) is not exact and $\dfrac{\frac{\partial N}{\partial x} - \frac{\partial M}{\partial y}}{M} = k(y)$ is a function of y only, then an integrating factor is given by $e^{\int k(y)dy}$.

Example 2.14 Solve $(xy^3 + y) dx + 2(x^2y^2 + x + y^4) dy = 0$.

This is not an exact differential equation. However

$$\frac{\frac{\partial N}{\partial x} - \frac{\partial M}{\partial y}}{M} = \frac{2(2xy^2 + 1) - (3xy^2 + 1)}{xy^3 + y} = \frac{xy^2 + 1}{y(xy^2 + 1)} = \frac{1}{y}$$

So an integrating factor by the above rule is $e^{\int (dy/y)} = y$.

Multiplying this on both the sides of the differential equation, we get $(xy^3 + y)ydx + 2y(x^2y^2 + x + y^4)dy = 0$, which is exact and by usual method its solution is given by $3x^2y^4 + 6xy^2 + 2y^6 = c$.

Rule 6: If Eq. (2.4) is not exact and can be put in the form $x^a y^b (mydx + nxdy) + x^r y^s (pydx + qxdy) = 0$, where a, b, m, n, r, s, p and q are constants, then an integrating factor is $x^h y^k$, where h and k are constants to be determined by applying the condition that after multiplication by $x^h y^k$, the equation must be exact.

Example 2.15 Solve $(3x + 2y^2) y \, dx + 2x (2x + 3y^2) \, dy = 0$.

This equation is not exact. But it can be put in the form

$$x(3y \, dx + 4x \, dy) + y^2(2y \, dx + 6x \, dy) = 0$$

By the above rule, $x^h y^k$ should be an integrating factor, and multiplying the equation by this, we get

$$(3x^{h+1}y^{k+1} + 2x^h y^{k+3}) \, dx + (4x^{h+2}y^k + 6x^{h+1}y^{k+2}) \, dy = 0$$

If this is exact, then we must have

$$3(k+1) x^{h+1}y^k + 2(k+3) x^h y^{k+2} = 4(h+2) x^{h+1}y^k + 6(h+1) x^h y^{k+2}$$

This is satisfied if $3(k + 1) = 4(h + 2)$ and $2(k + 3) = 6(h + 1)$.
Solving these, we get $h = 1, k = 3$.
So an integrating factor is xy^3; multiplying this with the given equation, we get

$$(3x + 2y^2) xy^4 dx + 2x^2y^3(2x + 3y^2) \, dy = 0$$

which is exact. Then by usual method, we get the solution

$$x^2y^4(x + y^2) = c$$

First Order Differential Equations

Exercises 2.4

Check if the following equations are exact. If an equation is exact, solve it; if not, then find an integrating factor to make it exact and then solve it.

1. $(2ax + by)y\, dx + (ax + 2by)x\, dy = 0$.

2. $x\, dx + y\, dy = \dfrac{a^2(x\, dy - y\, dx)}{x^2 + y^2}$

3. $(x^2 - ay)\, dx = (ax - y^2)\, dy$

4. $(e^y + 1)\cos x\, dx + e^y \sin x\, dy = 0$

5. $\cos x(\cos x - \sin \alpha \sin y)\, dx + \cos y(\cos y - \sin \alpha \sin x)\, dy = 0$

6. $\left\{y\left(1 + \dfrac{1}{x}\right) + \cos y\right\}dx + \{x + \ln x - x \sin y\}dy = 0$

7. $\left(1 + e^{x/y}\right)dx + e^{x/y}\left(1 - \dfrac{x}{y}\right)dy = 0$

8. $(\sin x \cos y + e^{2x})\, dx + (\cos x \sin y + \tan y)\, dy = 0$

9. $(x^2 - 2xy - y^2)\, dx - (x + y)^2\, dy = 0$

10. $ax^{a-1}y^b\, dx + bx^a y^{b-1}\, dy = 0$ for nonzero values of the real constants a and b.

11. $(y^2 e^x + 2xy)\, dx - x^2 \sin y\, dy = 0$

12. $y\, dx - x\, dy + (1 + x^2)\, dx + x^2 \sin y\, dy = 0$

13. $y(axy + e^x)\, dx - e^x\, dy = 0$

14. $(x^4 e^x - 2mxy^2)\, dx + 2mx^2 y\, dy = 0$

15. $2xy\, dx + (y^2 - x^2)\, dy = 0$

16. $y(xy + 2x^2 y^2)\, dx + x(xy - x^2 y^2)\, dy = 0$

17. $(x^2 y^2 + xy + 1)y\, dx + (x^2 y^2 - xy + 1)x\, dy = 0$

18. $(xy \sin xy + \cos xy)\, y\, dx + (xy \sin xy - \cos xy)\, x\, dy = 0$

19. $(x^3 y^3 + x^2 y^2 + xy + 1)y\, dx + (x^3 y^3 - x^2 y^2 - xy + 1)\, x\, dy = 0$

20. $y(2xy + 1)\, dx + x(1 + 2xy - x^3 y^3)\, dy = 0$

21. $(x^2 + y^2 + x)\, dx + xy\, dy = 0$

22. $(2x^3 y^2 + 4x^2 y + 2xy^2 + xy^4 + 2y)\, dx + 2(y^3 + x^2 y + x)\, dx = 0$

23. $\left(y + \dfrac{1}{3}y^3 + \dfrac{1}{2}x^2\right)dx + \dfrac{1}{4}(x + xy^2)\, dy = 0$

24. $(20x^2 + 8xy + 4y^2 + 3y^3)y\,dx + 4(x^2 + xy + y^2 + y^3)x\,dy = 0$

25. $(3x^2y^4 + 2xy)\,dx + (2x^3y^3 - x^2)\,dy = 0$

26. $(y^4 + 2y)\,dx + (xy^3 + 2y^4 - 4x)\,dy = 0$

27. $(xy^2 - x^2)\,dx + (3x^2y^2 + x^2y - 2x^3 + y^2)\,dy = 0$

28. $dy = (2xy - x)\,dx$

29. $y^2 dx + xy\,dy = 0$

30. $2xy\,dy = dx$

31. $(2xy^4 e^y + 2xy^3 + y)\,dx + (x^2 y^4 e^y - x^2 y^2 - 3x)\,dy = 0$

32. $x(2 + y^3)\,dy + 3y\,dx = 0$

33. $(2x^2 y - 3y^4)\,dx + (3x^3 + 2xy^3)\,dy = 0$

34. $(2y\,dx + 3x\,dy) + 2xy(3y\,dx + 4x\,dy) = 0$

35. $(y^2 + 2x^2 y)\,dx + (2x^3 - xy)\,dy = 0$

2.5 LINEAR EQUATIONS

We know a differential equation in which the dependent variable and its differential coefficients occur only in the first degree is called a *Linear Differential Equation* (refer to Section 1.2). Therefore, a first order linear differential equation is of the form

$$\frac{dy}{dx} + p(x)y = Q(x) \qquad (2.11)$$

where $p(x)$ and $Q(x)$ are functions of x only.

In order to solve Eq. (2.11), let us multiply both the sides by $e^{\int p\,dx}$, so that the left hand side of the new equation is an exact differential.

$$e^{\int p\,dx}\frac{dy}{dx} + p e^{\int p\,dx} y = e^{\int p\,dx} Q(x)$$

$$\Rightarrow \frac{d}{dx}\left(e^{\int p\,dx} y\right) = e^{\int p\,dx} Q(x)$$

$$\Rightarrow \int d\left(e^{\int p\,dx} y\right) = \int e^{\int p\,dx} Q(x)\,dx + c$$

$$\Rightarrow y e^{\int p\,dx} = \int e^{\int p\,dx} Q(x)\,dx + c$$

is the required solution of Eq. (2.11).

Example 2.16 Solve $(1 + x^2)\dfrac{dy}{dx} + y = e^{\tan^{-1} x}$.

Rewriting this equation, we get

$$\frac{dy}{dx} + \frac{1}{1+x^2} y = \frac{e^{\tan^{-1} x}}{1+x^2}$$

$$\int p\,dx = \int \frac{1}{1+x^2}\,dx = \tan^{-1} x$$

Thus the solution is

$$y e^{\tan^{-1} x} = \int e^{\tan^{-1} x} \frac{e^{\tan^{-1} x}}{1+x^2}\,dx + c$$

To integrate, put $\tan^{-1} x = t$. Then $\dfrac{dt}{dx} = \dfrac{1}{1+x^2}$

So

$$y e^t = \int e^{2t}\,dt + c = \frac{1}{2} e^{2t} + c$$

Therefore

$$y = \frac{1}{2} e^{\tan^{-1} x} + c e^{-\tan^{-1} x}$$

is the required solution.

Example 2.17 Solve $1 + y = (1 - y + xy)\dfrac{dy}{dx}$.

Sometimes a given differential equation becomes linear if we take x as the dependent variable and y as the independent variable. Though the equation is not linear, but it can be made linear by applying this technique. Thus, the equation can be rewritten as

$$\frac{dx}{dy} - \frac{y}{1+y} x = \frac{1-y}{1+y}$$

which is linear.

$$\int p\,dx = -\int \frac{y}{1+y}\,dy = -y + \ln(1+y)$$

$$= \ln e^{-y} + \ln(1+y)$$

$$= \ln e^{-y}(1+y)$$

So the solution is

$$x e^{\int p\,dy} = \int e^{\int p\,dy} Q\,dy + c$$

$$\Rightarrow x e^{\ln e^{-y}(1+y)} = \int e^{\ln e^{-y}(1+y)} \frac{1-y}{1+y}\,dy + c$$

$$\Rightarrow x e^{-y}(1+y) = \int e^{-y}(1-y)\,dy + c = y e^{-y} + c \quad \text{integrating by parts}$$

$$\Rightarrow x(1+y) = y + c e^{y} \quad \text{which is the required solution}$$

Exercises 2.5

Solve the following equations:

1. $(1 + y^2) \, dx = (\tan^{-1} y - x) \, dy$

2. $x \cos x \dfrac{dy}{dx} + y (x \sin x + \cos x) = 1$

3. $\sin 2x \dfrac{dy}{dx} - y = \tan x$

4. $x(x^2 + 1) \dfrac{dy}{dx} = y(1 - x^2) + x^3 \ln x$

5. $\sqrt{(a^2 + x^2)} \dfrac{dy}{dx} + y = \sqrt{(a^2 + x^2)} - x$

6. $\dfrac{dy}{dx} + \dfrac{y}{(1-x^2)^{3/2}} = \dfrac{x + \sqrt{(1-x^2)}}{(1-x^2)^2}$

7. $\sin x \dfrac{dy}{dx} + 3y = \cos x$

8. $\dfrac{dy}{dx} + y \tan x = \sec x$

9. $x(x - 1) \dfrac{dy}{dx} - y = x^2(x - 1)^2$

10. $(1 + y + x^2 y) \, dx + (x + x^3 y) \, dy = 0$

11. $x \dfrac{dy}{dx} + 2y = \sin x$

12. $\sec x \dfrac{dy}{dx} = y + \sin x$

13. $\dfrac{dy}{dx} = y \tan x - 2 \sin x$

14. $(1 + x^2) \dfrac{dy}{dx} + 2xy = \cos x$

15. $\dfrac{x \, dy}{dx} - y = 2x^2 \operatorname{cosec} 2x$

First Order Differential Equations

16. $(y - x)\dfrac{dy}{dx} = a^2$

17. $(2x - 10y^3)\dfrac{dy}{dx} + y = 0$

18. $(1 + x^2)\dfrac{dy}{dx} + 2yx - 4x^2 = 0$

19. $x(1 - x^2)\, dy + (2x^2 y - y - ax^3)\, dx = 0$

20. $\dfrac{dy}{dx} + \dfrac{y}{x} = x^2$ given that $y = 1$ when $x = 1$

21. $y' - 2xy = 2x e^{x^2}$

22. $y' \, x \ln x - y = 3x^3 \ln 2x$

23. $y' = \dfrac{y}{2y \ln y + y - x}$

24. $y' - y e^x = 2x e^x$

25. $y' + x e^x y = e^{(1-x)e^x}$

2.6 EQUATIONS REDUCIBLE TO KNOWN FORMS

We have seen, in Section 2.2, how a first order equation can be reduced to an equation in which the variables are separable (see Exercise 15) by an appropriate substitution. Further, we have also seen in Section 2.3 how a nonhomogeneous first order equation reduces to a homogeneous equation by an appropriate substitution (see Example 2.6). In this section we shall discuss certain types of differential equations, which can be reduced to one of the forms of the first order equation discussed earlier through a suitable substitution. These methods are explained through the following examples:

Example 2.18 The equation $\dfrac{dy}{dx} + P(x)y = Q(x)y^n$, known as *Bernoulli's Equation*, is linear when $n = 0$ or 1. Show that it can be reduced to a linear equation for any other value of n by the change of variable $z = y^{1-n}$.

Assume $n \neq 0$ or 1, and then divide both the sides by y^n.

$$\dfrac{1}{y^n}\dfrac{dy}{dx} + P(x)\dfrac{1}{y^{n-1}} = Q(x)$$

Now put $z = \dfrac{1}{y^{n-1}} \Rightarrow \dfrac{dz}{dx} = \dfrac{1-n}{y^n}\dfrac{dy}{dx} \Rightarrow \dfrac{1}{y^n}\dfrac{dy}{dx} = \dfrac{1}{1-n}\dfrac{dz}{dx}$

So that $$\frac{1}{1-n}\frac{dz}{dx} + P(x)z = Q(x)$$

i.e. $\frac{dz}{dx} + (1-n)P(x)z = (1-n)Q(x)$ is a linear equation and can be solved by applying the method explained in Section 2.5.

Example 2.19 Solve $x\frac{d^2y}{dx^2} + \frac{dy}{dx} = 0$.

This is a second order equation in which the variable y is missing. This can be reduced to a first order equation by substituting $y' = p$, and then $y'' = \frac{dp}{dx}$.

The equation becomes $$x\frac{dp}{dx} + p = 0$$

$$\Rightarrow \frac{dp}{dx} + \frac{1}{x}p = 0$$

which is a linear equation.
Hence the solution is

$$\frac{dy}{dx} = p = c_1 e^{-\int \frac{1}{x}dx} = c_1 e^{-\ln x} = \frac{c_1}{x}$$

$$\Rightarrow y = c_1 \ln x + c_2$$

which is the required solution.

Example 2.20 Solve $\frac{d^2y}{dx^2} + y\frac{dy}{dx} = 0$.

This is a second order equation in which the variable x is missing. This can be reduced to a first order equation by substituting $y' = p$ and then

$$\frac{d^2y}{dx^2} = \frac{dp}{dx} = \frac{dp}{dy}\frac{dy}{dx} = p\frac{dp}{dy}$$

So the equation becomes

$$p\frac{dp}{dy} + yp = 0$$

which is a first order equation in which variables can be separated.
Thus,

$$dp + y\,dy = 0 \Rightarrow p + \frac{y^2}{2} = c_1$$

$$\Rightarrow \frac{dy}{dx} + \frac{1}{2}y^2 = c_1 \Rightarrow 2dy = (k^2 - y^2)\,dx$$

Again, variables can be separated. Here
$$2c_1 = k^2$$
$$\Rightarrow \frac{2dy}{k^2 - y^2} = dx \Rightarrow \frac{2}{k}\tanh^{-1}\left(\frac{y}{k}\right) = x + c_2$$
$$\Rightarrow \frac{y}{k} = \tanh\left(\frac{kx + kc_2}{2}\right)$$
$$\Rightarrow y = k\tanh\left(\frac{kx + kc_2}{2}\right) = 2a\tanh(ax + c)$$

by renaming the constants $\frac{k}{2} = a, kc_2 = c$.

Example 2.21 Solve $f'(y)\dfrac{dy}{dx} + f(y)P(x) = Q(x)$ for $y(x)$.

Put
$$v = f(y) \Rightarrow \frac{dv}{dx} = f'(y)\frac{dy}{dx}$$
Then the differential equation reduces to
$$\frac{dv}{dx} + P(x)v = Q(x)$$
which is a linear equation with v as the dependent variable and can be solved by the method explained in Section 2.5.

Example 2.22 Solve $x \sin t\, dt + (x^3 - 2x^2 \cos t + \cos t)\, dx = 0$.

Rewriting this as $2 \cos t\, dx - \dfrac{x\sin t\, dt + \cos t\, dx}{x^2} = x\, dx$ and substituting $xy = \cos t$, $dy = -\dfrac{x\sin t\, dt + \cos t\, dx}{x^2}$, we have
$$dy + 2xy\, dx = x\, dx$$
$$\Rightarrow \frac{dy}{dx} + 2xy = x$$
which is a linear equation. Its solution by the method shown in Section 2.5 is
$$ye^{x^2} = \frac{\cos t}{x}e^{x^2} = \int e^{x^2} x\, dx = \frac{1}{2}e^{x^2} + c$$
i.e. $2\cos t = x + 2cxe^{-x^2}$ is the required solution.

Exercises 2.6

Solve the following equations:

1. $\dfrac{dy}{dx} + \dfrac{y}{x} = y^2$

2. $2\dfrac{dy}{dx} = \dfrac{y}{x} + \dfrac{y^2}{x^2}$

Introduction to Differential Equations

3. $(1-x^2)\dfrac{dy}{dx} + xy = xy^2$

4. $x\dfrac{dy}{dx} + y = y^2 \ln x$

5. $\dfrac{dy}{dx} + \dfrac{y}{x} = y^2 \sin x$

6. $(1+x^2)\dfrac{dy}{dx} = xy - y^2$

7. $2\dfrac{dy}{dx} - y \sec x = y^3 \tan x$

8. $x^2 y - x^3 \dfrac{dy}{dx} = y^4 \cos x$

9. $y(2xy + e^x)dx - e^x dy = 0$

10. $\cos x\, dy = y(\sin x - y)\, dx$

11. $\dfrac{d^2 y}{dx^2} + \dfrac{dy}{dx} = 0$

12. $\dfrac{d^2 y}{dx^2} + x\dfrac{dy}{dx} = 0$

13. $\dfrac{d^2 y}{dx^2} - y = 0$

14. $\dfrac{d^2 y}{dx^2} + a^2 y = 0$ where $a \neq 0$ is a constant

15. $y\dfrac{d^2 y}{dx^2} = \left(\dfrac{dy}{dx}\right)^2$

16. $y\dfrac{d^2 y}{dx^2} + \left(\dfrac{dy}{dx}\right)^2 - 2y\dfrac{dy}{dx} = 0$

17. $\dfrac{d^2 y}{dx^2} + 2x\left(\dfrac{dy}{dx}\right)^2 = 0$

First Order Differential Equations

18. $x^2 \dfrac{d^2 y}{dx^2} + \dfrac{x \, dy}{dx} = 1$

19. $(1 + x^2) \dfrac{d^2 y}{dx^2} + x \dfrac{dy}{dx} = 0$

20. $\dfrac{x \, d^2 y}{dx^2} - \dfrac{dy}{dx} = 3x^2$

21. $x \dfrac{dy}{dx} + y = y^2 \ln x$

22. $y \dfrac{dy}{dx} + \dfrac{x}{2} \sin 2y = x^3 \cos^2 y$

    (**Hint:** Divide by $\cos^2 y$.)

23. $\dfrac{dy}{dx} + \dfrac{1}{x} = \dfrac{e^y}{x^2}$

    (**Hint:** Divide by e^y.)

24. $\dfrac{dy}{dx} + \dfrac{y}{x} \ln y = \dfrac{y}{x^2} (\ln y)^2$

    (**Hint:** Divide by $y (\ln y)^2$.)

25. $\sin y \dfrac{dy}{dx} = \cos y \, (1 - x \cos y)$

CHAPTER 3

Linear Differential Equations of Higher Order

3.1 INTRODUCTION

So far we have discussed different methods of solving a few restricted types of first order differential equations and those second order differential equations which can be reduced to first order equations. But in practice we actually need to solve other types of differential equations. The simplest problem of the motion of a particle along a straight line, the motion of a pendulum, the problems of electrical circuits and of mechanical oscillations, etc. are governed by differential equations of second order. The problem of determining curves of a given perimeter which, under given conditions, enclose a maximum area (i.e. isoperimetric problem), depends on a differential equation of third order. The trajectories of rocket launching lead to differential equations of higher order depending on the conditions imposed. Therefore in this chapter we will discuss different methods of solving higher order differential equations of a very special class called *Linear Differential Equations* (see Eq. (1.15)). In order to discuss different methods of solving linear differential equations, we shall consider only the second order linear differential equations. The same method works for third and higher order linear differential equations.

3.2 SECOND ORDER LINEAR DIFFERENTIAL EQUATIONS

The general second order linear differential equation is given by

$$a_2(x)\frac{d^2y}{dx^2} + a_1(x)\frac{dy}{dx} + a_0(x)y = g(x) \tag{3.1}$$

where $a_2(x)$, $a_1(x)$, $a_0(x)$ and $g(x)$ are continuous real-valued functions defined on an interval I, and y is a twice-differentiable function defined on I.

If $g(x) = 0$ for all $x \in I$, then Eq. (3.1) reduces to the homogeneous equation

$$a_2(x)\frac{d^2y}{dx^2} + a_1(x)\frac{dy}{dx} + a_0(x)y = 0 \tag{3.2}$$

which is called the *associated homogeneous equation* of (3.1).

Equation (3.2) will be used to find the general solution of the equation (3.1). For, if y_c is the general solution of the equation (3.2), y_p is a particular solution of Eq. (3.1), and if y is the general solution of the Eq. (3.1), then $y - y_p$ is a solution of Eq. (3.2). This is because

$$a_2(x)(y-y_p)'' + a_1(x)(y-y_p)' + a_0(x)(y-y_p)$$
$$= \{a_2(x)y'' + a_1(x)y' + a_0(x)y\} - \{a_2(x)y_p'' + a_1(x)y_p' + a_0(x)y_p\} = g(x) - g(x) = 0$$

Thus we have $y - y_p = y_c \Rightarrow y = y_c + y_p$ and we have proved the following.

Theorem 3.1 If y_c is the general solution of Eq. (3.2) and y_p is a particular solution of Eq. (3.1), then the general solution y of Eq. (3.1) is given by

$$y = y_c + y_p$$

We shall see later that if y_c is known, then a formal procedure is available to find y_p. Thus the central problem in finding the general solution of Eq. (3.1) lies in finding the general solution of Eq. (3.2). Because of this reason, Eq. (3.2) is called the *associated homogeneous equation* and its general solution is called the *complementary function* of the nonhomogeneous equation (3.1).

Next the natural equation is, "Does Eq. (3.1) always have a solution? If no, then under what condition(s) does it have a solution? If yes, then is it unique?" The following theorem answers all these questions. We shall not prove this theorem as the proof is beyond the scope of this book.

Theorem 3.2 If $x_0 \in I$ is a given point and $y_0, y_1, ..., y_{n-1}$ are given numbers, then Eq. (3.1) has a unique solution $y(x)$, provided (3.1) is a normal equation, such that

$$y(x_0) = y_0, \ y'(x_0) = y_1, \ ..., \ y^{(n-1)}(x_0) = y_{n-1}$$

This theorem guarantees the existence of a solution for a normal differential equation (3.1) and assures a unique solution for the initial value problem

$$a_2(x)\frac{d^2y}{dx^2} + a_1(x)\frac{dy}{dx} + a_0(x)y = g(x) \qquad a_2(x) \neq 0 \ \forall \in I$$

$$y(x_0) = y_0 \qquad y'(x_0) = y_1 \qquad (3.3)$$

Let us understand the concepts in this theorem through the following examples:

Example 3.1 Show that $y = \sin x$ is a unique solution of the initial value problem $y'' + y = 0, y(0) = 0, y'(0) = 1$.

Here $a_2(x) = 1$, which is not zero in any interval, so the equation is a normal equation with continuous coefficients. Also, $x_0 = 0, y_0 = 0, y_1 = 1$ are given numbers.

We have $y = \sin x$, therefore $y'' = -\sin x$ and hence $y'' + y = 0$. Further, $y(0) = \sin 0 = 0, y'(0) = \cos 0 = 1$.

Thus $y = \sin x$ is a solution of the given initial value problem. Moreover, the given differential equation is a linear, normal differential equation with continuous coefficients in any interval containing $x = 0$. Hence, from Theorem 3.2, we conclude that the given function is the unique solution of the given initial value problem.

Example 3.2 Show that $y = x^2 \sin x$ and $y = 0$ are both solutions of the differential equation $x^2 y'' - 4xy' + (x^2 + 6)y = 0$, and that both satisfy the conditions $y(0) = 0$, and $y'(0) = 0$. Does this contradict Theorem 3.2? Justify your answer. $y = 0$ is a solution of the given initial value problem and so also $y = x^2 \sin x$. Check!

No, this does not contradict Theorem 3.2, because the given differential equation is not a normal differential equation in any interval containing the origin. Normality is a very impoortant requirement for an initial value problem to have a unique solution.

One thing is clear about the homogeneous equation (3.2), that the function $y(x)$ which is identically zero, that is, $y(x) = 0$ for all x, is always a solution. This is called the *trivial solution* and is usually of no interest. The basic facts about solutions of (3.2) are given in the following theorems.

Theorem 3.3 If Eq. (3.2) is normal on I, then the set of all its solutions form a two-dimensional subspace of $C^{(2)}(I)$.

Proof: Suppose x_0 is a fixed point in I. Then, by Theorem 3.2, we know that Eq. (3.2) has solutions $y_1(x)$ and $y_2(x)$, which respectively satisfy the initial conditions

$$y_1(x_0) = 1 \qquad y_1'(x_0) = 0$$
$$y_2(x_0) = 0 \qquad y_2'(x_0) = 0 \qquad (3.4)$$

We assert that these solutions form a basis for the solution space of (3.2).

Let $c_1 y_1(x) + c_2 y_2(x) = 0$ on I for some real numbers c_1, c_2. Then the system

$$c_1 y_1(x) + c_2 y_2(x) = 0$$
$$c_1 y_1'(x) + c_2 y_2'(x) = 0$$

imply that at $x = x_0$.

$$c_1 y_1(x_0) + c_2 y_2(x_0) = 0 \qquad \Rightarrow c_1 = 0$$
$$c_1 y_1'(x_0) + c_2 y_2'(x_0) = 0 \qquad \Rightarrow c_2 = 0$$

due to the initial conditions given in Eq. (3.4).

Thus $y_1(x)$ and $y_2(x)$ are linearly independent in $C^{(2)}(I)$. Next we show that every solution of Eq. (3.2) can be a linear combination of $y_1(x)$ and $y_2(x)$.

Let $y(x)$ be an arbitrary solution of Eq. (3.2) and $y(x_0) = b_0$ and $y'(x_0) = b_1$, where b_0, b_1 are real numbers. Then $y(x)$ is a unique solution of Eq. (3.2) by Theorem 3.2. But using the initial conditions in Eq. (3.4), we find that $b_0 y_1(x) + b_1 y_2(x)$ is also a solution of the initial value problem. Hence

$$y(x) = b_0 y_1(x) + b_1 y_2(x)$$

Thus $y_1(x), y_2(x)$ span the solution space of Eq. (3.2). This completes the proof.

This theorem leads to the following theorem, whose proof directly follows from the knowledge of Linear Algebra.

Theorem 3.4 If Eq. (3.2) is normal on I and $y_1(x)$ and $y_2(x)$ are two linearly independent solutions of (3.2), then

$$y_c(x) = c_1 y_1(x) + c_2 y_2(x)$$

Theorem 3.5 Suppose Eq. (3.2) is normal on I. Then a set of solutions $y_1(x)$ and $y_2(x)$ of Eq. (3.2) is linearly independent iff

$$W[y_1(x), y_2(x)] \neq 0 \quad \text{in } I.$$

where $W[y_1(x), y_2(x)]$ is the Wronskian of $y_1(x)$, $y_2(x)$.

Proof: We first assume that $y_1(x)$ and $y_2(x)$ are nontrivial solutions of Eq. (3.2). Then we show that either $W[y_1(x), y_2(x)] = 0$ for all $x \in I$ or the Wronskian is never zero on I.

$$W[y_1(x), y_2(x)] = \begin{vmatrix} y_1(x) & y_2(x) \\ y_1'(x) & y_2'(x) \end{vmatrix}$$

$$\Rightarrow W' = y_1(x) y_2''(x) - y_1''(x) y_2(x)$$

Since both $y_1(x)$ and $y_2(x)$ are solutions of Eq. (3.2), we have

$$a_2(x) y_1''(x) + a_1(x) y_1'(x) + a_0(x) y_1(x) = 0$$
$$a_2(x) y_2''(x) + a_1(x) y_2'(x) + a_0(x) y_2(x) = 0$$

Multiplying the first equation by $y_2(x)$ and the second equation by $y_1(x)$ and subtracting, we get

$$a_2(x)(y_1''(x) y_2(x) - y_1(x) y_2''(x)) + a_1(x)(y_1'(x) y_2(x) - y_1(x) y_2'(x)) = 0$$

$$\Rightarrow W'(x) + \frac{a_1(x)}{a_2(x)} W(x) = 0$$

$$\Rightarrow W(x) = c e^{-\int \frac{a_1}{a_2} dx}$$

where c is the constant of integration.

Since the exponential function never vanishes we find $W(x) = 0$ if $c = 0$, otherwise $W(x) \neq 0$ in I.

Now $W(x) = 0$

$\Leftrightarrow y_1(x) y_2'(x) - y_1'(x) y_2(x) = 0$

$\Leftrightarrow \dfrac{y_1(x) y_2'(x) - y_1'(x) y_2(x)}{y_1^2(x)} = 0 \quad$ since $y_1(x)$ is a nontrivial solution

$\Leftrightarrow \left(\dfrac{y_2(x)}{y_1(x)}\right)' = 0$

$\Leftrightarrow y_2(x) = k y_1(x)$

for some constant k and for all $x \in I$. (Here $y_1(x)$ may not be different from zero throughout I; but if $y_1(x) \neq 0$ for all $x \in (c, d) \in I$, then $y_2(x) = ky_1(x)$ for all $x \in (c, d)$, and by Theorem 3.2 we shall have $y_2(x) = ky_1(x)$ for all $x \in I$.)

Thus $W(x) = 0$ iff $y_1(x)$, $y_2(x)$ are linearly dependent. Hence the theorem follows.

An immediate consequence of this theorem is the following corollary.

Corollary 3.1 Suppose Eq. (3.2) is normal on I and $y_1(x)$, $y_2(x)$ are two solutions of Eq. (3.2). Then $y_1(x)$, $y_2(x)$ are linearly dependent iff $W(x_0) = 0$ for some $x_0 \in I$.

Since the solutions of a normal linear differential equation always exist, let us now discuss the methods of solving such equations. First we shall discuss the methods of solving homogeneous linear differential equations with constant coefficients, then we shall discuss the methods of solving homogeneous linear differential equations with variable coefficients, and finally we shall discuss the methods of solving nonhomogeneous linear differential equations.

3.3 METHOD OF SOLVING HOMOGENEOUS EQUATIONS WITH CONSTANT COEFFICIENTS

We shall now consider Eq. (3.2) with constant coefficients, i.e., the coefficients $a_2(x) = a_2$, $a_1(x) = a_1$, $a_0(x) = a_0$ are all constants and $a_2 \neq 0$. Thus, a second order normal homogeneous linear differential equation with constant coefficient is

$$a_2 \frac{d^2 y}{dx^2} + a_1 \frac{dy}{dx} + a_0 y = 0 \qquad a_2 \neq 0 \qquad a_2, a_1, a_0 \text{ are constants} \qquad (3.5)$$

Suppose $y = e^{mx}$ is a solution of Eq. (3.5), where m is a constant. The reason for making such an assumption is that the derivatives of this function are only constant multiples of this function, No other elementary function has such a property. Further, we have also seen that the linear first order equation

$$\frac{dy}{dx} + ay = 0$$

where a is a constant, has the solution $y = c_1 e^{-ax}$ on $(-\infty, +\infty)$. Therefore, it is natural to expect whether exponential solution exist on $(-\infty, +\infty)$ for higher order equations. Putting this in Eq. (3.5), we have

$$a_2 m^2 e^{mx} + a_1 m e^{mx} + a_0 e^{mx} = 0$$

$$\Rightarrow (a_2 m^2 + a_1 m + a_0) e^{mx} = 0$$

$$\Rightarrow a_2 m^2 + a_1 m + a_0 = 0 \qquad \text{since } e^{mx} \neq 0 \qquad (3.6)$$

Equation (3.6) is called the *auxiliary equation* and is obtained from (3.5) by replacing $d/dx = D$ by the constant m and suppressing y. Let the two values of m in the equation (3.6) be m_1 and m_2. Then there are three possibilities.

Case 1: If $m_1 \neq m_2$ and real, then Eq. (3.6) is equivalent to $(m - m_1)(m - m_2) = 0$ and Eq. (3.5) is equivalent to

$$(D - m_1)(D - m_2)y = 0 \qquad (3.7)$$

If we put
$$(D - m_2)y = u \qquad (3.8)$$

Then (3.7) becomes
$$(D - m_1)u = 0 \qquad (3.9)$$

which is a first order equation in which variables can be separated i.e.

$$\frac{du}{dx} - m_1 u = 0$$

$$\Rightarrow \frac{du}{u} = m_1 dx \Rightarrow \ln u = m_1 + \ln c_1$$

$$\Rightarrow u = c_1 e^{m_1 x}$$

The constant of integration can be taken as $\ln c_1$.
Substituting this value of u in (3.8), we get

$$(D - m_2)y = c_1 e^{m_1 x}$$

$$\Rightarrow \frac{dy}{dx} - m_2 y = c_1 e^{m_1 x}$$

which is a linear equation.

$$y e^{-\int m_2 dx} = \int e^{-\int m_2 dx} c_1 e^{m_1 x} dx + c_2 \qquad (3.10)$$

$$\Rightarrow y = \frac{c_1 e^{m_1 x}}{m_1 - m_2} + c_2 e^{m_2 x}$$

Since c_1 is an arbitrary constant, $\dfrac{c_1}{m_1 - m_2}$ is also an arbitrary constant and therefore the solution of Eq. (3.5) can be simply written as

$$y = c_1 e^{m_1 x} + c_2 e^{m_2 x} \quad \text{if } m_1 \neq m_2 \qquad (3.11)$$

Case 2: If $m_1 = m_2$, then the integrand on the right hand side of Eq. (3.10) is c_1. So (3.10) becomes

$$y = (c_1 x + c_2) e^{m_1 x} \qquad (3.12)$$

Case 3: Suppose that m_1 and m_2 are complex, i.e., say $m_1 = \alpha + i\beta$ and $m_2 = \alpha - i\beta$. Since $m_1 \neq m_2$, we apply Eq. (3.11) to get $y = c_1 e^{(\alpha + i\beta)x} + c_2 e^{(\alpha - i\beta)x}$, which is the solution in this case,

i.e.
$$y = e^{\alpha x}\left(c_1 e^{i\beta x} + c_2 e^{-i\beta x}\right)$$

$$= e^{\alpha x}\{c_1(\cos \beta x + i \sin \beta x) + c_2(\cos \beta x - i \sin \beta x)\}$$

$$= e^{\alpha x}\{(c_1 + c_2)\cos \beta x - i(c_1 - c_2)\sin \beta x\}$$

$$\Rightarrow y = e^{\alpha x}(k_1 \cos \beta x + k_2 \sin \beta x) \qquad (3.13)$$

Introduction to Differential Equations

where $k_1 = c_1 + c_2$ and $k_2 = i(c_1 - c_2)$ are arbitrary constants (complex), since c_1 and c_2 are arbitrary constants (real).

Example 3.3 Solve $y''' - 3y'' + 4y = 0$.

The auxiliary equation is
$$m^3 - 3m^2 + 4 = 0$$
$$\Rightarrow m^3 - 3m^2 + 4m - 4m + 4 = 0$$
$$\Rightarrow m^3 - 4m^2 + m^2 + 4m - 4m + 4 = 0$$
$$\Rightarrow (m^3 - 4m^2 + 4m) + (m^2 - 4m + 4) = 0$$
$$\Rightarrow m(m^2 - 4m + 4) + (m^2 - 4m + 4) = 0$$
$$\Rightarrow (m + 1)(m - 2)^2 = 0$$

Thus the roots are $m_1 = -1$, $m_2 = m_3 = 2$. Since the first root is real and distinct from others, gives $c_1 e^{-x}$ as part of the solution. Further, since the second and third roots are repeated, $(c_2 x + c_3)e^{2x}$ is part of the solution. The complete solution is then given by $y = c_1 e^{-x} + (c_2 x + c_3)e^{2x}$.

Example 3.4 Solve the equation $y'' - 6y' + 25y = 0$.

The auxiliary equation is $m^2 - 6m + 25 = 0 \Rightarrow m = 3 \pm i4$. So the solution of the given equation is $y(x) = e^{3x}(c_1 \cos 4x + c_2 \sin 4x)$.

Example 3.5 Solve $y''' + 5y'' + 7y' - 13 = 0$.

The auxiliary equation is
$$m^3 + 5m^2 + 7m - 13 = 0$$
$$\Rightarrow m^3 + 6m^2 + 13m - m^2 - 6m - 13 = 0$$
$$\Rightarrow (m - 1)(m^2 + 6m + 13) = 0$$

Thus the roots are $m_1 = 1$, $m_2 = -3 + 2i$, $m_3 = -3 - 2i$.

Since the first root is real and distinct from all others, gives $c_1 e^x$ as a part of the solution. Further, since other two roots are complex, $e^{-3x}(c_2 \cos 2x + c_3 \sin 2x)$ is part of the solution. The complete solution is given by
$$y = c_1 e^x + e^{-3x}(c_2 \cos 2x + c_3 \sin 2x)$$

Exercises 3.3

Solve the following equations:

1. $y'' - y' - 2y = 0$
2. $y'' + y' - 6y = 0$
3. $2y'' - 5y' + 2y = 0$
4. $y''' - 6y'' + 11y' - 6y = 0$

5. $(D^4 - 13D^2 + 36)y = 0$
6. $y'' - 6y' + 25y = 0$
7. $y'' + 9y = 0$
8. $y'' + 8y' + 25y = 0$
9. $y'' + 4y' + 8y = 0$
10. $y'' - 3y' + 4y = 0$
11. $y''' + y' = 0$
12. $y''' + 4y' = 0$
13. $(D^4 - 16)y = 0$
14. $y'' + 4y' + 4y = 0$
15. $(D^2 + 30D + 225)y = 0$
16. $y''' + 4y'' - 3y' - 18y = 0$
17. $(D^4 + 6D^3 + 5D^2 - 24D - 36)y = 0$
18. $(D^4 + 4D^2)y = 0$
19. $9y'' + 18y' - 16y = 0$
20. $y'' - 2y' - 3y = 0$ when $x = 0$ $y = 4$ and $y' = 0$

3.4 METHOD OF SOLVING HOMOGENEOUS EQUATIONS WITH VARIABLE COEFFICIENTS

In this section we shall discuss the methods of solving equations of the form (3.2) with $a_2(x) \neq 0$ for all $x \in I$. Thus Eq. (3.2) now can be written as

$$y'' + \frac{a_1(x)}{a_2(x)} y' + \frac{a_0(x)}{a_2(x)} y = 0$$

$$\Rightarrow y'' + P(x)y' + Q(x)y = 0 \qquad (3.14)$$

where $P(x) = \dfrac{a_1(x)}{a_2(x)}$ and $Q(x) = \dfrac{a_0(x)}{a_2(x)}$ are continuous functions of x on I.

If $y_1(x)$ is a solution of (3.14), then we shall find a second solution $y_2(x)$ of (3.14) such that $y_1(x)$ and $y_2(x)$ are linearly independent on I.

Let $y_2(x) = v(x)y_1(x)$, where $v(x)$ is an unknown function to be determined.

Thus $y'(x) = v'(x)y_1(x) + v(x)y_1'(x)$

and $y_2''(x) = v''(x)y_1(x) + 2v'(x)y_1'(x) + v(x)y_1''(x)$

Putting all these in Eq. (3.14), we have

$v''(x)y_1(x) + 2v'(x)y_1'(x) + v(x)y_1''(x) + P(x)v'(x)y_1(x) + P(x)v(x)y_1'(x)$
$+ Q(x)v(x)y_1(x) = 0$

$\Rightarrow v''(x)y_1(x) + (2y_1'(x) + P(x)y_1(x))v'(x) + (y_1''(x) + P(x)y_1'(x) + Q(x)y_1(x))v(x) = 0$

$\Rightarrow v''(x)y_1(x) + (2y_1'(x) + P(x)y_1(x))v'(x) = 0$ since $y_1(x)$ is solution of Eq. (3.14)

$\Rightarrow \dfrac{v''(x)}{v'(x)} + \dfrac{2y_1'(x)}{y_1(x)} + P(x) = 0$

$\Rightarrow \ln v'(x) + \ln y_1^2(x) + \int P(x)\,dx = 0$ integrating

Here the constant of integration is not taken since we are looking for one function $v(x)$.

$\Rightarrow \ln v'(x)y_1^2(x) + \int P(x)\,dx = 0$

$\Rightarrow v(x) = \int \dfrac{e^{-\int P(x)dx}}{y_1^2(x)}\,dx$

Thus $y_2(x) = v(x)y_1(x) = y_1(x)\displaystyle\int \dfrac{e^{-\int P(x)dx}}{y_1^2(x)}\,dx$ (3.15)

Are $y_1(x)$ and $y_2(x)$ actually linearly independent? Let us check by finding their Wronskian

$W[y_1(x), y_2(x)] = \begin{vmatrix} y_1(x) & y_2(x) \\ y_1'(x) & y_2'(x) \end{vmatrix} = \begin{vmatrix} y_1(x) & y_1(x)\int \dfrac{e^{-\int P(x)dx}}{y_1^2(x)}dx \\ y_1'(x) & y_1'(x)\int \dfrac{e^{-\int P(x)dx}}{y_1^2(x)}dx + \dfrac{e^{-\int P(x)dx}}{y_1(x)} \end{vmatrix}$

$= e^{-\int P(x)dx}$

which being an exponential function never varishes in any interval.
Thus, by Theorem 3.5, $y_1(x)$ and $y_2(x)$ are linearly independent solutions of (3.14). Consequently the general solution of (3.14) is given by

$$y(x) = c_1 y_1(x) + c_2 y_2(x) \qquad (3.16)$$

where c_1 and c_2 are two arbitrary constants.

The next question is how to find $y_1(x)$ for (3.14)? There are no standard methods available to find a solution for a given equation (3.14). However, the following are some tips to find a solution. If this works, then $y_1(x)$ for Eq. (3.14) is known, otherwise trial and error method is the only way to obtain a solution $y_1(x)$ for (3.14).

Rule 1: If $m(m-1) + mx P(x) + x^2 Q(x) = 0$ for some real number m, then $y_1(x) = x^m$ is a solution for Eq. (3.14).

Rule 2: If $a^2 + aP(x) + Q(x) = 0$ for some real number a, then $y_1(x) = e^{ax}$ is a solution for Eq. (3.14). Note that if $1 + P(x) + Q(x) = 0$ (i.e. the sum of the coefficients of y'', y', y is zero), then $y_1(x) = e^x$ is a solution of Eq. (3.14).

Linear Differential Equations of Higher Order   47

Example 3.6 Solve the equation $xy'' - (2x + 1)y' + (x + 1)y = 0$ in any interval I not containing the origin.

The equation can be reduced to

$$y'' - \frac{2x+1}{x}y' + \frac{(x+1)}{x}y = 0$$

since $x \neq 0$ in I.

Here $P(x) = -\frac{2x+1}{x}$, $Q(x) = \frac{x+1}{x}$ and $1 + P(x) + Q(x) = 0$.

Hence as per Rule 2, $y_1(x) = e^x$ is a solution of the given differential equation.

$$y_2(x) = y_1(x) \int \frac{e^{-\int P(x)dx}}{y_1^2(x)} dx = e^x \int \frac{e^{\int \frac{2x+1}{x}dx}}{e^{2x}} dx = e^x \int \frac{e^{2x}x}{e^{2x}} dx = \frac{x^2 e^x}{2}$$

So the general solution of the equation is given by

$$y(x) = c_1 y_1(x) + c_2 y_2(x) = c_1 e^x + c_2 \frac{x^2 e^x}{2}$$

Example 3.7 Solve the differential equation $(x^2 - 1)y'' - 2xy' + 2y = 0$ in the interval $I = (-1, 1)$.

Here $P(x) = \frac{-2x}{x^2-1}$, $Q(x) = \frac{2}{x^2-1}$ and

$$m(m-1) + mx P(x) + x^2 Q(x) = m(m-1) + mx\left(\frac{-2x}{x^2-1}\right) + x^2\left(\frac{2}{x^2-1}\right) = 0$$

for $m = 1$

So by Rule 1, $y_1(x) = x$ is a solution of the equation.

$$y_2(x) = y_1(x) \int \frac{e^{-\int P(x)dx}}{y_1^2(x)} dx = x \int \frac{e^{\int \frac{2xdx}{x^2-1}}}{x^2} dx = x \int \frac{e^{\ln(x^2-1)}}{x^2} dx$$

$$= x \int \frac{x^2-1}{x^2} dx = x\left(x + \frac{1}{x}\right) = x^2 + 1$$

So the general solution of the equation is given by

$$y(x) = c_1 y_1(x) + c_2 y_2(x) = c_1 x + c_2 (x^2 + 1)$$

Example 3.8 Solve the following differential equation in the interval $I = (0, +\infty)$:

$$x^2 y'' + x^3 y' - 2(1 + x^2)y = 0$$

48 Introduction to Differential Equations

The given equation can be rewritten as

$$y'' + \frac{x^3}{x^2}y' - 2\left(\frac{1+x^2}{x^2}\right)y = 0$$

Then

$$m(m-1) + mx(x) + x^2\left(\frac{-2(1+x^2)}{x^2}\right) = 0 \quad \text{if } m = 2$$

Thus $y_1(x) = x^2$ is a solution of the given equation.

$$y_2(x) = y_1(x)\int \frac{e^{-\int P(x)dx}}{y_1^2(x)}dx = x^2\int \frac{e^{-\int x\,dx}}{x^2}dx = x^2\int \frac{e^{(-x^2/2)}}{x^2}dx$$

Hence the general solution of the given equation is

$$y(x) = c_1 x^2 + c_2 x^2 \int \frac{e^{(-x^2/2)}}{x^2}dx$$

The following example illustrates the method of solving a class of homogeneous equations with variable coefficients, called either *Euler's equations, Cauchy's equations* or *equidimensional equations*, by making an appropriate substitution. An equation of the form

$$a_n x^n \frac{d^n y}{dx^n} + a_{n-1}x^{n-1}\frac{d^{n-1}y}{dx^{n-1}} + \cdots + a_1 x \frac{dy}{dx} + a_0 y = 0 \quad x > 0 \quad (3.17)$$

where $a_n \neq 0, a_{n-1}, \ldots, a_0$ are constants, is called either a *Euler's equation, Cauchy's equation*, or *equidimensional equation*.

If we put $z = \ln x$, then Eq. (3.17) will be reduced to an equation with constant coefficients. This is explained in the following example.

Example 3.9 Solve the Euler equation $a_3 x^3 y''' + a_2 x^2 y'' + a_1 xy' + a_0 y = 0$, where a_3, a_2, a_1, a_0 are constants and $x > 0$.

This is not a linear homogeneous equation with constant coefficients. However, with a suitable substitution, this can be reduced to one with constant coefficients.

Put $z = \ln x$, so $x = e^z$. Then by chain rule

$$\frac{dy}{dx} = \frac{dy}{dz}\cdot\frac{dz}{dx} = \frac{dy}{dz}\cdot\frac{1}{x}$$

$$\Rightarrow x\frac{dy}{dx} = \frac{dy}{dz}$$

Linear Differential Equations of Higher Order

$$\frac{d^2y}{dx^2} = \frac{d}{dx}\left(\frac{1}{x}\frac{dy}{dz}\right) = -\frac{1}{x^2}\frac{dy}{dz} + \frac{1}{x^2}\frac{d^2y}{dz^2}$$

$$\Rightarrow x^2\frac{d^2y}{dx^2} = \frac{d^2y}{dz^2} - \frac{dy}{dz} = \frac{d}{dz}\left(\frac{d}{dz}-1\right)y$$

Similarly, we can find that

$$x^3\frac{d^3y}{dx^3} = \frac{d}{dz}\left(\frac{d}{dz}-1\right)\left(\frac{d}{dz}-2\right)y$$

Thus the given equation reduces to

$$\left\{a_3\frac{d}{dz}\left(\frac{d}{dz}-1\right)\left(\frac{d}{dz}-2\right) + a_2\frac{d}{dz}\left(\frac{d}{dz}-1\right) + a_1\frac{d}{dz} + a_0\right\}y = 0$$

$$\Rightarrow \left\{a_3D^3 + (a_2 - 3a_3)D^2 + (a_1 - a_2 + 2a_3)D + a_0\right\}y = 0$$

where $D = \frac{d}{dz}$; this is a linear differential equation with constant coefficients.

If we take $a_3 = 1$, $a_2 = 2$, $a_1 = 4$ and $a_0 = -4$, then we have the Euler equation $x^3y''' + 2x^2y'' + 4xy' - 4y = 0$, which reduces by the above method to $(D^3 - D^2 + 4D - 4)y = 0$.

The auxiliary equation is $m^3 - m^2 + 4m - 4 = 0$, i.e., $(m^2 + 4)(m - 1) = 0$, whose roots are $m_1 = 1$, $m_2 = 2i$, $m_3 = -2i$.

So the solution of the linear equation with constant coefficient is $y = c_1e^z + c_2 \cos 2z + c_3 \sin 2z$.

Hence the solution of the required equation is

$$y = c_1x + c_2 \cos(2 \ln x) + c_3 \sin(2 \ln x)$$

$$\Rightarrow y = c_1x + c_2 \cos(\ln x^2) + c_3 \sin(\ln x^2)$$

Example 3.10 Show that the equation $y'' + P(x)y' + Q(x)y = 0$ where $Q(x) > 0$, can be reduced to a differential equation with constant coefficients by substituting

$$z(x) = \int\sqrt{Q(x)}\,dx \text{ iff } \frac{Q'(x) + 2P(x)Q(x)}{(Q(x))^{3/2}} \text{ is constant.}$$

We have

$$z(x) = \int\sqrt{Q(x)}\,dx \Rightarrow \frac{dz}{dx} = \sqrt{Q(x)}$$

$$\frac{dy}{dx} = \frac{dy}{dz}\cdot\frac{dz}{dx} = \sqrt{Q(x)}\,\frac{dy}{dz}$$

$$\frac{d^2y}{dx^2} = \frac{d}{dx}\left(\frac{dy}{dx}\right)$$

$$= \frac{d}{dx}\left(\sqrt{Q(x)}\frac{dy}{dz}\right)$$

$$= \frac{d\sqrt{Q(x)}}{dx}\frac{dy}{dz} + \sqrt{Q(x)}\frac{d^2y}{dz^2}\cdot\frac{dz}{dx}$$

$$= \frac{Q'(x)}{2\sqrt{Q(x)}}\frac{dy}{dz} + Q(x)\frac{d^2y}{dz^2}$$

Putting these values in the given differential equation, we obtain

$$Q(x)\frac{d^2y}{dz^2} + \frac{Q'(x)}{2\sqrt{Q(x)}}\frac{dy}{dz} + P(x)\sqrt{Q(x)}\frac{dy}{dz} + Q(x)y = 0$$

$$\Rightarrow \frac{d^2y}{dz^2} + \left(\frac{Q'(x) + 2P(x)Q(x)}{2(Q(x))^{3/2}}\right)\frac{dy}{dz} + y = 0$$

which is an equation with constants coefficients iff $\dfrac{Q'(x) + 2P(x)Q(x)}{(Q(x))^{3/2}}$ is constant.

Example 3.11 Solve $(1 - x^2)y'' - xy' + k^2y = 0$ for $-1 < x < 1$.

This equation can be rewritten as $y'' - \dfrac{x}{1-x^2}y' + \dfrac{k^2}{1-x^2}y = 0$.

Here $Q(x) = \dfrac{k^2}{1-x^2} > 0$, and $\dfrac{Q'(x) + 2P(x)Q(x)}{(Q(x))^{3/2}} = 0$ is a constant.

Thus, due to Example 3.10, this equation will be reduced to an equation with constant coefficients if we make $z(x) = \displaystyle\int \dfrac{k\,dx}{\sqrt{1-x^2}} = k\sin^{-1}x$. So the given equation reduces to $\dfrac{d^2y}{dz^2} + y = 0$, whose solution is given by

$$y(z) = c_1 \cos z + c_2 \sin z \Rightarrow y(x) = c_1 \cos(k\sin^{-1}x) + c_2 \sin(k\sin^{-1}x)$$

is the required solution.

Exercises 3.4

Solve the following equations:

1. $x^2y'' + 2xy' - 2y = 0$ in $I = (-\infty, 0) \cup (0, +\infty)$
2. $x^2y'' - x(x+2)y' + (x+2)y = 0$ in $I = (0, +\infty)$

3. $(x-1)y'' - xy' + y = 0$ for $x > 1$
4. $(1-x^2)y'' - 2xy' + 2y = 0$ for $I = (-1, 1)$
5. $x^2y'' + xy' - 4y = 0$ for $x > 0$
6. $x^2y'' - 2xy' + 2y = 0$ for $x > 0$
7. $x^2y'' + xy' - y = 0$ for $x > 0$
8. $xy'' - (x+1)y' + y = 0$ for $x < 0$
9. $x^2y'' + 3xy' = 0$ for $x > 0$
10. $y'' - xf(x)y' + f(x)y = 0$
11. $y'' - f(x)y' + [f(x) - 1]y = 0$
12. $(3x+2)^2 y'' + 3(3x+2)y' - 36y = 0$ for $3x+2 > 0$
13. $(x+a)^2 y'' - 4(x+a)y' + 6y = 0$ for $x+a < 0$
14. $(1+x)^2 y'' + (1+x)y' + y = 0$ for $x+1 < 0$
15. $x^2y'' - 6y = 0$ in $I = (0, +\infty)$

3.5 METHOD OF SOLVING NONHOMOGENEOUS EQUATIONS

In Section 3.2, we learnt that to find the general solution of the equation

$$a_2(x)\frac{d^2y}{dx^2} + a_1(x)\frac{dy}{dx} + a_0(x)y = g(x) \qquad (3.18)$$

where $a_2(x) \neq 0$ and $a_1(x)$, $a_0(x)$, and $g(x)$ are continuous functions on an interval I, it is necessary to find the general solution y_c of the associated homogeneous equation

$$a_2(x)\frac{d^2y}{dx^2} + a_1(x)\frac{dy}{dx} + a_0(x)y = 0 \qquad (3.19)$$

and then add to it any particular solution y_p of (3.18). In the preceding sections we discussed the methods of finding y_c for (3.19). But how do we find y_p for Eq. (3.18)? We shall now discuss a method of finding a particular solution y_p for Eq. (3.18).

Several methods exist for finding a particular solution y_p for a nonhomogeneous linear differential equation (3.18). In this section we shall develop a method called *variation of parameters* which can be applied to all linear differential equations regardless of the nature of the coefficients and the right hand side function $g(x)$, provided y_c for the equation is known. In this respect this method is more powerful than any other existing method.

First let us rewrite both Eqs. (3.18) and (3.19) respectively as

$$\frac{d^2y}{dx^2} + \frac{a_1(x)}{a_2(x)}\frac{dy}{dx} + \frac{a_0(x)}{a_2(x)}y = \frac{g(x)}{a_2(x)} \qquad \text{since } a_2(x) \neq 0 \text{ in } I$$

52 Introduction to Differential Equations

or
$$\frac{d^2y}{dx^2} + P(x)\frac{dy}{dx} + Q(x)y = R(x) \qquad (3.20)$$

and
$$\frac{d^2y}{dx^2} + P(x)\frac{dy}{dx} + Q(x)y = 0 \qquad (3.21)$$

where
$$P(x) = \frac{a_1(x)}{a_2(x)}, \quad Q(x) = \frac{a_0(x)}{a_2(x)}, \quad R(x) = \frac{g(x)}{a_2(x)}$$

Let the complementary function y_c for Eq. (3.20) be $y_c = c_1 y_1(x) + c_2 y_2(x)$, where c_1 and c_2 are arbitrary constants and $y_1(x)$ and $y_2(x)$, are two linearly independent solutions of (3.21).

Let $y_p = v_1(x)y_1(x) + v_2(x)y_2(x)$ be a particular solution of (3.20), where $v_1(x)$ and $v_2(x)$ are two twice-differentiable functions to be determined. There is no justification for this assumption. This is just a wild guess. But this guess works! Since c_1 and c_2 in y_c are replaced by variables to get y_p, this method is called *variation of parameters*.

$$y'p = v_1(x)y_1'(x) + v_2(x)y_2'(x) + v_1'(x)y_1(x) + v_2'(x)y_2(x)$$

Since two functions $v_1(x)$ and $v_2(x)$ are to be determined, we need two conditions. One condition will be obtained when we put the value of y_p in Eq. (3.20), and for the second condition let us assume

$$v_1'(x)y_1(x) + v_2'(x)y_2(x) = 0 \qquad (3.22)$$

This is just to make the calculations simpler. Then y_p' is given by

$$y_p' = v_1(x)y_1'(x) + v_2(x)y_2'(x)$$

and
$$y_p'' = v_1(x)y_1''(x) + v_2(x)y_2''(x) + v_1'(x)y_1'(x) + v_2'(x)y_2'(x)$$

Putting these values of y_p, y_p', y_p'' in Eq. (3.20), we have

$$[v_1(x)y_1''(x) + v_2(x)y_2''(x) + v_1'(x)y_1'(x) + v_2'(x)y_2'(x)]$$
$$+ P(x)[v_1(x)y_1'(x) + v_2(x)y_2'(x)] + Q(x)[v_1(x)y_1(x) + v_2(x)y_2(x)] = R(x)$$
$$\Rightarrow v_1(x)[y_1''(x) + P(x)y_1'(x) + Q(x)y_1(x)] + v_2(x)[y_2''(x) + P(x)y_2'(x) + Q(x)y_2(x)]$$
$$+ v_1'(x)y_1'(x) + v_2'(x)y_2'(x) = R(x) \qquad (3.23)$$

Since $y_1(x)$ and $y_2(x)$ are solutions of (3.21), we have

$$y_1''(x) + P(x)y_1'(x) + Q(x)y_1(x) = 0$$

and
$$y_2''(x) + P(x)y_2'(x) + Q(x)y_2(x) = 0$$

Consequently Eq. (3.23) becomes

$$v_1'(x)y_1'(x) + v_2'(x)y_2'(x) = R(x) \qquad (3.24)$$

Thus we get two conditions (3.22) and (3.24) to determine $v_1(x)$ and $v_2(x)$. Putting these two equations together, we have

$$v_1'(x)y_1(x) + v_2'(x)y_2(x) = 0$$

$$v_1'(x)y_1'(x) + v_2'(x)y_2'(x) = R(x)$$

Solving for $v_1'(x)$ and $v_2'(x)$ by Cramer's rule, we get

$$v_1'(x) = \frac{\begin{vmatrix} 0 & y_2(x) \\ R(x) & y_2'(x) \end{vmatrix}}{\begin{vmatrix} y_1(x) & y_2(x) \\ y_1'(x) & y_2'(x) \end{vmatrix}} = \frac{-y_2(x)R(x)}{y_1(x)y_2'(x) - y_1'(x)y_2(x)} \qquad (3.25)$$

$$v_2'(x) = \frac{\begin{vmatrix} y_1(x) & 0 \\ y_1'(x) & R(x) \end{vmatrix}}{\begin{vmatrix} y_1(x) & y_2(x) \\ y_1'(x) & y_2'(x) \end{vmatrix}} = \frac{y_1(x)R(x)}{y_1(x)y_2'(x) - y_1'(x)y_2(x)} \qquad (3.26)$$

Integrating (3.25) and (3.26), we obtain $v_1(x)$ and $v_2(x)$ respectively. All constants of integration can be disregarded in these integrations since we are looking only for a particular solution of (3.20). Thus the general solution of (3.20) is given by

$$y = c_1 y_1(x) + c_2 y_2(x) + v_1(x)y_1(x) + v_2(x)y_2(x)$$

where $v_1(x)$ and $v_2(x)$ are obtained from (3.25) and (3.26) respectively.

Here a word of caution for working out numerical problems: one should go through the process instead of using (3.25) and (3.26) as formulae.

Example 3.12 Solve $y'' - y' - 2y = \sin 2x$.

The associated homogeneous equation is $y'' - y' - 2y = 0$. The auxiliary equation is $m^2 - m - 2 = 0$, whose roots are $m_1 = -1$ and $m_2 = 2$.

The complementary function y_c of the given equation is

$$y_c = c_1 e^{-x} + c_2 e^{2x}$$

So, let
$$y_p = v_1(x)e^{-x} + v_2(x)e^{2x}$$

$$y_p' = -v_1(x)e^{-x} + 2v_2(x)e^{2x} + v_1'(x)e^{-x} + v_2'(x)e^{2x}$$

Putting $v_1'(x)e^{-x} + v_2'(x)e^{2x} = 0$, we get

54 ▶ *Introduction to Differential Equations*

$$y'_p = -v_1(x)e^{-x} + 2v_2(x)e^{2x}$$

$$y''_p = v_1(x)e^{-x} + 4v_2(x)e^{2x} - v'_1(x)e^{-x} + 2v'_2(x)e^{2x}$$

Now $y''_p - y'_p - 2y_p = \sin 2x$. Putting the values, we get

$$v_1(x)e^{-x} + 4v_2(x)e^{2x} - v'_1(x)e^{-x} + 2v'_2(x)e^{2x} + v_1(x)e^{-x}$$

$$-2v_2(x)e^{2x} - 2v_1(x)e^{-x} - 2v_2(x)e^{2x} = \sin 2x$$

$$\Rightarrow -v'_1(x)e^{-x} + 2v'_2(x)e^{2x} = \sin 2x$$

Thus the two equations to determine $v_1(x)$ and $v_2(x)$ are

$$v'_1(x)e^{-x} + v'_2(x)e^{2x} = 0$$

$$-v'_1(x)e^{-x} + 2v'_2(x)e^{2x} = \sin 2x$$

$$v'_1(x) = \frac{\begin{vmatrix} 0 & e^{2x} \\ \sin 2x & 2e^{2x} \end{vmatrix}}{\begin{vmatrix} e^{-x} & e^{2x} \\ -e^{-x} & 2e^{2x} \end{vmatrix}} = \frac{-e^{2x}\sin 2x}{2e^x + e^x} = \frac{-e^x \sin 2x}{3}$$

$$v'_2(x) = \frac{\begin{vmatrix} e^{-x} & 0 \\ -e^{-x} & \sin 2x \end{vmatrix}}{\begin{vmatrix} e^{-x} & e^{2x} \\ -e^{-x} & 2e^{2x} \end{vmatrix}} = \frac{e^{-x}\sin 2x}{3e^x} = \frac{e^{-2x}\sin 2x}{3}$$

Now, on integrating we have

$$v_1(x) = -\int \frac{e^x \sin 2x}{3}dx = -\frac{1}{15}e^x(\sin 2x - 2\cos 2x),$$

and

$$v_2(x) = \int \frac{e^{-2x}\sin 2x}{3}dx = -\frac{1}{12}e^{-2x}(\sin 2x + \cos 2x)$$

Hence the general solution of the given equation is

$$y = c_1 e^{-x} + c_2 e^{2x} + v_1(x)e^{-x} + v_2(x)e^{2x}$$

$$= c_1 e^{-x} + c_2 e^{2x} - \frac{1}{15}(\sin 2x - 2\cos 2x) - \frac{1}{12}(\sin 2x + \cos 2x)$$

$$y = c_1 e^{-x} + c_2 e^{2x} - \frac{3}{20}\sin 2x + \frac{1}{20}\cos 2x$$

Example 3.13 Solve the equation $(D^4 + D^2)y = x$.

The associated homogeneous equation is $(D^4 + D^2)y = 0$.

The auxiliary equation is $m^4 + m^2 = 0$ and its roots are $m_1 = m_2 = 0$, $m_3 = +i$, $m_4 = -i$. So the complementary function y_c of the given differential equation is $y_c = c_1 + c_2 x + c_3 \cos x + c_4 \sin x$.

Let $y_p = v_1(x) + v_2(x)x + v_3(x)\cos x + v_4(x)\sin x$. Here we have to determine four functions, so we need four conditions. We proceed exactly the same way and impose conditions to simplify the calculations.

$$y_p' = v_2(x) - v_3(x)\sin x + v_4(x)\cos x$$

putting $v_1'(x) + v_2'(x)x + v_3'(x)\cos x + v_4'(x)\sin x = 0$ (3.27)

$$y_p'' = -v_3(x)\cos x - v_4(x)\sin x$$

putting $v_2'(x) - v_3'(x)\sin x + v_4'(x)\cos x = 0$ (3.28)

$$y_p''' = v_3(x)\sin x - v_4(x)\cos x$$

putting $v_3'(x)(-\cos x) + v_4'(x)(-\sin x) = 0$ (3.29)

$$y_p^{(iv)} = v_3(x)\cos x + v_4(x)\sin x + v_3'(x)\sin x - v_4'(x)\cos x$$

Putting these values of $y_p^{(iv)}$ and y_p'' in the given differential, we get

$$v_3(x)\cos x + v_4(x)\sin x + v_3'(x)\sin x - v_4'(x)\cos x - v_3(x)\cos x - v_4(x)\sin x = x$$

$$\Rightarrow v_3'(x)\sin x + v_4'(-\cos x) = x \qquad (3.30)$$

Solving Eqs. (3.27) to (3.30) for $v_1(x)$, $v_2(x)$, $v_3(x)$ and $v_4(x)$ by Cramer's rule, we have

$$v_1'(x) = \frac{\begin{vmatrix} 0 & x & \cos x & \sin x \\ 0 & 1 & -\sin x & \cos x \\ 0 & 0 & -\cos x & -\sin x \\ x & 0 & \sin x & -\cos x \end{vmatrix}}{\begin{vmatrix} 1 & x & \cos x & \sin x \\ 0 & 1 & -\sin x & \cos x \\ 0 & 0 & -\cos x & -\sin x \\ 0 & 0 & \sin x & -\cos x \end{vmatrix}} = \frac{-x^2}{1}$$

$$v_2'(x) = \frac{\begin{vmatrix} 1 & 0 & \cos x & \sin x \\ 0 & 0 & -\sin x & \cos x \\ 0 & 0 & -\cos x & -\sin x \\ 0 & x & \sin x & -\cos x \end{vmatrix}}{1} = x$$

$$v_3' = \begin{vmatrix} 1 & x & 0 & \sin x \\ 0 & 1 & 0 & \cos x \\ 0 & 0 & 0 & -\sin x \\ 0 & 0 & x & -\cos x \end{vmatrix} = x \sin x$$

$$v_4' = \begin{vmatrix} 1 & x & \cos x & 0 \\ 0 & 1 & -\sin x & 0 \\ 0 & 0 & -\cos x & 0 \\ 0 & 0 & \sin x & x \end{vmatrix} = -x \cos x$$

Thus, by integrating,

$$v_1(x) = -\int x^2 \, dx = \frac{-x^3}{3}$$

$$v_2(x) = \int x \, dx = \frac{x^2}{2}$$

$$v_3(x) = \int x \sin x \, dx = -x \cos x + \sin x$$

$$v_4(x) = -\int x \cos x \, dx = -x \sin x - \cos x$$

Hence the general solution of the given differential equation is

$$y = c_1 + c_2 x + c_3 \cos x + c_4 \sin x - \frac{x^3}{3} + \frac{x^3}{2} + (\sin x - x \cos x) \cos x - (x \sin x + \cos x) \sin x$$

$$\Rightarrow y = c_1 + c_2 x + c_3 \cos x + c_4 \sin x + \frac{x^3}{6} - x$$

Example 3.14 Solve the differential equation

$$(x^2 - 1)y'' - 2xy' + 2y = (x^2 - 1)^2 \text{ in } (-1, 1)$$

The complementary function $y_c = c_1 x + c_2(x^2 + 1)$ for this equation was obtained in Example 3.7.

$$v_1'(x) = \frac{\begin{vmatrix} 0 & 1 + x^2 \\ x^2 - 1 & 2x \end{vmatrix}}{\begin{vmatrix} x & 1 + x^2 \\ 1 & 2x \end{vmatrix}} = \frac{-(x^2 - 1)(x^2 + 1)}{(x^2 - 1)} \quad \text{from (3.25)}$$

$$v_1(x) = -\int (x^2 + 1) \, dx = -\frac{x^3}{3} - x$$

$$v_2'(x) = \frac{\begin{vmatrix} x & 0 \\ 1 & x^2-1 \\ x & x^2+1 \\ 1 & 2x \end{vmatrix}}{\begin{vmatrix} x & x^2-1 \\ 1 & 2x \end{vmatrix}} = \frac{x(x^2-1)}{(x^2-1)} \qquad \text{from (3.26)}$$

$$v_2(x) = \int x \, dx = \frac{x^2}{2}$$

Thus the general solution of the given equation is

$$y(x) = c_1 x + c_2(x^2+1) + \left(\frac{-x^3}{3} - x\right)x + \frac{x^2}{2}(x^2+1)$$

$$= c_1 x + c_2(x^2+1) - \frac{x^4}{3} - x^2 + \frac{x^4}{2} + \frac{x^2}{2}$$

$$= c_1 x + c_2(x^2+1) + \frac{x^4}{6} - \frac{x^2}{2}$$

Exercises 3.5

Solve the following differential equations:

1. $y'' - 2y' + y = \dfrac{e^x}{x}$
2. $y'' - 2y' + y = \dfrac{e^x}{x^2}$
3. $y'' - 2y' + y = \dfrac{e^x}{x^3}$
4. $y'' - y' - 2y = 4x^2$
5. $y'' + 25y = 2\sin 2x$
6. $y'' + y' = x$
7. $y'' + y = \tan x$
8. $y'' + y = \sin x$
9. $y'' + 2y' + y = e^x$
10. $y'' + 2y + y = e^{-x}$
11. $y'' - y = x$
12. $y'' - y = e^x$
13. $y'' - y = \sin x$
14. $y'' + 4y' + 5y = 10$
15. $y'' + 4y' + 5y = x + 2$
16. $y''' + y' = \sec x$
17. $y''' + y' = \csc x$
18. $(D^4 - 9D^2)y = 54x^2$
19. $(D^4 + D^2)y = e^x$
20. $(D^5 - 4D^3)y = 32e^{2x}$
21. $x^2 y'' + 2xy' - 12y = x^4$ for $x > 0$
22. $(1 - x^2)y'' - 2xy' + 2y = (x^2 - 1)^2$ in $(-1, 1)$
23. $x^3 y''' + x^2 y'' - 2xy' + 2y = x \ln x$ for $x > 0$

24. $(x-1)y'' - xy' + y = (x-1)^2$ for $x > 1$

25. $xy'' - (x+1)y' + y = x^2 e^x$ for I not containing the origin

3.6 REDUCTION OF ORDER

If one nontrivial solution of Eq. (3.21) is known, then we can solve the nonhomogeneous equation (3.20) by the method of reduction of order and obtain both a particular solution of (3.20) and a second independent solution of (3.21) directly. In other words, if p independent solutions of the associated homogeneous equation of an nth order nonhomogeneous equation are some way known, then the method of reduction of order can be used to reduce the order of the equation to one of order $(n - p)$. This will help to solve the given differential equation. This fact is particularly interesting when $n = 2$ as the reduced first order equation can always be solved by the methods discussed in Chapter 2. Let us now see how this method works for a second order linear equation (3.20).

Let $y_1(x)$ be a solution of the associated homogeneous equation (3.21).

Let $y(x) = v(x)y_1(x)$ be a solution of (3.20), where $v(x)$ is a twice-differentiable function to be determined. Putting this in Eq. (3.20), we have

$$v''(x)y_1(x) + 2v'(x)y_1'(x) + v(x)y_1''(x) + P(x)\{v'(x)y_1(x) + v(x)y_1'(x)\}$$
$$+ Q(x)v(x)y_1(x) = R(x)$$

$$\Rightarrow v''(x)y_1(x) + v'(x)\{2y_1'(x) + P(x)y_1(x)\}$$
$$+ v(x)\{y_1''(x) + P(x)y_1'(x) + Q(x)y_1(x)\} = R(x)$$

Since $y_1(x)$ is a solution of (3.21), we have

$$v''(x)y_1(x) + v'(x)\{2y_1'(x) + P(x)y_1(x)\} = R(x)$$

Put $v'(x) = u(x)$; then

$$u'(x)y_1(x) + \{2y_1'(x) + P(x)y_1(x)\}u(x) = R(x) \tag{3.31}$$

This is a first order linear differential equation, which can be solved by the method discussed in Section 2.5. Then integrating $u(x) = v'(x)$, we get the general solution of Eq. (3.20) by putting this value of $v(x)$ in $y(x) = v(x)y_1(x)$. We shall illustrate this in the following example.

Example 3.15 Solve, by method of reduction of order, the equation

$$xy'' - 2(x+1)y' + (x+2)y = e^x \text{ for } x \neq 0.$$

The associated homogeneous equation of the given equation is

$$xy'' - 2(x+1)y' + (x+2)y = 0.$$

By Rule 2 of Section 3.4, this homogeneous equation has a solution $y_1(x) = e^x$.

Let $y(x) = v(x)e^x$ be a solution of the given nonhomogeneous equation. Putting this in the equation, we get

$$x[v''(x)e^x + 2v'(x)e^x + v(x)e^x] - 2(x+1)[v'(x)e^x + v(x)e^x]$$
$$+ (x+2)v(x)e^x = e^x$$
$$\Rightarrow v''(x)x + [2x - 2(x+1)]v'(x) + [x - 2(x+1) + (x+2)]v(x) = 1 \quad (\because e^x \neq 0)$$
$$\Rightarrow v''(x) - \frac{2}{x}v'(x) = \frac{1}{x}$$
$$\Rightarrow u'(x) - \frac{2}{x}u(x) = \frac{1}{x} \quad \text{putting } v'(x) = u(x)$$
$$\Rightarrow u(x)e^{-\int \frac{2}{x}dx} = \int e^{-\int \frac{2}{x}dx} \frac{1}{x} dx + c_1 \quad \text{by Eq. (2.5)}$$
$$\Rightarrow u(x)\frac{1}{x^2} = \int \frac{dx}{x^3} + c_1 = -\frac{1}{2x^2} + c_1$$
$$\Rightarrow v'(x) = c_1 x^2 - \frac{1}{2} \Rightarrow v(x) = c_1 \frac{x^3}{3} - \frac{x}{2} + c_2$$

This yields the general solution of the given equation as

$$y(x) = v(x)e^x = c_1 \frac{x^3 e^x}{3} - \frac{xe^x}{2} + c_2 e^x$$

In this solution, $y_c(x) = c_1 \frac{x^3 e^x}{3} + c_2 e^x$ and $y_p(x) = -\frac{xe^x}{2}$. These are obtained directly. This is one advantage of this method.
And now some exercises for you.

Exercises 3.6

Solve the following differential equations by using reduction of order method.

1. $x^2 y'' - xy' + y = \sqrt{x}$ for $0 < x < \infty$
2. $xy'' - (2x - 1)y' + (x - 1)y = e^x$ for $x > 0$
3. $4x^2 y'' - 8xy' + 9y = 0$ for $x > 0$
4. $x^2 y'' - 2xy' + 2y = 4x^2$ for $x > 0$
5. $x^2 y'' + 5xy' - 5y = x^{-\frac{1}{2}}$ for $0 < x < \infty$
6. $(x - 1)y'' - xy' + y = xe^x$ for $x > 1$
7. $xy'' - (x + 2)y' + 2y = x^3$ for $x \neq 0$
8. $x^2 y'' + 3xy' + y = \sin x$ for $x \neq 0$
9. $x^2 y'' + x^3 y' - 2(1 + x^2)y = x$ for $x > 0$
10. $xy'' + (2 + x)y' + y = e^{-x}$ for $x > 0$

3.7 SERIES SOLUTION METHOD

The methods studied so far give solutions in terms of known elementary functions. There are linear differential equations whose solutions connot be expressed in such simple form, viz. $y'' + xy = 0$. Thus there is a need of other means of expression for the solutions of these equations. One such means of expression is furnished by power series representations, and in terms of those functions which can have power series representation. This section is devoted to methods of obtaining solutions in power series form.

So the natural equation is, "Under what conditions can we be certain that the differential equation (3.20) or (3.21) actually does have a power series solution?" This is a question of considerable importance; for it would be quite absurd to actually try to find a power series solution if there were really no such solution to be found for the equation! In order to answer this important question concerning the existence of a power series solution, we shall first introduce certain basic definitions.

A function f is said to be *analytic* at a point x_0 iff it has a Taylor series expansion $\sum_{n=0}^{\infty} c_n(x - x_0)^n$ in $|x - x_0| < \delta$ for some $\delta > 0$, which converges to $f(x)$ in $|x - x_0| < \delta$.

Polynomial functions, rational functions, trigonometric functions, exponential functions, logarithmic functions are a few examples of analytic functions at every point of their domain of definitions. For example, a rational function defined by $f(x) = x/(x^2 - 3x + 2)$ is analytic everywhere except at those values of x at which its denominator is zero, i.e. at $x = 1$ and $x = 2$.

We are now in a position to state a theorem concerning the existence of power series solution for Eq. (3.20).

Theorem 3.6 If $P(x)$, $Q(x)$ and $R(x)$ are analytic at x_0, then Eq. (3.20) has a unique solution $f(x)$ such that $f(x)$ is analytic at x_0, $f(x_0) = a$ and $f'(x_0) = b$, where a and b are given constants. Further, all solutions of (3.20) are analytic at x_0, and, if the Taylor series for the coefficients converge in $|x - x_0| < \delta$ for some $\delta > 0$, then the Taylor series for the solutions also converge in $|x - x_0| < \delta$.

This theorem gives a sufficient condition for the existence of power series solutions of the differential equation (3.20). We shall omit the proof of this important theorem. However, we shall explain the importance of this theorem through the following:

Example 3.16 Solve the equation $(1 - x^2)y'' - 2xy' + \alpha(\alpha + 1)y = 0$, where α is a constant.

Here $P(x) = \dfrac{-2x}{1 - x^2}$ and $Q(x) = \dfrac{\alpha(\alpha + 1)}{1 - x^2}$ are analytic at every point except at $x = 1$ and $x = -1$.

Linear Differential Equations of Higher Order

Thus Theorem 3.6 guarantees the existence of power series solutions at every point except at $x = 1$ and $x = -1$. Let us explain the method of obtaining power series solutions for the given equation at $x = 0$. But what happens to solutions at $x = 1$ and $x = -1$? We shall discuss them separately later.

Since Theorem 3.6 guarantees the existence of a power series solution at $x = 0$, let us assume that

$$y(x) = \sum_{n=0}^{\infty} c_n x^n$$

is a solution. This assumption is due to Theorem 3.6. Our objective is to determine the constants c_n.

Then

$$y'(x) = \sum_{n=1}^{\infty} n c_n x^{n-1}$$

$$y''(x) = \sum_{n=2}^{\infty} n(n-1) c_n x^{n-2}$$

Putting all these in the given differential equation, we have

$$(1 - x^2) \sum_{n=2}^{\infty} n(n-1) c_n x^{n-2} - 2x \sum_{n=1}^{\infty} n c_n x^{n-1} + \alpha(\alpha + 1) \sum_{n=0}^{\infty} c_n x^n = 0$$

$$\Rightarrow \sum_{n=2}^{\infty} n(n-1) c_n x^{n-2} - \sum_{n=2}^{\infty} n(n-1) c_n x^n - \sum_{n=1}^{\infty} 2n c_n x^n$$

$$+ \sum_{n=0}^{\infty} \alpha(\alpha + 1) c_n x^n = 0$$

$$\Rightarrow \sum_{n=0}^{\infty} (n+2)(n+1) c_{n+2} x^n - \sum_{n=2}^{\infty} n(n-1) c_n x^n - \sum_{n=1}^{\infty} 2n c_n x^n$$

$$+ \sum_{n=0}^{\infty} \alpha(\alpha + 1) c_n x^n = 0$$

$$\Rightarrow [2 \cdot 1 c_2 + \alpha(\alpha + 1) c_0] + [3 \cdot 2 c_3 - 2 c_1 + \alpha(\alpha + 1) c_1] x$$

$$+ \sum_{n=2}^{\infty} [(n+2)(n+1) c_{n+2} - n(n-1) c_n - 2n c_n + \alpha(\alpha + 1) c_n] x^n = 0$$

Note the changes in the first summation of the previous step; here n is replaced by $n + 2$ and accordingly the whole thing is adjusted. Since n is a dummy suffix, there will be no loss of generality. This is an important step for getting the final solution.

Introduction to Differential Equations

The terms for $n = 0$ and $n = 1$ are separated from the summation, and $n = 2$ onwards are put together under a single summation in the next step. This step is also equally important.

For the power series to be equal to zero, we must have the coefficients of different powers of x to be equal to zero. So we have

$$2 \cdot 1 c_2 + \alpha(\alpha + 1) c_0 = 0 \quad \Rightarrow \quad c_2 = -\frac{\alpha(\alpha+1)}{2 \cdot 1} c_0$$

$$3 \cdot 2 c_3 - 2 c_1 + \alpha(\alpha + 1) c_1 = 0 \quad \Rightarrow \quad c_3 = -\frac{(\alpha-1)(\alpha+2)}{3 \cdot 2} c_1$$

$$(n+2)(n+1)c_{n+2} - n(n-1)c_n - 2n c_n + \alpha(\alpha+1) c_n = 0 \qquad n \geq 2$$

$$\Rightarrow c_{n+2} = \frac{n(n-1) + 2n - \alpha(\alpha+1)}{(n+2)(n+1)} c_n = -\frac{(\alpha-n)(\alpha+n+1)}{(n+2)(n+1)} c_n \qquad n \geq 2$$

Thus we get

$$c_{2n} = (-1)^n \frac{\alpha(\alpha-2) \cdots (\alpha - 2n+2)(\alpha+1)(\alpha+3) \cdots (\alpha+2n-1)}{(2n)!} c_0$$

and $\quad c_{2n+1} = (-1)^n \dfrac{(\alpha-1) \cdots (\alpha-3) \cdots (\alpha-2n+1)(\alpha+2)(\alpha+4) \cdots (\alpha+2n)}{(2n+1)!} c_1$

where c_0 and c_1 are arbitrary constants.
Hence the solution is given by

$$y(x) = c_0 \left[1 + \sum_{n=1}^{\infty} (-1)^n \frac{\alpha(\alpha-2) \cdots (\alpha - 2n+2)(\alpha+1)(\alpha+3) \cdots (\alpha+2n-1)}{(2n)!} x^{2n} \right]$$

$$+ c_1 \left[x + \sum_{n=1}^{\infty} (-1)^n \frac{(\alpha-1)(\alpha-3) \cdots (a-2n+1)(\alpha+2)(\alpha+4) \cdots (\alpha+2n)}{(2n+1)!} x^{2n+1} \right]$$

This is the general solution of the given differential equation as c_0 and c_1 are arbitrary constants and the power series in both the brackets are not constant multiple of each other, i.e., they are independent. It may be noted that if α is a nonnegative integer one of these power series terminates, i.e., it becomes a polynomial.

The point where both $P(x)$ and $Q(x)$ are analytic is called a *ordinary point* for the differential equation (3.20); otherwise the point is called a *singular point*.

On the other hand a singular point $x_0 \in I$ is called a *regular singular point* for the differential equation iff $P(x)(x - x_0)$ and $Q(x)(x - x_0)^2$ are both analytic at x_0; otherwise the point x_0 is called an *irregular singular point* for the differential equation (3.20).

In Example 3.16, every point except $x = 1$ and $x = -1$ in $(-\infty, +\infty)$ is an ordinary point for the differential equation, whereas $x = 1$ and $x = -1$ are regular singular points for the differential equation.

Linear Differential Equations of Higher Order

On the other hand, for the differential equation $x^2(1-x^2)y'' - (1-x)^2 y' + y = 0$ in $(-2, 2)$, the singular points are $x = 0, 1, -1$. All other points in $(-2, 2)$ are ordinary points for this equation. Here $x = 1, -1$ are regular singular points (check!) and $x = 0$ is an irregular singular point, for $\dfrac{-(1-x)^2}{x^2(1-x^2)} \cdot x = \dfrac{x-1}{x(1+x)}$ is not analytic at $x = 0$, though $\dfrac{1}{x^2(1-x^2)} \cdot x^2 = \dfrac{1}{1-x^2}$ is analytic at $x = 0$.

Thus the method explained in Example 3.16 is applicable to obtain power series solution at any ordinary point x_0 for a differential equation. Now, what if $a_2(x)$, $a_1(x)$, and $a_0(x)$ for a differential equation are analytic in an interval I and $a_2(x_0) = 0$ at some point x_0 in I, then can such a differential equation have an analytic solution at x_0? In other words, can a nonnormal differential equation have analytic solutions? Let us examine some examples.

Example 3.17 Can the equation $x^2 y'' + (3x - 1)y' + y = 0$ have an analytic solution at $x = 0$?

Let, if possible, $y(x) = \sum\limits_{n=0}^{\infty} c_n x^n$ be a solution of the given differential equation. Then following the steps explained in Example 3.16, we have

$$\sum_{n=2}^{\infty} n(n-1)c_n x^n + \sum_{n=1}^{\infty} 3n c_n x^n - \sum_{n=1}^{\infty} n c_n x^{n-1} + \sum_{n=0}^{\infty} c_n x^n = 0$$

$$\Rightarrow \sum_{n=2}^{\infty} n(n-1)c_n x^n + \sum_{n=1}^{\infty} 3n c_n x^n - \sum_{n=0}^{\infty} (n+1)c_{n+1} x^n + \sum_{n=0}^{\infty} c_n x^n = 0$$

$$\Rightarrow (c_0 - c_1) + (3c_1 - 2c_2 + c_1)x + \sum_{n=2}^{\infty} [n(n-1)c_n + 3n c_n - (n+1)c_{n+1} + c_n]x^n = 0$$

Equating the coefficients of different powers of x from both the sides, we get

$$c_0 - c_1 = 0 \Rightarrow c_0 = c_1$$

$$3c_1 - 2c_2 + c_1 = 0 \Rightarrow c_2 = 2c_1 = 2c_0$$

$$n(n-1)c_n + 3n c_n - (n+1)c_{n+1} + c_n = 0 \Rightarrow c_{n+1} = (n+1)c_n \quad n \geq 2$$

So that $c_n = n! c_0$ for all n.

But $\sum\limits_{n=0}^{\infty} n! x^n$ does not converge for any $x \neq 0$, as the ratio test suggests.

Hence the given equation does not have an analytic solution at $x = 0$.

Example 3.18 Show that the only solution, analytic at $x = 0$, for the equation $x^3 y'' + xy' - y = 0$ is $c_1 x$.

Proceeding the way we have done in Example 3.17, we arrive at $c_0 = 0$, $0.c_1 = 0$, $c_2 = 0$, $c_n = -(n-2)c_{n-1}$ for ≥ 3.
Thus the only analytic solution is $c_1 x$.

We shall later discuss how to find the second solution independent of the first and then the general solution.

These examples show that the condition of normality, i.e., non-vanishing of the leading coefficient is necessary in general, for the existence of an analytic solution, or for all the solutions to be analytic. If all the three analytic coefficients vanish at a point x_0, then each of $\dfrac{a_i(x)}{x-x_0}$, $i = 0, 1, 2$ is again analytic at x_0. Such a point is a singular point for the differential equation. The above discussions suggest that at a singular point the differential equation may or may not have an analytic solution. Clearly we must seek a different type of solution in such a case, but what type of solution can we expect? For this, we shall not discuss solutions of a differential equation at an irregular singular point, nor shall we tackle all the problems arising in connection with the solutions at a regular singular point. Our aim is to introduce the solutions of some important second order linear differential equations at regular singular points, which often arise in applications. So we discuss the theory just enough for our purpose.

The simplest example of an equation with a regular singular point at x_0 is the Euler's equation

$$(x-x_0)^2 a_2 y'' + (x-x_0) a_1 y' + a_0 y = 0$$

where a_0, a_1, a_2 are constants and $a_2 \neq 0$. Without any loss of generality, we may assume $x_0 = 0$ and the above equation reduces to

$$x^2 a_2 y'' + x a_1 y' + a_0 y = 0 \tag{3.32}$$

We know by making the substitution $z = \ln x$, for $x > 0$ and $z = \ln(-x)$ for $x < 0$, Eq. (3.32) can be reduced to a differential equation with constant coefficients and the solution therefore will be a linear combination of $|x|^\alpha$, $|x|^\alpha \ln|x|$, $|x|^\alpha \cos \beta \ln|x|$, $|x|^\alpha \sin \beta \ln|x|$, where α and β are real numbers (refer to Example 3.11). This suggests that for a regular singular point in general, we should expect a solution of the form $x^\alpha \sum_{n=0}^{\infty} c_n x^n$, where α is a real number and $c_0 \neq 0$.

Such a series is called a *quasi-power series* or *Frobenius series*. If α is a nonnegative integer, then a Frobenius series is a power series; if α is a negative integer, then a Frobenius series is a Laurent series. The method of finding solutions of a differential equation at a regular singular point in the Frobenius form is called the *Frobenius method*.

The next question is in connection with the existence of one or more Frobenius series solutions for a differential equation at a regular singular point. With regard to this, let us consider the general equation with a regular singular point at $x = 0$.

$$x^2 y'' + x p(x) y' + q(x) y = 0 \tag{3.33}$$

where $p(x)$ and $q(x)$ are analytic at $x = 0$, i.e., we can write them as

Linear Differential Equations of Higher Order 65

$$p(x) = \sum_{n=0}^{\infty} p_n x^n \qquad q(x) = \sum_{n=0}^{\infty} q_n x^n$$

where p_n's and q_n's are constants.

Let $y(x) = x^\alpha \sum_{n=0}^{\infty} c_n x^n, c_0 \neq 0$, be a solution, as per our earlier conclusion.

Then
$$y'(x) = \sum_{n=0}^{\infty} (n+\alpha) c_n x^{n+\alpha-1}$$

and
$$y''(x) = \sum_{n=0}^{\infty} (n+\alpha)(n+\alpha-1) c_n x^{n+\alpha-2}$$

Putting these in Eq. (3.33), we get

$$\sum_{n=0}^{\infty} (n+\alpha)(n+\alpha-1) c_n x^{n+\alpha} + p(x) \sum_{n=0}^{\infty} (n+\alpha) c_n x^{n+\alpha} + q(x) \sum_{n=0}^{\infty} c_n x^{n+\alpha} = 0$$

$$\Rightarrow \sum_{n=0}^{\infty} (n+\alpha)(n+\alpha-1) c_n x^{n+\alpha} + \left(\sum_{n=0}^{\infty} p_n x^n\right) \sum_{n=0}^{\infty} (n+\alpha) c_n x^{n+\alpha}$$

$$+ \left(\sum_{n=0}^{\infty} q_n x^n\right) \sum_{n=0}^{\infty} c_n x^{n+\alpha} = 0$$

Equating the coefficients of x^α from both the sides yields the equation
$$\alpha(\alpha-1)c_0 + p_0 \alpha c_0 + q_0 c_0 = 0$$
$$\Rightarrow \alpha^2 + (p_0 - 1)\alpha + q_0 = 0 \qquad \text{since } c_0 \neq 0$$
$$\Rightarrow \alpha^2 + (p_0 - 1)\alpha + q_0 = 0 \qquad (3.34)$$

Thus, α must satisfy Eq. (3.34) for the existence of a Frobenius series solution.

Equation (3.34) is called the *indicial equation* of the differential equation (3.33). If this equation has complex roots, then the Frobenius series solution does not exist. Some modifications are needed for the existence of Frobenius series solution, which we shall not discuss here. We shall only discuss those differential equations of the form (3.33) for which the indicial equation has real roots. We have the following theorem to this effect:

Theorem 3.7 If $\alpha_1 \geq \alpha_2$ are two real roots of the indicial equation (3.34), then Eq. (3.33) has a solution of the form $x^{\alpha_1} \sum_{n=0}^{\infty} b_n x^n, b_0 \neq 0$, on the interval $0 < x < R$. Further, if $\alpha_1 - \alpha_2$ is not an integer, then Eq. (3.33) has a second solution of the form $x^{\alpha_2} \sum_{n=0}^{\infty} c_n x^n$, $c_0 \neq 0$, on the interval $0 < x < R$, which is independent of the first one, where $R > 0$ is the minimum of the radii of convergence of the power series of $p(x)$ and $q(x)$.

We shall not prove this theorem. The theorem guarantees the existence of two independent Frobenius series solutions when $\alpha_1 - \alpha_2$ is not an integer, and guarantees the existence of one Frobenius series solution whenever $\alpha_1 = \alpha_2$ or whenever $\alpha_1 - \alpha_2$ is an integer. However, the theorem does not say anything regarding the existence of a second independent Frobenius series solution whenever $\alpha_1 - \alpha_2$ is an integer; at least it does not rule out the possibility.

Let us examine some examples.

Example 3.19 If possible, find two Frobenius series solutions for the equation $2x^2 y'' + x(1 + 2x)y' - y = 0$ at regular singular point $x = 0$.

First rewrite the equation in the form of Eq. (3.33), i.e.

$$x^2 y'' + x\left(\frac{1+2x}{2}\right) y' + \left(\frac{-1}{2}\right) y = 0$$

Thus $p(x) = \dfrac{1+2x}{2}$ and $q(x) = -\dfrac{1}{2}$; so the indicial equation is

$$\alpha^2 + (p(0) - 1)\alpha + q(0) = 0$$

$$\Rightarrow \alpha^2 + \left(\frac{1}{2} - 1\right)\alpha - \frac{1}{2} = 0 \Rightarrow 2\alpha^2 - \alpha - 1 = 0$$

The roots of the indicial equation are $\alpha_1 = 1$ and $\alpha_2 = -\dfrac{1}{2}$. Since $\alpha_1 - \alpha_2 = \dfrac{3}{2}$, which is not an integer, the given equation has two independent Frobenius series solutions according to Theorem 3.7. Let us find them.

Let $y_1(x) = x \sum_{n=0}^{\infty} b_n x^n = \sum_{n=0}^{\infty} b_n x^{n+1}$, $b_0 \neq 0$, be a solution since Theorem 3.4 guarantees such a solution. Our task is to determine all the coefficients.

Putting this solution in the given differential equation, we get

$$\sum_{n=0}^{\infty} 2(n+1)n b_n x^{n+1} + \sum_{n=0}^{\infty} 2(n+1) b_n x^{n+2} + \sum_{n=0}^{\infty} (n+1) b_n x^{n+1} - \sum_{n=0}^{\infty} b_n x^{n+1} = 0$$

$$\Rightarrow \sum_{n=0}^{\infty} 2(n+1)n b_n x^{n+1} + \sum_{n=1}^{\infty} 2n b_{n-1} x^{n+1} + \sum_{n=0}^{\infty} (n+1) b_n x^{n+1} - \sum_{n=0}^{\infty} b_n x^{n+1} = 0$$

Equating different powers of x^{n+1} to zero, we have $2.0 x b_0 + b_0 x - b_0 x = 0$, which implies that $b_0 \neq 0$ is arbitrary.

$$\{2(n+1)n + (n+1) - 1\}b_n = -2nb_{n-1} \qquad n \geq 1$$

$$\Rightarrow b_n = \frac{-2}{2n+3}b_{n-1} \qquad n \geq 1$$

$$= (-1)^n \frac{2^n}{5 \cdot 7 \cdot 9 \cdots (2n+3)} b_0 \qquad n \geq 1$$

Thus one solution is

$$y_1(x) = b_0 \left(x + x \sum_{n=1}^{\infty} \frac{(-1)^n 2^n}{5 \cdot 7 \cdot 9 \cdots (2n+3)} x^n \right)$$

Let $y_2(x) = x^{-\frac{1}{2}} \sum_{n=0}^{\infty} c_n x^n$, $c_0 \neq 0$, be the second solution of the equation, due to Theorem 3.7. Proceeding as above by putting $y_2(x)$ in the given equation, we arrive at the second solution, which is

$$y_2(x) = x^{-\frac{1}{2}} c_0 \sum_{n=0}^{\infty} \frac{(-1)^n}{n!} x^n = c_0 x^{-\frac{1}{2}} e^{-x}$$

This yields the general solution of the given differential equation at $x = 0$ as

$$y(x) = b_0 x \left(1 + \sum_{n=1}^{\infty} \frac{(-1)^n 2^n x^n}{5 \cdot 7 \cdot 9 \cdots (2n+3)} \right) + c_0 x^{-\frac{1}{2}} e^{-x}$$

Example 3.20 Show that the differential equation

$$4x^2 y'' - 8x^2 y' + (4x^2 + 1) y = 0$$

has one Frobenius series solution at a regular singular point $x = 0$, and find the series solution.

For this equation

$$p(x) = \frac{-8x}{4} \text{ and } q(x) = \frac{4x^2 + 1}{4}$$

so the indicial equation is $\alpha^2 + (p(0) - 1)\alpha + q(0) = 0$

i.e. $\alpha^2 - \alpha + \frac{1}{4} = 0 \Rightarrow \alpha_1 = \alpha_2 = \frac{1}{2}$

Thus, according to Theorem 3.7, the given equation can have only one Frobenius series solution. Let $y_1(x) = x^{\frac{1}{2}} \sum_{n=0}^{\infty} b_n x^n$ be a solution with $b_0 \neq 0$.

Putting this in the given equation results in

$$\sum_{n=0}^{\infty} 4\left(n+\frac{1}{2}\right)\left(n-\frac{1}{2}\right)b_n x^{n+\frac{1}{2}} - \sum_{n=0}^{\infty}\left(n+\frac{1}{2}\right)b_n x^{n+\frac{3}{2}}$$

$$+ \sum_{n=0}^{\infty} 4b_n x^{n+\frac{5}{2}} + \sum_{n=0}^{\infty} b_n x^{n+\frac{1}{2}} = 0$$

$$\Rightarrow \sum_{n=0}^{\infty} 4\left(n+\frac{1}{2}\right)\left(n-\frac{1}{2}\right)b_n x^{n+\frac{1}{2}} - \sum_{n=1}^{\infty}\left(n-\frac{1}{2}\right)b_{n-1} x^{n+\frac{1}{2}}$$

$$+ \sum_{n=2}^{\infty} 4b_{n-2} x^{n+\frac{1}{2}} + \sum_{n=0}^{\infty} b_n x^{n+\frac{1}{2}} = 0$$

replacing n by $n-1$ in the second summand and n by $n-2$ in the third summand

$$\Rightarrow \left\{4\left(\frac{1}{2}\right)\left(-\frac{1}{2}\right)b_0 + b_0\right\} x^{\frac{1}{2}} + \left\{4\left(\frac{3}{2}\right)\left(\frac{1}{2}\right)b_1 - \left(\frac{1}{2}\right)b_0 + b_1\right\} x^{1+\frac{1}{2}}$$

$$+ \sum_{n=2}^{\infty}\left[4\left(n+\frac{1}{2}\right)\left(n-\frac{1}{2}\right)b_n - \left(n-\frac{1}{2}\right)b_{n-1} + 4b_{n-2} + b_n\right] x^{n+\frac{1}{2}} = 0$$

Equating the coefficients of different powers of x to zero, we obtain ($b_0 \neq 0$ is arbitrary)

$$4b_1 = \frac{1}{2}b_0 \Rightarrow b_1 = \frac{1}{2\cdot 4}b_0 = \frac{1}{8}b_0$$

$$\left\{4\left(n+\frac{1}{2}\right)\left(n-\frac{1}{2}\right)+1\right\} b_n = \left(n-\frac{1}{2}\right)b_{n-1} - 4b_{n-2} \quad \text{for } n \geq 2$$

$$\Rightarrow b_n = \frac{\left(n-\frac{1}{2}\right)b_{n-1} - 4b_{n-2}}{4n^2} \quad \text{for } n \geq 2$$

$$b_2 = \frac{\frac{3}{2}b_1 - 4b_0}{16} = \frac{\frac{3}{16} - 4}{16} b_0 = -\frac{61}{16^2} b_0 = -\frac{61}{256} b_0$$

$$b_3 = \frac{\frac{5}{2}b_2 - 4b_1}{4\cdot 3^2} = \frac{-\frac{5\cdot 61}{2\cdot 16^2} - \frac{1}{2}}{4\cdot 3^2} b_0 = \frac{5\cdot 61 - 16^2}{2\cdot 4\cdot 3^2 \cdot 16^2} b_0 = -\frac{305 - 256}{72 \times 256} b_0 = -\frac{49}{18432} b_0$$

Thus the Frobenius series solution is given by

$$y_1(x) = x^{\frac{1}{2}} b_0 \left\{ 1 + \frac{1}{8}x - \frac{61}{256}x^2 - \frac{49}{18432}x^2 + \cdots \right\}$$

Example 3.21 Find, if possible, two independent Frobenius series solutions of the differential equation

$$x^2 y'' + x(1+x)y' - y = 0$$

at a regular singular point $x = 0$.

For the given equation $p(x) = 1 + x$ and $q(x) = -1$. So the indicial equation is

$$\alpha^2 + (p(0) - 1)\alpha + q(0) = 0$$

i.e.
$$\alpha^2 - 1 = 0$$

$$\Rightarrow \alpha_1 = 1 \text{ and } \alpha_2 = -1$$

Here $\alpha_1 - \alpha_2 = 2$. According to Theorem 3.7, nothing can be said about whether the second Frobenius series solution exists or not. So let us do it the following way.

If possible, let $y(x) = x^{-1} \sum_{n=0}^{\infty} b_n x^n$, $b_0 \neq 0$, be a solution of the given equation.

Putting this in the given equation, we obtain

$$\sum_{n=0}^{\infty} (n-1)(n-2)b_n x^{n-1} + \sum_{n=0}^{\infty} (n-1)b_n x^n + \sum_{n=0}^{\infty} (n-1)b_n x^{n-1} - \sum_{n=0}^{\infty} b_n x^{n-1} = 0$$

$$\Rightarrow \sum_{n=0}^{\infty} (n-1)(n-2)b_n x^{n-1} + \sum_{n=1}^{\infty} (n-2)b_{n-1} x^{n-1}$$

$$+ \sum_{n=0}^{\infty} (n-1)b_n x^{n-1} - \sum_{n=0}^{\infty} b_n x^{n-1} = 0$$

(Replacing n by $n - 1$ in the second summand)

$$\Rightarrow (2b_0 - b_0 - b_0)x^{-1} + \sum_{n=1}^{\infty} \left[(n-1)(n-2)b_n + (n-2)b_{n-1} + (n-1)b_n - b_n \right] x^{n-1} = 0$$

Equating the coefficients of different powers of x to zero, we obtain

$$(2-2)b_0 = 0 \Rightarrow b_0 \neq 0 \text{ is arbitrary}$$

$$[(n-1)(n-2) + (n-1) - 1]b_n = -(n-2)b_{n-1} \quad \text{for } n \geq 1$$

$$\Rightarrow n(n-2)b_n = -(n-2)b_{n-1} \quad \text{for } n \geq 1 \tag{3.35}$$

For $n = 1$ $\quad b_1 = b_0$
For $n = 2$ $\quad 0b_2 = 0b_1$
$\quad\quad\quad\quad \Rightarrow b_2 \neq 0$ is arbitrary.

For $n \geq 3$
$$b_3 = -\frac{1}{3}b_2$$
$$b_4 = -\frac{1}{4}b_3 = \frac{(-1)^2}{3 \cdot 4}b_2$$
$$\cdots\cdots\cdots\cdots\cdots\cdots$$
$$\cdots\cdots\cdots\cdots\cdots\cdots$$
$$b_n = \frac{(-1)^{n-2}}{3 \cdot 4 \cdots n}b_2 = \frac{(-1)^{n-2} 2}{n!}b_2$$

This yields
$$y(x) = \sum_{n=0}^{\infty} b_n x^{n-1} = (b_0 x^{-1} - b_0) + \left(b_2 x - \frac{2x^2}{3!}b_2 + \frac{2x^3}{4!}b_2 - \cdots\right)$$

$$= b_0 \left(x^{-1} - 1\right) + \frac{2b_2}{x}\left(\frac{x^2}{2!} - \frac{x^3}{3!} + \frac{x^4}{4!} - \cdots\right)$$

$$= b_0 \left(\frac{1}{x} - 1\right) + \frac{2b_2}{x}(e^{-x} - 1 + x)$$

This is the general solution of the given equation, because
$$y_1(x) = \frac{1}{x} - 1$$
$$y_2(x) = \frac{2}{x}(e^{-x} - 1 + x)$$

are independent of each other, and b_0 and b_2 are arbitrary constants. They are also both Frobenius series as per Theorem 3.7.

Note that the given equation has two Frobenius series solutions in spite of the fact that $\alpha_1 - \alpha_2$ is an integer.

Example 3.22 Find, if possible, two Frobenius series solutions of the following differential equation at its regular singular point:
$$x^2 y'' + xy' + (x^2 - 1)y = 0$$

Clearly $x = 0$ is the only regular singular point for the given differential equation (check!).

Here $p(x) = 1$, $q(x) = x^2 - 1$, and so the indicial equation is $\alpha^2 - 1 = 0$, which implies that $\alpha_1 = 1$, $\alpha_2 = -1$ and $\alpha_1 - \alpha_2 = 2$.

Like the previous example, let us check if the second Frobenius series solution exists or not. For that, let $y_2(x) = x^{-1} \sum_{n=0}^{\infty} b_n x^n$, $b_0 \neq 0$ be a solution. Putting this in the given differential equation and proceeding exactly the way we got the final step in Examples 3.21, we obtain on equating the coefficients of different powers of x to zero

$$(-1)(-2)b_0 - b_0 - b_0 = 0 \Rightarrow b_0 \neq 0 \text{ is arbitrary}$$
$$0 \cdot b_1 + 0 \cdot b_1 - b_1 = 0 \Rightarrow b_1 = 0$$
$$(n-1)(n-2)b_n + (n-1)b_n + b_{n-2} - b_n = 0 \quad n \geq 2$$
$$\Rightarrow n(n-2)b_n = -b_{n-2} \quad n \geq 2 \quad (3.36)$$

For $n = 2$, $0 \cdot b_2 = -b_0$. This suggests that b_2 cannot be determined. Hence the second Frobenius series solution for the given differential equation does not exist.

This equation has one Frobenius series solution at its regular singular point in the form

$$y_1(x) = x \sum_{n=0}^{\infty} c_n x^n \quad c_0 \neq 0$$

This can be determined by the usual method and is left as an exercise.

Compare Examples 3.21 and 3.22. In both these examples, the roots of the indicial equation differ by an integer. In one case, the second Frobenius series solution exists and in the case of the other the second Frobenius series solution does not exist. Then how do we decide when does it exist and when not? The answer is given in (3.35) and (3.36). For $n = 2$, in (3.35) both the coefficients of b_n and b_{n-1} are zeros. In such a situation, b_2 is arbitrary, that is, for any value of b_2 the relation (3.35) will be valid. On the other hand, in (3.36) for $n = 2$ the coefficient of b_n is zero, whereas the coefficient of b_{n-2} is not zero. Thus b_2 is equal to some nonzero quantity divided by zero, which cannot be determined. Hence the tip is that whenever for some n the coefficients from both the sides are zero, the Frobenius series exists and if the coefficient on the left hand side is zero and that on the right hand side nonzero for some value of n, then the Frobenius series does not exist.

Whenever the second Frobenius series solution exists, the general solution of the given differential equation is obtained by taking the linear combination of the two Frobenius series solutions. If the second Frobenius series solution does not exist then the following theorem gives the method of finding the second linearly independent solution and hence the general solution.

Theorem 3.8 Let $\alpha_1 \geq \alpha_2$ be two real roots of the indicial equation (3.34) of the differential equation (3.33), which has a regular singular point at $x = 0$.

(i) If $\alpha_1 - \alpha_2$ is not an integer, then there exist two linearly independent solutions of (3.33) at $x = 0$ of the form

$$y_1(x) = x^{\alpha_1} \sum_{n=0}^{\infty} b_n x^n \quad b_0 \neq 0$$

$$y_2(x) = x^{\alpha_2} \sum_{n=0}^{\infty} c_n x^n \quad c_0 \neq 0$$

(ii) If $\alpha_1 = \alpha_2$, then there exist two linearly independent solutions of (3.33) at $x = 0$ of the form

$$y_1(x) = x^{\alpha_1} \sum_{n=0}^{\infty} b_n x^n \qquad b_0 \neq 0$$

$$y_2(x) = y_1(x) \ln x + x^{\alpha_1} \sum_{n=1}^{\infty} c_n x^n$$

(iii) If $\alpha_1 - \alpha_2$ is a positive integer, then there exist two linearly independent solutions of (3.33) at $x = 0$ of the form

$$y_1(x) = x^{\alpha_1} \sum_{n=0}^{\infty} b_n x^n \qquad b_0 \neq 0$$

$$y_2(x) = y_1(x) \ln x + x^{\alpha_2} \sum_{n=0}^{\infty} c_n x^n \qquad c_0 \neq 0$$

We shall neither prove this theorem nor illustrate it through an example. The proof is routine and the reader should work out some examples himself.

Now try the following exercises.

Exercises 3.7

Find power series solutions in powers of x of each of the differential equations in questions 1–5.

1. $y'' + xy' + y = 0$
2. $y'' + xy' + (2x^2 + 1)y = 0$
3. $y'' + xy' + (3x + 2)y = 0$
4. $(x^2 + 1)y'' + xy' + xy = 0$
5. $(x^3 - 1)y'' + x^2 y' + xy = 0$

Find the power series solution in powers of x of each of the initial value problems in Exercises 6 and 7.

6. $y'' - xy' - y = 0 \qquad y(0) = 1 \qquad y'(0) = 0$
7. $(x^2 + 1)y'' + xy' + 2xy = 0 \qquad y(0) = 2 \qquad y'(0) = 3$
8. Find power series solutions in powers of $(x - 1)$ of the equation
   $x^2 y'' + xy' + y = 0$
9. Find the power series solutions in powers of $(x - 1)$ of the initial value problem
   $xy'' + y' + 2y = 0 \qquad y(0) = 2 \qquad y'(1) = 4$
10. The solutions of Airy's equation $y'' + xy = 0$ are called Airy functions.
    (a) Find the Airy functions in the form of power series and verify directly that these series converge for all x.

(b) Find the general solution of $y'' - xy = 0$ by using part (a) without calculation.

11. Find two independent power series solutions valid for $|x| < 1$ of the Chebyshev's equation $(1 - x^2)y'' - xy' + p^2y = 0$, where p is a constant.

12. Find two independent power series solutions in powers of x, valid for all x of the Hermite equation $y'' - 2xy' + 2py = 0$, where p is a constant.

Locate and classify the singular points of the differential equations in Exercises 13–17.

13. $(x^2 - 3x)y'' + (x + 2)y' + y = 0$

14. $(x^4 - 2x^3 + x^2)y'' + 2(x - 1)y' + x^2y = 0$

15. $(3x + 1)xy'' - (x + 1)y' + 2y = 0$

16. $x^3(x - 1)y'' - 2(x - 1)y' + 3xy = 0$

17. $x^2(x^2 - 1)^2 y'' - x(1 - x)y' + 2y = 0$

Use the method of Frobenius to find solutions near $x = 0$ of each of the differential equations in Exercises 18–30.

18. $2x^2y'' + xy' + (x^2 - 1)y = 0$

19. $x^2y'' - xy' + \left(x^2 + \dfrac{8}{9}\right)y = 0$

20. $x^2y'' + xy' + \left(x^2 - \dfrac{1}{9}\right)y = 0$

21. $3xy'' - (x - 2)y' - 2y = 0$

22. $x^2\dfrac{d^2y}{dx^2} + x\dfrac{dy}{dx} + \left(x^2 - \dfrac{1}{4}\right)y = 0$

23. $xy'' - (x^2 + 2)y' + xy = 0$

24. $(2x^2 - x)y'' + (2x - 2)y' + (-2x^2 + 3x - 2)y = 0$

25. $x^2y'' + xy' + (x - 1)y = 0$

26. $x^2y'' - xy' + 8(x^2 - 1)y = 0$

27. $xy'' + y' + 2y = 0$

28. $x^2y'' - xy' + (x^2 + 1)y = 0$

29. $x^2y'' - 3xy' + (4x + 4)y = 0$

30. $4x^2y'' - 8x^2y' + (4x^2 + 1)y = 0$

CHAPTER 4

Properties of Solutions

4.1 INTRODUCTION

In the previous chapter, our main aim was to develop methods of finding explicit solutions of the second order, normal, homogeneous linear equation

$$y'' + P(x)y' + Q(x)y = 0 \qquad (4.1)$$

We have developed different methods and solved the equation. Unfortunately only rarely do we got solutions in terms of familiar elementary functions. Therefore to understand the nature and properties of the solutions of Eq. (4.1) we have to directly analyse the equation itself. In this chapter, we shall discuss the essential characteristics of the solutions of (4.1) by studying the equation without obtaining the formal expressions for these solutions. It is interesting to know how much useful information can be gained in this way.

4.2 ZEROS OF A SOLUTION

The zeros of a function play a very important role in determining the characteristics of the function.

If $f(x_0) = 0$ for a function f, then x_0 is called a *zero* of the function f.

If x_0 is a zero of a function f such that $f(x_0) = 0$, $f'(x_0) = 0$, $\cdots$, $f^{(n-1)}(x_0) = 0$ and $f^{(n)}(x_0) \neq 0$, then x_0 is called a *zero of order n*. If x_0 is a zero of order 1, then it is called a *simple zero*.

Example 4.1 $f(x) = \sin x$ has a simple zero at 0, $\pm\pi$, $\pm 2\pi$, $\pm 3\pi$, ...

Example 4.2 $f(x) = x^n$ has a zero of order n at 0.

Example 4.3 $f(x) = e^x$ has no zeros.

What can we say about the zeros of a nontrivial solution of Eq. (4.1), without actually obtaining an expression for the solution? Can we know whether a

Properties of Solutions

nontrivial solution of Eq. (4.1) has distinct zeros or not, finite number of zeros or not? Can two linearly independent solutions of Eq. (4.1) have the same zeros? The following theorems answer all these questions and many more.

Theorem 4.1 A nontrivial solution of Eq. (4.1) has only simple zeros.

Proof: Let $y_1(x)$ be a nontrivial solution of Eq. (4.1) and x_0 be a zero of $y_1(x)$ (not a simple zero). Then

$$y_1(x_0) = y_1'(x_0) = 0 = y_1''(x_0) = \cdots$$

Thus we found that $y_1(x)$ is a solution of the initial value problem

$$y'' + P(x)y' + Q(x)y = 0 \qquad (4.2)$$

$$y(x_0) = y'(x_0) = 0$$

We know that the trivial solution $y_2(x) = 0$ for all x also satisfies the initial value problem (4.2), i.e., a solution of (4.2).

But we know the initial value problem (4.2) has a unique solution by Theorem 3.2.

Hence $y_1(x) = y_2(x)$, i.e., $y_1(x)$ is not a nontrivial solution of Eq. (4.1), a contradiction!

Therefore, $y_1(x_0) = 0$ and $y_1'(x_0) \neq 0$, which means that x_0 is a simple zero of $y_1(x)$. This completes the proof.

This theorem implies that the graph of a nontrivial solution $y_1(x)$ of Eq. (4.1) can never touch the x-axis but always intersects it. Therefore the zeros of $y_1(x)$ are all distinct. How many zeros can $y_1(x)$ have, finite or infinite?

If $x_1, x_2, x_3, \ldots$ are zeros of $y_1(x)$, then due to the above Theorem 4.1 the graph of $y_1(x)$ is alternately positive and negative or vice versa in the intervals (x_1, x_2), (x_2, x_3), ...

A function is said to *oscillate* in an interval if it changes its signs infinitely many times. If a function oscillates in an interval, then we say the function is an *oscillating function* in that interval. A function is said to oscillate slowly or rapidly depending on whether the distance between successive zeros is more or less.

Example 4.4 The trigonometric functions $\sin x$ and $\cos x$ are oscillating functions in the interval $(-\infty, +\infty)$, whereas polynomial functions and exponential functions are not oscillating functions in $(-\infty, +\infty)$.

Example 4.5 The function $f(x) = \sin \theta x$ oscillates, in $(-\infty, +\infty)$, slowly if θ is very small and rapidly if θ is very large.

The following theorem suggests that a nontrivial solution of Eq. (4.1) is not an oscillating function in any finite interval.

Theorem 4.2 A nontrivial solution of Eq. (4.1) has only finitely many zeros in any finite interval $[a, b]$.

Proof: Let $y_1(x)$ be a nontrivial solution of Eq. (4.1) and let it have an infinite number of zeros in $[a, b]$. It follows from this that there exist in $[a, b]$ a point x_0

and a sequence of zeros $x_n \neq x_0$ such that $x_n \to x_0$. This is due to the Bolzano–Weierstrass theorem of advanced calculus, which is beyond the scope of this book.

Since $y_1(x)$ is continuous and differentiable at x_0, we have

$$y_1(x_0) = \lim_{x_n \to x_0} y_1(x_n) = 0$$

and

$$y_1'(x_0) = \lim_{x_n \to x_0} \frac{y_1(x_n) - y_1(x_0)}{x_n - x_0} = 0$$

This implies, due to Theorem 3.2, that $y_1(x)$ is the trivial solution of (4.1), and this contradiction completes the proof.

The following example illustrates the concepts of this theorem.

Example 4.6 Consider the equation $y''(x) + cy(x) = 0$, where c is a constant.

If $c < 0$, say $c = -\alpha^2$, where α is any real number, then a nontrivial solution of this equation is given by

$$y = Ae^{\alpha x} + Be^{-\alpha x}$$

where A, B are arbitrary constants with at least one of A and B different from zero. Let us suppose that $A \neq 0$. To find the zeros of this solution, let

$$Ae^{\alpha x} + Be^{-\alpha x} = 0$$

$$\Rightarrow e^{2\alpha x} = -\frac{B}{A}$$

$$\Rightarrow 2\alpha x = \ln\left(\frac{-B}{A}\right) \Rightarrow x = \frac{1}{2\alpha} \ln\left(\frac{-B}{A}\right)$$

Thus y can have at most one zero.

If $c = 0$, then the equation reduces to $y'' = 0 \Rightarrow y = Ax + B$, with at least one of A and B different from zero, is a nontrivial solution, which has exactly one zero.

If $c > 0$, say $c = \alpha^2$, where α is any real number, then a nontrivial solution of the given equation is given by

$$y = A \cos \alpha x + B \sin \alpha x$$

where at least one of A and B is different from zero, has infinitely many zeros in $(-\infty, +\infty)$, but has finitely many zeros in any finite interval.

Further, as c increases, the number of zeros increases in the sense that any finite interval contains more and more zeros.

Next, let us discuss the zeros of the two linearly independent solutions of Eq. (4.1). The following theorem says they cannot have the same zeros.

Theorem 4.3 Two linearly independent solutions of Eq. (4.1) cannot vanish at the same point.

Proof: Let $y_1(x)$ and $y_2(x)$ be two linearly independent solutions of Eq. (4.1), and let us assume the contrary, that is, there is a point x_0 such that $y_1(x_0) = y_2(x_0) = 0$.

Then the Wronskian of $y_1(x)$, $y_2(x)$ at x_0

$$W[x_0; y_1(x), y_2(x)] = 0$$

which contradicts the fact that $y_1(x)$ and $y_2(x)$ are linearly independent. This contradiction completes the proof of the theorem.

Example 4.7 Consider the equation $y'' + y = 0$.

We know that the two linearly independent solutions of this equation are $\sin x$ and $\cos x$.

The zeros of $\sin x$ are $0, \pm\pi, \pm 2\pi, \pm 3\pi, \ldots$

The zeros of $\cos x$ are $\pm\dfrac{\pi}{2}, \pm\dfrac{3\pi}{2}, \pm\dfrac{5\pi}{2}, \ldots$

These are distinct from eachother. Not only the zeros of $\sin x$ are different from those of $\cos x$, but they also occur alternately. Though these are peculiar to these functions, they can be generalised for beyond these particular functions. The following theorem in this direction is called the *Sturm Separation Theorem* in honour of the Swiss mathematician Jacques-Charles Francois Sturm (1803–1855).

Theorem 4.4 (Sturm Separation Theorem) Let $y_1(x)$ and $y_2(x)$ be two linearly independent solutions of Eq. (4.1). Then $y_1(x)$ vanishes exactly once between any two consecutive zeros of $y_2(x)$ and vice versa.

Proof: Let x_1 and x_2 be two consecutive zeros of $y_2(x)$, so $y_2(x_1) = y_2(x_2) = 0$. Then, $W[x_1; y_1, y_2] = y_1(x_1) y_2'(x_1) - y_1'(x_1) y_2(x_1) = y_1(x_1) y_2'(x_1)$

and $W[x_2; y_1, y_2] = y_1(x_2) y_2'(x_2) - y_1'(x_2) y_2(x_2) = y_1(x_2) y_2'(x_2)$

are of the same sign, otherwise $W[x; y_1, y_2]$ will vanish in $[x_1, x_2]$.

So, let us assume that $y_1(x_1) y_2'(x_1) > 0$ and $y_1(x_2) y_2'(x_2) > 0$.

But on the other hand $y_2'(x_1)$ and $y_2'(x_2)$ are of opposite signs because, by Rolle's theorem, there exists a point $x_0 \in (x_1, x_2)$ such that $y_2'(x_0) = 0$.

Let us suppose that $y_2'(x_1) > 0$ and $y_2'(x_2) < 0$. Then from the above we must have $y_1(x_1) > 0$ and $y_1(x_2) < 0$. Since $y_1(x)$ is a continuous function, this implies that there exists a point $x_3 \in (x_1, x_2)$ such that $y_1(x_3) = 0$.

Thus, between any two consecutive zeros of $y_2(x)$, there is a zero of $y_1(x)$. Similarly, we can prove that between any two consecutive zeros of $y_1(x)$ there is a zero of $y_2(x)$ by repeating the above argument with interchanging $y_1(x)$ and $y_2(x)$.

To prove that there is exactly one zero x_3 of $y_1(x)$ between two consecutive zeros x_1 and x_2 of $y_2(x)$, let us assume the contrary, i.e., let us suppose there is another zero $x_4 \neq x_3$ of $y_1(x)$ such that $x_4 \in (x_1, x_2)$.

If there is no other zero of $y_1(x)$ in (x_3, x_4), then x_3 and x_4 are two consecutive zeros of $y_1(x)$, and by the above argument there is a zero x_5 of $y_2(x)$ such that $x_5 \in (x_3, x_4) \subset (x_1, x_2)$, which contradicts the fact that x_1 and x_2 are two consecutive zeros of $y_2(x)$. Hence there is exactly one zero of $y_1(x)$ between any two consecutive zeros of $y_2(x)$ and vice versa. This completes the proof of the theorem.

Introduction to Differential Equations

This theorem implies that all solutions of Eq. (4.1) oscillate with essentially the same rapidity. This means that on a given interval the number of zeros of any solution cannot differ by more than 1 from the number of zeros of any other solution.

Example 4.8 Show that the zeros of the functions $a \sin x + b \cos x$ and $c \sin x + d \cos x$ are distinct and occur alternately whenever $ad - bc \neq 0$.

Let
$$y_1(x) = a \sin x + b \cos x$$
$$y_2(x) = c \sin x + d \cos x$$

Clearly $y_1(x)$ and $y_2(x)$ are two nontrivial solutions of the equation $y''(x) + y(x) = 0$.
Are they linearly independent? Yes, because

$$\alpha y_1(x) + \beta y_2(x) = 0$$
$$\Rightarrow \alpha a \sin x + \alpha b \cos x + \beta c \sin x + \beta d \cos x = 0$$
$$\Rightarrow (\alpha a + \beta c) \sin x + (\alpha b + \beta d) \cos x = 0$$
$$\Rightarrow \alpha a + \beta c = 0 \quad \text{or} \quad \alpha b + \beta d = 0 \quad \text{as } \sin x \text{ and } \cos x \text{ are linearly independent}$$
$$\Rightarrow \alpha = 0 = \beta \quad \text{provided } ad - bc \neq 0$$

Thus $y_1(x)$ and $y_2(x)$ are two linearly independent solutions of the equation $y''(x) + y(x) = 0$.
Hence by Theorem 4.4, the zeros of $y_1(x)$ and $y_2(x)$ are distinct and occur alternately.

It is clear from Example 4.6 that the solutions of $y'' + 4y = 0$ oscillate more rapidly than those of $y'' + y = 0$ in any given interval. The following theorem, known as *Sturm Comparison Theorem*, shows that the solutions of the equation

$$u''(x) + p(x) u(x) = 0 \qquad (4.3)$$

oscillate more rapidly when $p(x)$ is increased.

Before discussing the Sturm comparison theorem, let us look at the form of Eq. (4.3) in which the term involving $u'(x)$ is missing.

It is customary to refer to (4.1) as the *standard form* and to (4.3) as the *normal form* of a homogeneous, normal, second order linear differential equation. Further, every equation (4.1) can be reduced to the form (4.3) by putting in (4.1)

$$y(x) = f(x)u(x)$$

and then
$$y'(x) = f'(x) u(x) + f(x)u'(x)$$
$$y''(x) = f''(x)u(x) + 2f'(x)u'(x) + f(x)u''(x)$$

Then Eq. (4.1) becomes

$$u''(x) f(x) + 2f'(x)u'(x) + u(x)f''(x) + P(x)f'(x)u(x) + P(x)f(x)u'(x) + Q(x)f(x)u(x) = 0$$
$$\Rightarrow u''(x)f(x) + u'(x)\{2f'(x) + P(x)f(x)\} + u(x)\{f''(x) + P(x)f'(x) + Q(x)f(x)\} = 0$$

Since we do not want a term involving $u'(x)$, we set its coefficient equal to zero. Then we have

$$2f'(x) + P(x)f(x) = 0 \Rightarrow \frac{f'(x)}{f(x)} = \frac{-P(x)}{2}$$

$$\Rightarrow f(x) = e^{-\frac{1}{2}\int P(x)dx} \qquad (4.4)$$

Thus

$$y(x) = \left(e^{-\frac{1}{2}\int P(x)dx}\right)u(x) \qquad (4.5)$$

and Eq. (4.1) becomes

$$u''(x)e^{-\frac{1}{2}\int Pdx} + u(x)\left\{Q(x) - \frac{1}{4}P^2(x) - \frac{1}{2}P'(x)\right\}e^{-\frac{1}{2}\int Pdx} = 0$$

$$\Rightarrow u''(x) + \left\{Q(x) - \frac{1}{4}P^2(x) - \frac{1}{2}P'(x)\right\}u(x) = 0$$

Since $e^{-\frac{1}{2}\int Pdx} \neq 0$

Here, if we put

$$p(x) = Q(x) - \frac{1}{4}P^2(x) - \frac{1}{2}P'(x) \qquad (4.6)$$

Then we get the normal form $u''(x) + p(x)u(x) = 0$

From (4.5) it is quite evident that the zeros of the solutions of (4.1) are exactly the same as those of the solutions of (4.3). Therefore the above transformation does not affect the oscillation behaviour of the solutions of (4.1). Since it is convenient to deal with equations of the form (4.3) to discuss the oscillation behaviour of the solutions, we shall henceforth consider the equations of the form (4.3).

The following theorem suggests how the oscillation of solutions of Eq. (4.3) depends on the sign of $p(x)$ in an interval.

Theorem 4.5 Let $u(x)$ be a nontrivial solution of (4.3). If

(i) $p(x) \leq 0$, then $u(x)$ has at most one zero;

(ii) $p(x) > 0$ for all $x > 0$ and $\int_1^\infty p(x)dx = \infty$, then $u(x)$ has infinitely many zeros in $(0, +\infty)$.

Proof: (i) If $p(x) = 0$, then we know that $u(x)$ has exactly one zero. So let us assume that $p(x) < 0$ and x_0 is a zero of $u(x)$, so that $u(x_0) = 0$, and by Theorem 4.1, $u'(x_0) \neq 0$. Let us assume $u'(x_0) > 0$, so that $u(x)$ is positive over some interval $(x_0, x_0 + a)$, $a > 0$.

Thus

$u''(x) = -p(x)u(x)$ is a positive function on $(x_0, x_0 + a)$ as $p(x) < 0$.

$\Rightarrow u'(x)$ is an increasing function in $(x_0, x_0 + a)$

$\Rightarrow u(x) \neq 0$ in $(x_0, x_0 + a)$

Similarly, $u(x) \neq 0$ in $(x_0 - a, x_0)$.

A similar argument holds good when $u'(x_0) < 0$. Thus $u(x)$ has at most one zero.

(ii) Let us assume the contrary, namely, $u(x)$ vanishes at most finite number of times in $(0, \infty)$, so that there exists a point $x_0 > 1$ such that $u(x) \neq 0$ for $x > x_0$.

Since $u(x)$ has no zeros for $x > x_0$, $u(x)$ is either positive or negative in (x_0, ∞). Let us assume that $u(x) > 0$ in (x_0, ∞). Then $u''(x) = -p(x)u(x)$ is negative, which implies that the graph of $u(x)$ is above the x-axis, concave down and the slope $u'(x)$ is decreasing. Does $u'(x)$ become negative? Yes, because if we put

$$V(x) = -\frac{u'(x)}{u(x)} \text{ for } x > x_0, \text{ then}$$

$$V'(x) = \frac{-u''(x)u(x) + (u'(x))^2}{(u(x))^2} = p(x) + (V(x))^2$$

Integrating from x_0 to x, we get

$$V(x) - V(x_0) = \int_{x_0}^{x} p(x)dx + \int_{x_0}^{x} (V(x))^2 dx$$

$$\Rightarrow V(x) = V(x_0) + \int_{x_0}^{x} p(x)dx + \int_{x_0}^{x} (V(x))^2 dx$$

This implies that $V(x)$ is positive for large values of x as $\int_{1}^{\infty} p(x)dx = \infty$,

and hence $\int_{x_0}^{\infty} p(x)dx = \infty$.

This in turn implies that $V(x) = -\frac{u'(x)}{u(x)} > 0$, and hence $u'(x) < 0$ as $u(x) > 0$ for large values of x.

$u'(x) < 0$ implies that the graph of $u(x)$ crosses the x-axis in (x_0, a), $a > 0$. This in turn implies that $u(x)$ has a zero for $x > x_0$. This contradiction completes the proof of the theorem.

Remark 4.1 The hypothesis $\int_{1}^{\infty} p(x)dx = \infty$ in part (ii) of Theorem 4.5 is a must.

Properties of Solutions

The conclusion of the theorem may not be true without this. The following example explains this idea.

Example 4.9 Consider the Euler equation $y'' + \left(\dfrac{k}{x^2}\right) y = 0, x > 0$. Every nontrivial solution of this equation has an infinite number of positive zeros if $k > \dfrac{1}{4}$ and only a finite number if $k \leq \dfrac{1}{4}$. But the hypothesis is not satisfied, as

$$\int_1^\infty \frac{k}{x^2} \, dx = k \neq \infty$$

To solve the differential equation we follow the method discussed in Section 3.4.

Putting $z = \ln x$ in the given equation, we get

$$\frac{d^2 y}{dz^2} - \frac{dy}{dz} + ky = 0$$

Then the auxiliary equation is
$$m^2 - m + ky = 0$$
whose roots are $m = \dfrac{1 \pm \sqrt{1 - 4k}}{2}$.

These roots are real or complex depending on $k \leq \dfrac{1}{4}$ or $k > \dfrac{1}{4}$. If $k < \dfrac{1}{4}$, then the roots are real and distinct, say m_1 and m_2, and the solution is given by

$$y = A e^{m_1 z} + B e^{-m_2 z} = A x^{m_1} + B x^{-m_2}$$

If $k = \dfrac{1}{4}$, then the roots are equal, say $m_1 = m_2$, and the solution is given by

$$y = (Az + B) e^{m_1 z} = (A \ln x + B) x^{m_1}$$

In both the above cases, the solution of the given equation has finitely many zeros in $(0, \infty)$.

If $k > \dfrac{1}{4}$, then the roots are complex, say $\alpha \pm i\beta$ and the solution is given by
$$y = e^{\alpha z} (A \cos \beta z + B \sin \beta z)$$
$$= x^\alpha (A \cos \beta \ln x + B \sin \beta \ln x)$$

which has infinitely many zeros in $(0, \infty)$.

4.3 STURM COMPARISON THEOREM

Now we state and prove the Sturm comparison theorem, which has many applications and provides information about the zeros of special functions, to be discussed in the next chapter.

Theorem 4.6 (Sturm Comparison Theorem) Let $u(x)$ and $v(x)$ be nontrivial solutions of

$$u''(x) + p(x)u(x) = 0$$

and

$$v''(x) + q(x)v(x) = 0$$

where $p(x) > q(x)$ in an interval I. Then between any two consecutive zeros of $v(x)$, there is a zero of $u(x)$ in I.

Proof: Let x_1 and x_2 be two consecutive zeros of $v(x)$ in the interval I.
Then $v(x_1) = v(x_2) = 0$, and $v(x)$ does not vanish in (x_1, x_2). So $v(x)$ is either positive or negative in (x_1, x_2). Let us suppose that $v(x) > 0$ in (x_1, x_2).
If $u(x)$ does not vanish in (x_1, x_2), then let us suppose that $u(x) > 0$ in (x_1, x_2). Consider the Wronskian $W[x; u(x), v(x)]$ of the functions $u(x), v(x)$ at x.

$$W[x; u(x), v(x)] = u(x)v'(x) - u'(x)v(x)$$
$$\Rightarrow W'[x; u(x), v(x)] = u(x)v''(x) - u''(x)v(x)$$
$$\Rightarrow W'(x) = u(x)(-q(x)v(x)) - v(x)(-p(x)u(x))$$
$$\Rightarrow W'(x) = (p(x) - q(x))u(x)v(x) > 0 \quad \text{on} \quad (x_1, x_2)$$

Integrating both the sides from x_1 to x_2, we get

$$W(x_2) - W(x_1) > 0 \Rightarrow W(x_2) > W(x_1) \tag{4.7}$$

But

$$W(x_1) = W[x_1; u(x_1), v(x_1)] = u(x_1)v'(x_1) - u'(x_1)v(x_1) = u(x_1)v'(x_1) \geq 0$$

as $v(x)$ is positive in (x_1, x_2), $v'(x_1)$ must be increasing at x_1 and decreasing at x_2, so $u(x_1) \geq 0$ and $v'(x_1) > 0$
and

$$W(x_2) = W[x_2; u(x_2), v(x_2)] = u(x_2)v'(x_2) - u'(x_2)v(x_2) = u(x_2)v'(x_2) \leq 0$$

by the above argument.
This contradicts (4.7) and proves the theorem that $u(x)$ has at least one zero in (x_1, x_2).

We conclude from the above theorem that: (i) If $v(x)$ has infinitely many zeros in any interval I, then $u(x)$ also has infinitely many zeros in I; in other words, if $v(x)$ oscillates in I then $u(x)$ also does so.

(ii) If $u(x)$ has finitely many zeros in any interval I, then $v(x)$ also has finitely many zeros in I, i.e., if $u(x)$ is a nonoscillating function in I, then so is $v(x)$.

Let us now discuss a few examples to understand the concepts discussed above.

Example 4.10 Show that any Airy function (a solution of Airy's equation $y'' + xy = 0$) has infinitely many positive zeros and at most one negative zero. Further, an Airy function vanishes at least once in any interval of length π which is included in $(1, \infty)$.

If $x < 0$, then by Theorem 4.5, an Airy function has at most one zero. If $x > 0$ and

$$\int_1^\infty x\, dx = \infty,$$ then again, by Theorem 4.5, an Airy function has infinitely many positive zeros.

Properties of Solutions 83

To show that an Airy function vanishes at least once in an interval of length π contained in $(1, \infty)$, let us apply Theorem 4.6 by taking two equations

$$y'' + xy = 0 \qquad (4.8)$$

and
$$v'' + v = 0 \qquad (4.9)$$

We know the general solution of (4.9) is given by

$$v(x) = A \sin x + B \cos x$$

where A and B are arbitrary constants. We can choose them as $A = \alpha \cos \beta$, $B = \alpha \sin \beta$, for any α, β. Then

$$v(x) = \alpha \sin x \cos \beta + \alpha \sin \beta \cos x = \alpha \sin(x + \beta)$$
$$\Rightarrow v(x) = 0 \text{ iff } x + \beta = n\pi \qquad \text{for } n = 0, 1, 2, 3, \ldots$$

For $x > 1$, an Airy function vanishes at least once between any two consecutive zeros $n\pi$, $(n+1)\pi$ of the solution of Eq. (4.9), by Theorem 4.6.

Clearly, the length of the interval $[n\pi, (n+1)\pi]$ is π. This completes the proof of the example.

Example 4.11 Prove that every nontrivial solution of the Hermite equation $y'' - 2xy' + 2py = 0$, $p \geq 0$, has at most finitely many zeros on $(-\infty, +\infty)$.

First let us find the normal form of the Hermite equation.

Putting $y = uf$, $y' = u'f + uf'$, $y'' = u''f + 2u'f' + uf''$ in the Hermite equation we get

$$u''f + 2u'f' + uf'' - 2xu'f - 2xuf' + 2puf = 0$$
$$\Rightarrow u''f + u'(2f' - 2xf) + u(f'' - 2xf' + 2pf) = 0 \qquad (4.10)$$

Putting $2f' - 2xf = 0$, we get $f = e^{x^2/2}$. Then $f' = xe^{x^2/2}$ and $f'' = e^{x^2/2} + x^2 e^{x^2/2}$.

So the normal form of the Hermite equation is obtained by putting these values in (4.10) as follows:

$$u'' + \{1 - x^2 + 2p\}u = 0 \qquad (4.11)$$

By Theorem 4.5, if $1 - x^2 + 2p \leq 0$, then $u(x)$, a nontrivial solution of (4.11) has at most one zero in $(-\infty, -\sqrt{1+2p}] \cup [\sqrt{1+2p}, +\infty)$, since $x^2 \geq 1 + 2p$.

If
$$1 - x^2 + 2p > 0$$
$$\Rightarrow x^2 < 1 + 2p$$
$$\Rightarrow -\sqrt{1+2p} < x < \sqrt{1+2p}$$

then $u(x)$ has finitely many zeros in a finite interval $(-\sqrt{1+2p}, \sqrt{1+2p})$ by Theorem 4.2.

Thus $u(x)$ and hence $y(x) = e^{x^2/2} u(x)$ has finitely many zeros in $(-\infty, +\infty)$.

Example 4.12 Show that every nontrivial solution of Bessel's equation

$$x^2y'' + xy' + (x^2 - p^2)y = 0 \quad p \geq 0 \quad (4.12)$$

has infinitely many positive zeros.

Putting $y = uf$, $y' = u'f + uf'$, $y'' = u''f + 2u'f' + uf''$ in Eq. (4.12), we get

$$x^2u''f + x^22u'f' + x^2uf'' + xu'f + xuf' + (x^2 - p^2)uf = 0$$

$$\Rightarrow x^2fu'' + u'(2x^2f' + xf) + u(x^2f'' + xf' + (x^2 - p^2)f) = 0 \quad (4.13)$$

Now put $2x^2f' + xf = 0 \Rightarrow f = \dfrac{1}{\sqrt{x}}$

Then $f' = \dfrac{-1}{2x^{3/2}}$ and $f'' = \dfrac{3}{4x^{5/2}}$. With this, Eq. (4.13) becomes

$$x^{3/2}u'' + \frac{1 + 4(x^2 - p^2)}{4\sqrt{x}} u = 0$$

$$\Rightarrow u'' + \left(1 + \frac{1 - p^2}{4x^2}\right)u = 0 \quad (4.14)$$

is the normal form of Bessel's equation (4.12).

Here, we can choose $x_0 > 0$ such that

$$1 + \frac{1 - p^2}{4x^2} > 0 \quad \text{for all } x > x_0$$

and

$$\int_1^\infty \left(1 + \frac{1 - p^2}{4x^2}\right) dx = \infty$$

Hence by Theorem 4.5, $u(x)$, the nontrivial solution of (4.14), has infinitely many positive zeros.

Therefore, for $x > 0$, $y(x) = \dfrac{u(x)}{\sqrt{x}}$, the nontrivial solution of (4.12), has infinitely many positive zeros.

These examples suggest the oscillation behaviour of some of the special functions like Hermite function (non-trivial solution of Hermite equation) and Bessel's function (nontrivial solution of Bessel's equation). We shall discuss special functions and equations in more detail in the next chapter.

Exercises 4.3

1. Find the normal form of the equation $xy'' + 2y' + xy = 0$, $x > 0$. Can two differential equations have the same normal form? Justify.
2. Find the normal form of the following equations:

(i) $\cos^2 x y'' + \sin 2x\, y' + \sin x\, y = 0 \qquad 0 < x < \dfrac{\pi}{2}$

(ii) $x^2 y'' - 4xy' + (x^4 + 6)y = 0 \qquad x > 0$

(iii) $4x^2 y'' + 4xy' + y = 0 \qquad x > 0$

(iv) $4x^2 y'' + 4x^3 y' + \{(x^2 + 1)^2 + 1\}y = 0 \qquad x > 0$

3. Show that every nontrivial solution of $y'' + (\sin^2 x + 1)y = 0$ has an infinite number of positive zeros.

4. Show that every nontrivial solution of the equation $y'' + (\sinh x)y = 0$ has at most one negative zero and infinitely many positive zeros.

5. Show that $\sin k_1 x$ has at least one zero between any two zeros of $\sin k_2 x$ whenever $k_1 > k_2 > 0$.

6. Show that $\cos k_1 x$ has at least one zero between any two zeros of $\cos k_2 x$ whenever $k_1 > k_2 > 0$.

7. Show that the nontrivial solutions of $y'' + \dfrac{y}{x \ln x} = 0,\ x > 1$ are oscillating functions.

8. If $y(x)$ is a nontrivial solution of $y'' + q(x)y = 0$, then show that $y(x)$ has an infinite number of positive zeros if $q(x) > k/x^2$ for some $k > 1/4$, and only a finite number if $q(x) < 1/4x^2$, $x > 0$.

9. If $1 \le f(x) \le 4$ for $c \le x \le c + \pi$, then show that any nontrivial solution of $y'' + f(x)y = 0$ has one or two zeros in $(c, c + \pi)$.

10. Does every nontrivial solution of the differential equation
$$x^2 y'' + 2x^3 y' + \{(x^2 + 1)^2 - 2\}y = 0$$
oscillate on the positive x-axis? Justify your answer.

CHAPTER 5

Special Functions and Equations

5.1 INTRODUCTION

In the solution of a great many types of problems in applied mathematics, we are led to the solution of linear differential equations or sets of linear differential equations. If the solutions to some of these equations are given in terms of elementary functions like polynomials, trigonometric functions, hyperbolic functions, exponential functions, logarithmic functions etc. or an algebraic combination of these functions, then the problem is considered to be solved. But unfortunately there are many equations which cannot be solved in terms of these elementary functions. Solutions of such equations motivate the introduction of many special functions. They are well-tabulated like the elementary functions. They also occur in many applications in Engineering, Physics and various branches of Mathematics. In this chapter we propose to study some of them in view of their practical importance. However, we start with two functions which cannot be introduced through differential equations, but often arise in the solution of physical problems and are of great importance in various branches of mathematical analysis.

5.2 GAMMA AND BETA FUNCTIONS

These two functions need not have to be studied independently because they are closely related. The Gamma function arose while solving an interpolation problem. The problem was to find a monotonic analytic function defined on $(1, \infty)$ which took the value $n!$ at n. It can be solved by using the improper integral $\int_0^\infty t^{x-1} e^{-t} \, dt$, which converges for $x > 0$.

5.3 GAMMA FUNCTION

The *Gamma function*, denoted by $\Gamma(x)$, is defined by

$$\Gamma(x) = \int_0^\infty t^{x-1} e^{-t} dt \quad \text{for } x > 0 \tag{5.1}$$

It is easy to see that $\Gamma(1) = 1$, and substitution of $x + 1$ for x in (5.1) gives

$$\Gamma(x+1) = \int_0^\infty t^x e^{-t} dt$$

$$= -t^x e^{-t}\Big]_0^\infty + \int_0^\infty xt^{x-1} e^{-t} dt \quad \text{integrating by parts}$$

$$= x\int_0^\infty t^{x-1} e^{-t} dt = x\Gamma(x)$$

Thus the recurrence formula for the Gamma function is

$$\Gamma(x+1) = x\,\Gamma(x) \tag{5.2}$$

In particular, if $x = n$, a positive integer, then

$$\Gamma(n+1) = n\Gamma(n) = n!$$

That is why sometimes $\Gamma(x)$ is called the *factorial function*.

5.4 DOMAIN OF GAMMA FUNCTION

Though the Gamma function was invented for solving a specific problem, its applicability made it desirable that its domain be extended as much as possible. It is natural to do so by using the relation (5.2) as $\Gamma(x) = \dfrac{\Gamma(x+1)}{x}$.

Thus the domain is first extended to $(-1, 0)$ then to $(-2, -1)$, and so on. For example

$$\Gamma\left(-\frac{7}{3}\right) = \frac{\Gamma\left(-\frac{4}{3}\right)}{-\frac{7}{3}} = \frac{\Gamma\left(-\frac{1}{3}\right)}{\left(-\frac{7}{3}\right)\left(-\frac{4}{3}\right)} = \frac{\Gamma\left(\frac{2}{3}\right)}{\left(-\frac{7}{3}\right)\left(-\frac{4}{3}\right)\left(-\frac{1}{3}\right)} = \frac{-27}{28}\Gamma\left(\frac{2}{3}\right)$$

Hence the Gamma function is defined on the whole of real line, provided we assume that $\Gamma(x) = \infty$ for $x = 0, -1, -2, \ldots$.

Introduction to Differential Equations

Before we discuss further the properties of Gamma function, let us define the Beta function.

5.5 BETA FUNCTION

Beta function, denoted by $\beta(x, y)$, is defined by

$$\beta(x, y) = \int_0^1 t^{x-1}(1-t)^{y-1} dt \qquad (5.3)$$

which is convergent for $x > 0$, $y > 0$.

It is easy to see by letting $t = 1 - s$ that

$$\beta(x, y) = \int_0^1 t^{x-1}(1-t)^{y-1} dt = \int_0^1 s^{y-1}(1-s)^{x-1} ds = \beta(y, x) \qquad (5.4)$$

If we let $t = \sin^2 \theta$ in (5.3), then we get

$$\beta(x, y) = 2 \int_0^{\pi/2} (\sin \theta)^{2x-1} (\cos \theta)^{2y-1} d\theta \qquad (5.5)$$

The substitution $t = s/a$ in (5.3) gives

$$\beta(x, y) = \frac{1}{a^{x+y-1}} \int_0^a s^{x-1}(a-s)^{y-1} ds \qquad (5.6)$$

If $t = \dfrac{s}{1+s}$ in (5.3), then we obtain

$$\beta(x, y) = \int_0^\infty \frac{s^{y-1} ds}{(1+s)^{x+y}} \qquad (5.7)$$

Replacing t by $\dfrac{s+1}{2}$ in (5.3), we get yet another form for

$$\beta(x, y) = \frac{1}{2^{x+y-1}} \int_{-1}^1 (1+t)^{x-1}(1-t)^{y-1} dt \qquad (5.8)$$

These are the more common forms of the Beta function.

5.6 RELATIONSHIP BETWEEN GAMMA AND BETA FUNCTIONS

Consider the Gamma function as given in (5.1),

$$\Gamma(x) = \int_0^\infty t^{x-1} e^{-t}\, dt$$

Changing the variable t by the substitution $t = u^2$ we have

$$\Gamma(x) = 2 \int_0^\infty u^{2x-1} e^{-u^2}\, du \qquad (5.9)$$

We may also write $\Gamma(y) = 2 \int_0^\infty v^{2y-1} e^{-v^2}\, dv \qquad (5.10)$

Thus from (5.9) and (5.10), we obtain

$$\Gamma(x)\,\Gamma(y) = 4 \left(\int_0^\infty u^{2x-1} e^{-u^2}\, du \right)\left(\int_0^\infty v^{2y-1} e^{-v^2}\, dv \right)$$

$$= 4 \int_0^\infty \int_0^\infty u^{2x-1} v^{2y-1} e^{-(u^2+v^2)}\, du\, dv \qquad (5.11)$$

This can be visualised as a surface integral in the uv-plane. Changing the uv-coordinate system to $r\theta$-coordinate system by $u = r \cos\theta$ and $v = \sin\theta$, (5.11) becomes

$$\Gamma(x)\Gamma(y) = 4 \int_0^\infty \int_0^{\pi/2} r^{2(x+y)-2} (\sin\theta)^{2x-1}(\cos\theta)^{2y-1} e^{-r^2} r\, dr\, d\theta$$

$$= \left(2\int_0^\infty r^{2(x+y)-1} e^{-r^2}\, dr \right)\left(2\int_0^{\pi/2} (\sin\theta)^{2x-1}(\cos\theta)^{2y-1}\, d\theta \right)$$

$$= \Gamma(x+y)\,\beta(x,y) \qquad \text{by (5.9) and (5.5)}$$

Hence we have the relation

$$\beta(x, y) = \frac{\Gamma(x)\,\Gamma(y)}{\Gamma(x+y)} \qquad (5.12)$$

In particular, $\left\{\Gamma\left(\dfrac{1}{2}\right)\right\}^2 = \beta\left(\dfrac{1}{2}, \dfrac{1}{2}\right) = \int_0^{\pi/2} 2\, d\theta = \pi$ from (5.5).

Hence $\Gamma\left(\dfrac{1}{2}\right) = \sqrt{\pi}$, since the integrand in (5.1) is positive, $\Gamma(x) > 0$. Thus the Gamma function can now be evaluated easily for integral and half-integral values of x.

The formula (5.12) is very useful for the evaluation of certain classes of definite integrals. For example, from (5.5) and (5.12) we obtain

$$\int_0^{\pi/2} (\sin\theta)^{2x-1}(\cos\theta)^{2y-1} d\theta = \frac{\Gamma(x)\Gamma(y)}{2\Gamma(x+y)} \tag{5.13}$$

We conclude this section by introducing another important function called the *Digamma function*, which is obtained from the Gamma function by differentiating it.

5.7 DIGAMMA FUNCTION

Differentiating (5.1) with respect to x, we get $\Gamma'(x) = \int_0^\infty t^{x-1} e^{-t} \ln t\, dt$, using logarithmic differentiation for the integrand on the right hand side of (5.1).

Again differentiating (5.2), we have

$$\Gamma'(x+1) = x\Gamma'(x) + \Gamma(x)$$

Dividing both the sides of this by (5.2), we have

$$\frac{\Gamma'(x+1)}{\Gamma(x+1)} = \frac{\Gamma'(x)}{\Gamma(x)} + \frac{1}{x}$$

If we put $\psi(x) = \dfrac{\Gamma'(x)}{\Gamma(x)}$ (5.14)

then we have $\psi(x+1) = \psi(x) + \dfrac{1}{x}$.

Equation (5.14) defines the Digamma function $\psi(x)$.

In particular, for integral values n of x we have

$$\psi(n+1) = \psi(n) + \frac{1}{n} = \psi(n-1) + \frac{1}{n-1} + \frac{1}{n}$$

or

$$\psi(n+1) = \psi(1) + \sum_{k=1}^{n} \frac{1}{k} \tag{5.15}$$

Now $\psi(1) = \dfrac{\Gamma'(1)}{\Gamma(1)} = \int_0^\infty e^{-t} \ln t\, dt$, and it can be shown that its value is $-\gamma$,

Special Functions and Equations

where γ, known as Euler's constant, is defined as $\lim\limits_{n\to\infty}\left\{\sum\limits_{k=1}^{n}\dfrac{1}{k} - \ln n\right\}$. This number considered important as e or π, is approximately 0.5772156649.

Example 5.1 Evaluate $\int_0^{\pi/2} \sqrt{\cot\theta}\, d\theta$.

$$\int_0^{\pi/2} \sqrt{\cot\theta}\, d\theta = \int_0^{\pi/2} \cos^{\frac{1}{2}}\theta \sin^{-\frac{1}{2}}\theta\, d\theta = \frac{\Gamma\left(\frac{1}{4}\right)\Gamma\left(\frac{3}{4}\right)}{2\Gamma\left(\frac{1}{4}+\frac{3}{4}\right)} \quad \text{by (5.13)}$$

$$= \frac{1}{2}\Gamma\left(\frac{1}{4}\right)\Gamma\left(\frac{3}{4}\right)$$

$$= \frac{1}{2}\frac{\pi}{\sin\dfrac{\pi}{4}} \quad \text{by (5.19) and Example 5.5}$$

$$= \frac{\pi}{\sqrt{2}}$$

Example 5.2 Prove that $\Gamma(x) = \int_0^1 \left(\ln\dfrac{1}{y}\right)^{x-1} dy$.

Putting $t = \ln\dfrac{1}{y} = -\ln y$ on the right hand side, we obtain

$$-\int_\infty^0 t^{x-1} e^{-t}\, dt = \int_0^\infty t^{x-1} e^{-t}\, dt = \Gamma(x)$$

Example 5.3 Express the following integral in terms of the Beta function:

$$\int_a^b (b-z)^{x-1}(z-a)^{y-1} dz \qquad b > a \quad x > 0 \quad y > 0$$

Putting $t = \dfrac{z-a}{b-a}$ in the left hand side, we obtain

$$\int_a^b (b-z)^{x-1}(z-a)^{y-1} dz = \int_0^1 (b-a)^{x-1}(1-t)^{x-1}(b-a)^{y-1}t^{y-1}(b-a)\, dt$$

$$= (b-a)^{x+y-1} \int_0^1 t^{y-1}(1-t)^{x-1} dt = (b-a)^{x+y-1} \beta(y,x)$$

$$= (b-a)^{x+y-1} \beta(x,y) \qquad \text{by (5.4)}$$

Example 5.4 Prove that $\beta(x, x) = \dfrac{\sqrt{\pi}\,\Gamma(x)}{2^{2x-1}\,\Gamma\left(x + \dfrac{1}{2}\right)}$.

We have

$$\beta(x,x) = \int_0^1 t^{x-1}(1-t)^{x-1}\,dt$$

Putting $t = \sin^2\theta$ in this, we get

$$\beta(x,x) = \int_0^{\pi/2} 2\sin^{2x-2}\theta \cos^{2x-2}\theta \sin\theta \cos\theta\,d\theta$$

$$= \int_0^{\pi/2} \frac{(\sin 2\theta)^{2x-2} \sin 2\theta\,d\theta}{2^{2x-2}}$$

$$= \int_0^{\pi/2} \frac{(\sin 2\theta)^{2x-1}}{2^{2x-2}}\,d\theta$$

$$= \frac{1}{2^{2x-1}} \int_0^{\pi} (\sin\phi)^{2x-1}(\cos\phi)^{2\frac{1}{2}-1}\,d\phi \qquad \text{putting } 2\theta = \phi$$

$$= \frac{2}{2^{2x-1}} \int_0^{\pi/2} (\sin\phi)^{2x-1}(\cos\phi)^{2\frac{1}{2}-1}\,d\phi$$

$$= \frac{1}{2^{2x-1}}\beta\left(x, \frac{1}{2}\right) = \frac{1}{2^{2x-1}} \frac{\Gamma(x)\,\Gamma\left(\dfrac{1}{2}\right)}{\Gamma\left(x + \dfrac{1}{2}\right)} \qquad \text{by (5.12) and (5.13)}$$

$$= \frac{\sqrt{\pi}\,\Gamma(x)}{2^{2x-1}\,\Gamma\left(x + \dfrac{1}{2}\right)}$$

Example 5.5 Prove that $\Gamma(x)\,\Gamma(1 - x) = \dfrac{\pi}{\sin \pi x}$ for non-integral values of x. This is called the *Reflection Formula*.

Substituting (5.7) into the relation (5.12), we obtain

$$\int_0^{\infty} \frac{s^{x-1}\,ds}{(1+s)^{x+y}} = \frac{\Gamma(x)\,\Gamma(y)}{\Gamma(x+y)} \qquad x > 0 \quad y > 0 \qquad (5.16)$$

If we now let $y = 1 - x$, $0 < x < 1$ in (5.16), we obtain

$$\int_0^\infty \frac{s^{x-1}ds}{(1+s)} = \frac{\Gamma(x)\Gamma(1-x)}{\Gamma(1)} \tag{5.17}$$

We know from complex variables by integrating around a branch point that

$$\int_0^\infty \frac{s^{x-1}}{1+s} ds = \frac{\pi}{\sin \pi x} \qquad 0 < x < 1 \tag{5.18}$$

Now, from (5.17) and (5.18) and using the fact that $\Gamma(1) = 1$, we have

$$\Gamma(x)\Gamma(1-x) = \frac{\pi}{\sin \pi x} \quad \text{for } 0 < x < 1$$

We now remove the restriction $0 < x < 1$. Suppose that $x = t + n$, where n is an integer and $0 < t < 1$.
Then
$$\Gamma(x)\Gamma(1-x) = \Gamma(t+n)\Gamma(1-t-n)$$
$$= (t+n-1)(t+n-2) \cdots t\Gamma(t) \cdot \left(\frac{1}{1-t-n}\right)\left(\frac{1}{2-t-n}\right)\cdots$$
$$\left(\frac{1}{-t}\right)\Gamma(1-t) \qquad \text{by (5.2)}$$
$$= (-1)^n \Gamma(t)\Gamma(1-t)$$
$$= (-1)^n \frac{\pi}{\sin \pi t} \qquad \text{by (5.18)}$$
$$\Rightarrow \Gamma(x)\Gamma(1-x) = \frac{\pi}{\sin(n\pi + \pi t)} = \frac{\pi}{\sin \pi x} \tag{5.19}$$

Exercises 5.7

1. Evaluate $\Gamma\left(-\frac{1}{2}\right)$ and $\Gamma\left(-\frac{7}{2}\right)$.

2. Show that $\Gamma\left(\frac{2n+1}{2}\right) = \frac{1 \cdot 3 \cdot 5 \cdots (2n-1)}{2^n}\sqrt{\pi}$.

3. Show that $\int_0^1 \frac{dx}{\sqrt{(1-x^n)}} = \frac{\sqrt{\pi}}{n} \frac{\Gamma(1/n)}{\Gamma\left(\frac{1}{n}+\frac{1}{2}\right)}$.

4. Show that $\int_0^1 \frac{x^n dx}{\sqrt{(1-x^2)}} = \begin{cases} \dfrac{1 \cdot 3 \cdot 5 \cdots (n-1)}{2 \cdot 4 \cdot 6 \cdots n} \cdot \dfrac{\pi}{2} & \text{if } n \text{ is an even positive integer} \\ \dfrac{2 \cdot 4 \cdot 6 \cdots (n-1)}{1 \cdot 3 \cdot 5 \cdots n} & \text{if } n \text{ is an odd positive integer} \end{cases}$

5. Evaluate (a) $\int_0^\infty e^{-x^4} dx$ (b) $\int_0^\infty \dfrac{e^{-x}}{\sqrt{x}} dx$ (c) $\int_0^\infty e^{-\sqrt{x}} dx$.

6. Show that $\beta(x, y) = \beta(x + 1, y) + \beta(x, y + 1)$.

7. Prove that $\int_0^{\pi/2} \tan^n \theta\, d\theta = \dfrac{1}{2}\Gamma\left(\dfrac{n+1}{2}\right)\Gamma\left(\dfrac{1-n}{2}\right)$ if $|n| < 1$.

8. Prove that $\int_0^{\pi/2} \sin^n \theta\, d\theta = \int_0^{\pi/2} \cos^n \theta\, d\theta = \dfrac{\sqrt{\pi}}{2} \dfrac{\Gamma\left(\dfrac{n+1}{2}\right)}{\Gamma\left(\dfrac{n+2}{2}\right)}$.

9. Show that $\dfrac{d^n \Gamma(x)}{dx^n} = \int_0^\infty t^{x-1} e^{-t} (\ln t)^n\, dt$.

10. Show that $2\int_0^{\pi/2} \sqrt{\tan \theta}\, d\theta = \Gamma\left(\dfrac{1}{4}\right)\Gamma\left(\dfrac{3}{4}\right) = 4\int_0^\infty \dfrac{x^2 dx}{1+x^4} = \pi\sqrt{2}$.

11. For $x, y > 0$, prove that $\int_0^\infty e^{-yt} t^{x-1} dt = y^{-x} \Gamma(x)$.

12. Show that if $c > 1$, $\displaystyle\int_0^\infty \dfrac{x^c}{c^x}\, dx = \dfrac{\Gamma(c+1)}{(\ln c)^{c+1}}$.

13. Show that if $n > -1$,

$$\int_0^\infty x^n e^{-k^2 x^2}\, dx = \dfrac{1}{k^{n+1}} \Gamma\left(\dfrac{n+1}{2}\right).$$

Hence or otherwise evaluate $\displaystyle\int_{-\infty}^{+\infty} e^{-k^2 x^2}\, dx$.

14. Express $\displaystyle\int_0^1 x^m (1-x^n)^p\, dx$ in terms of the Beta function and hence

evaluate $\displaystyle\int_0^1 x^5 (1-x^3)^{10}\, dx$.

15. Prove that $\int_0^\infty e^{-at} t^x dt = \dfrac{1}{a^{x+1}} \Gamma(x+1)$.

16. Prove that $\int_0^\infty x^m e^{-x^n} dx = \dfrac{1}{n} \Gamma\left(\dfrac{m+1}{n}\right)$, where $m > -1$, $n > 0$.

17. Show that $\Gamma(x)\Gamma(-x) = \dfrac{-\pi}{x \sin \pi x}$.

18. Show that $\Gamma\left(\dfrac{1}{2} + x\right) \Gamma\left(\dfrac{1}{2} - x\right) = \dfrac{\pi}{\cos \pi x}$.

19. Prove that $\int_a^\infty \exp(2ax - x^2) dx = \dfrac{1}{2}\sqrt{\pi} \exp(a^2)$.

20. Prove that $\Gamma(2x) = \dfrac{2^{2x-1}}{\sqrt{\pi}} \Gamma(x) \Gamma\left(x + \dfrac{1}{2}\right)$. This is called *Duplication Formula*. (**Hint:** Use Example 5.4.)

5.8 HYPERGEOMETRIC FUNCTIONS

We now consider a second order linear differential equation which has two regular singular points which cannot be solved with the help of elementary functions.

The differential equation

$$x(1-x)y'' + [c - (a+b+1)x]y' - aby = 0 \tag{5.20}$$

where a, b, c are fixed parameters, is called the *Hypergeometric Equation* with parameters a, b, c.

This equation has regular singular points at 0 and 1. First let us find a Frobenius series solution to Eq. (5.20) about the regular singular point $x = 0$. For that, we have

$$xP(x) = \dfrac{c - (a+b+1)x}{(1-x)} = [c - (a+b+1)x](1 + x + x^2 + \cdots)$$

and

$$x^2 Q(x) = \dfrac{-abx}{1-x} = (-abx)(1 + x + x^2 + \cdots)$$

which suggest that $p_0 = c$ and $q_0 = 0$, so that the indicial equation is $m^2 + m(c-1) = 0$, whose roots are $m_1 = 0$ and $m_2 = 1 - c$. If $1 - c \leq 0$, then we know (see Chapter 3) that Eq. (5.20) has a power series solution of the form

$$y_1 = x^0 \sum_{n=0}^\infty a_n x^n \qquad a_0 \neq 0 \tag{5.21}$$

Introduction to Differential Equations

The usual calculation shows that

$$a_{n+1} = \frac{(a+n)(b+n)}{(n+1)(c+n)} a_n \quad \text{for } n = 0, 1, 2, \ldots$$

This "explains" rather unnatural coefficients occuring in Eq. (5.20). From the above recurrence relation of the coefficients, we observe that all the coefficients in (5.21) depend on $a_0 \neq 0$. If we set $a_0 = 1$, then (5.21) becomes

$$y_1 = 1 + \sum_{n=1}^{\infty} \frac{a(a+1) \cdots (a+n-1) b(b+1) \cdots (b+n-1)}{n! \, c(c+1) \cdots (c+n-1)} x^n \tag{5.22}$$

This is known as the *Hypergeometric Series*, for if we put $a = 1$ and $b = c$, then (5.22) becomes

$$1 + x + x^2 + \cdots = \frac{1}{1-x}$$

the familiar geometric series. This also "justifies" the name given to Eq. (5.20).

If a or b is either zero or a negative integer, then the series (5.22) terminates; otherwise it is easy to see that the series converges for $|x| < 1$. Further, if c is neither zero nor a negative integer, then (5.22) is an analytic function. Thus we have the following:

Definition The analytic function defined by (5.22) when $c \neq 0, -1, -2, \ldots$ is called the *Hypergeometric Function* with parameters a, b, c on the interval $(-1, 1)$ and is denoted by $F(a, b; c; x)$. Thus

$$F(a, b; c; x) = 1 + \sum_{n=1}^{\infty} \frac{a(a+1) \cdots (a+n-1) b(b+1) \cdots (b+n-1)}{n! \, c(c+1) \cdots (c+n-1)} x^n \tag{5.23}$$

To obtain a second linearly independent solution to (5.20) about $x = 0$, we use the fact that when c is not an integer, then (5.20) has a Frobenius series solution of the form

$$y_2 = x^{1-c} \Sigma b_n x^n, \quad \text{where } b_0 \neq 0. \tag{5.24}$$

Instead of calculating the coefficients in the series (5.24) by the usual method, we change the dependent variable by $y = x^{1-c} z$. Now the corresponding solution for z should be a series. The equation satisfied by z turns out to be

$$x(1-x)z'' + [(2-c) - (a+b-2c+3)x]z' - (a-c+1)(b-c+1)z = 0 \tag{5.25}$$

Since $2 - c$ is not an integer, we know that the series solution to (5.25) by (5.22) must be $F(a - c + 1, b - c + 1; 2 - c; x)$.

Accordingly, (5.24) becomes

$$y_2 = x^{1-c} F(a - c + 1, b - c + 1; 2 - c; x) \tag{5.26}$$

Hence the general solution of (5.20), when c is not an integer, about $x = 0$, can be written as

$$y = c_1 F(a, b; c; x) + c_2 x^{1-c} F(a - c + 1, b - c + 1; 2 - c; x) \quad (5.27)$$

If c is an integer, then either one of (5.22) and (5.26) does not exist, depending on the smaller of the roots 0 or $1 - c$. We then use the techniques of Chapter 3 to obtain the second linearly independent solution.

The solution to (5.20) about the singular point $x = 1$ can be obtained either by using similar conditions or by changing the independent variable from x to $t = 1 - x$. This makes $x = 1$ correspond to $t = 0$, and transforms Eq. (5.20) to

$$t(1-t)\frac{d^2 y}{dt^2} + [(a+b-c+1) - (a+b+1)t]\frac{dy}{dt} - aby = 0 \quad (5.28)$$

which is a hypergeometric equation with parameters a, b, and $a + b - c + 1$. If $c - a - b$ is not an integer, then by (5.27) the general solution of (5.28) about $t = 0$ is given by

$$y = c_1 F(a, b; a + b - c + 1; t) + c_2 t^{c-a-b} F(c - b, c - a; c - a - b + 1; t)$$

Hence the general solution of (5.20) about $x = 1$ is given by

$$y = c_1 F(a, b; a + b - c + 1; 1 - x) + c_2 (1 - x)^{c-a-b}$$
$$F(c - b, c - a; c - a - b + 1; 1 - x) \quad (5.29)$$

5.9 PROPERTIES OF HYPERGEOMETRIC FUNCTION

1. It follows from (5.23) that

$$F(a, b; c; x) = F(b, a; c; x) \quad (5.30)$$

2. Differentiating both the sides of (5.23), we find that the derivative of a hypergeometric function is another hypergeometric function, i.e.

$$F'(a,b;c;x) = \frac{ab}{c} F(a+1, b+1; c+1; x) \quad (5.31)$$

3. The way we arrive at (5.22) clearly shows that for $c \geq 1$ any solution of (5.20) which is bounded at $x = 0$ must be a constant multiple of $F(a, b; c; x)$.

4. By using the relation (5.2) of the Gamma function, we can easily express the hypergeometric function as

$$F(a, b; c; x) = \frac{\Gamma(c)}{\Gamma(a)\Gamma(b)} \sum_{n=0}^{\infty} \frac{\Gamma(a+n)\Gamma(b+n)}{n!\,\Gamma(c+n)} x^n \quad (5.32)$$

5. There are many linear relations which connect $F(a, b; c; x)$ with $F(a \pm 1, b \pm 1; c \pm 1; x)$, e.g.

$$(c - a)F(a - 1, b; c; x) + a(x - 1) F(a + 1, b; c; x)$$
$$= [(c - 2a) + (a - b)x] F(a, b; c; x) \quad (5.33)$$

6. For $c > b > 0$, we have the integral representation

$$F(a, b; c; x) = \frac{\Gamma(c)}{\Gamma(b)\Gamma(c-b)} \int_0^1 t^{b-1}(1-t)^{c-b-1}(1 - xt)^{-a} dt \quad (5.34)$$

We have from (5.32)

$$F(a,b;c;x) = \frac{\Gamma(c)}{\Gamma(a)\Gamma(b)} \sum_{n=0}^{\infty} \frac{\Gamma(a+n)\Gamma(b+n)}{n!\Gamma(c+n)} x^n$$

$$= \frac{\Gamma(c)}{\Gamma(a)\Gamma(b)\Gamma(c-b)} \sum_{n=0}^{\infty} \Gamma(a+n) \frac{\Gamma(c-b)\Gamma(b+n)}{\Gamma(c+n)} \frac{x^n}{n!}$$

$$= \frac{\Gamma(c)}{\Gamma(a)\Gamma(b)\Gamma(c-b)} \sum_{n=0}^{\infty} \Gamma(a+n)\beta(c-b,b+n) \frac{x^n}{n!}$$

by the relation (5.12)

$$= \frac{\Gamma(c)}{\Gamma(a)\Gamma(b)\Gamma(c-b)} \sum_{n=0}^{\infty} \Gamma(a+n) \int_0^1 t^{n+b-1}(1-t)^{c-b-1} dt \frac{x^n}{n!}$$

by Definition 5.3, which is valid for $c - b > 0$ and $b + n > 0$

$$= \frac{\Gamma(c)}{\Gamma(b)\Gamma(c-b)} \sum_{n=0}^{\infty} \int_0^1 (1-t)^{c-b-1} t^{b-1} \frac{\Gamma(a+n)}{\Gamma(a)} \frac{(xt)^n}{n!} dt$$

$$= \frac{\Gamma(c)}{\Gamma(b)\Gamma(c-b)} \int_0^1 (1-t)^{c-b-1} t^{b-1} \sum_{n=0}^{\infty} \frac{\Gamma(a+n)}{\Gamma(a)} \frac{(xt)^n}{n!} dt$$

$$= \frac{\Gamma(c)}{\Gamma(b)\Gamma(c-b)} \int_0^1 t^{b-1}(1-t)^{c-b-1}(1-xt)^{-a} dt$$

by using binomial theorem.

7. If $c \neq 0, -1, -2, \ldots;\ c > a + b;$ and $a, b > 0$, then

$$F(a,b;c;1) = \frac{\Gamma(c-a-b)\Gamma(c)}{\Gamma(c-a)\Gamma(c-b)} \qquad (5.35)$$

By (5.34), we have on putting $x = 1$

$$F(a,b;c;1) = \frac{\Gamma(c)}{\Gamma(b)\Gamma(c-b)} \int_0^1 t^{b-1}(1-t)^{c-a-b-1} dt$$

$$= \frac{\Gamma(c)}{\Gamma(b)\Gamma(c-b)} \beta(b, c-a-b) \qquad \text{by Definition 5.3}$$

$$= \frac{\Gamma(c)}{\Gamma(b)\Gamma(c-b)} \frac{\Gamma(b)\Gamma(c-a-b)}{\Gamma(c-a)} \qquad \text{using the relation (5.12)}$$

$$= \frac{\Gamma(c)\,\Gamma(c-a-b)}{\Gamma(c-a)\,\Gamma(c-b)}$$

8. Most of the elementary functions can be expressed in terms of hypergeometric functions in a rather simple way, as the following list shows:

(i) $(1 - x)^a = F(-a, b; b; x)$

(ii) $\ln(1 + x) = x\, F(1, 1; 2; -x)$

(iii) $\ln\left(\dfrac{1+x}{1-x}\right) = 2x\, F\left(\dfrac{1}{2}, 1; \dfrac{3}{2}; x^2\right)$

(iv) $\sin^{-1} x = xF\left(\dfrac{1}{2}, \dfrac{1}{2}; \dfrac{3}{2}; x^2\right)$

(v) $\tan^{-1} x = xF\left(\dfrac{1}{2}, 1; \dfrac{3}{2}; -x^2\right)$

(vi) $e^x = \lim\limits_{a \to \infty} F\left(a, b; b; \dfrac{x}{a}\right)$

(vii) $\sin x = x\left[\lim\limits_{a \to \infty} F\left(a, a; \dfrac{3}{2}; \dfrac{-x^2}{4a^2}\right)\right]$

(viii) $\cos x = \lim\limits_{a \to \infty} F\left(a, a; \dfrac{1}{2}; \dfrac{-x^2}{4a^2}\right)$

Many other functions, which we shall come across later on, can also be expressed in terms of hypergeometric functions. This shows the powerful unifying influence of the hypergeometric function.

Example 5.6 Transform the following differential equation into a hypergeometric equation and then find its general solution:

$$(x - \alpha)(x - \beta)y'' + (\gamma + \delta x)y' + \varepsilon y = 0 \qquad (5.36)$$

where $\alpha \neq \beta$.

If we compare Equations (5.20) and (5.36), then we notice the following common features in both the equations; the coefficients of y'', y' and y are polynomials of degrees 2, 1 and 0, and the first of these polynomials has distinct real zeros. This suggests by a suitable change of independent variable that Eq. (5.36) can be transformed to the hypergeometric form (5.20).

Let us suppose $t = \dfrac{x - \alpha}{\beta - \alpha}$. Then $x = \alpha$ corresponds to $t = 0$ and $x = \beta$ to $t = 1$.

Now $\quad x - \alpha = (\beta - \alpha)t \qquad x - \beta = (\beta - \alpha)(t - 1)$

$$\dfrac{dy}{dx} = \dfrac{1}{\beta - \alpha}\dfrac{dy}{dt} \qquad \dfrac{d^2y}{dx^2} = \dfrac{1}{(\beta - \alpha)^2}\dfrac{d^2y}{dt^2}$$

and Eq. (5.36) reduces to

$$t(t-1)\frac{d^2y}{dt^2} + [\gamma + \delta\alpha + \delta(\beta-\alpha)t]\frac{1}{\beta-\alpha}\frac{dy}{dt} + \varepsilon y = 0$$

or

$$t(1-t)\frac{d^2y}{dt^2} + \left[-\frac{\gamma+\delta\alpha}{\beta-\alpha} - \delta t\right]\frac{dy}{dt} - \varepsilon y = 0$$

which is a hypergeometric equation with parameters a, b and c, given by

$$a+b+1 = \delta, \quad ab = \varepsilon \quad \text{and} \quad c = -\frac{\gamma+\delta\alpha}{\beta-\alpha}$$

Once we obtain the values of the parameters a, b, c, we can get the general solution of (5.36) in terms of the hypergeometric functions by (5.27). For instance, if we take $\alpha = -1$, $\beta = 1$, $\gamma = 4$, $\delta = 5$, and $\varepsilon = 4$, then we have

$$a+b+1 = 5, \quad ab = 4, \quad c = -\frac{4+5(-1)}{2} = \frac{1}{2} \quad \text{and} \quad a = b = 2,$$

and hence the general solution is given by

$$y = c_1 F\left(2, 2; \frac{1}{2}; \frac{x+1}{2}\right) + c_2 \left(\frac{x+1}{2}\right)^{1/2} F\left(\frac{5}{2}, \frac{5}{2}; \frac{3}{2}; \frac{x+1}{2}\right)$$

Example 5.7 Find the general solution of

$$(1-e^x)y'' + \frac{1}{2}y' + e^x y = 0$$

about the singular point $x = 0$.

Some differential equations are of the hypergeometric type even if they do not appear to be so. This is an example of such a differential equation.

Let us put $t = e^x$. Then the above equation takes the form

$$(1-t)\left[t^2\frac{d^2y}{dt^2} + t\frac{dy}{dt}\right] + \frac{1}{2}t\frac{dy}{dt} + ty = 0$$

$$\Rightarrow t(1-t)\frac{d^2y}{dt^2} + \left(\frac{3}{2} - t\right)\frac{dy}{dt} + y = 0$$

which is a hypergeometric equation with parameters $a = 1$, $b = -1$ and $c = \frac{3}{2}$.

Since $x = 0$ corresponds to $t = 1$, the general solution of the hypergeometric equation about $t = 1$ by (5.29) is given by

$$y = c_1 F\left(1, -1; -\frac{1}{2}; 1-t\right) + c_2(1-t)^{3/2} F\left(\frac{5}{2}, \frac{1}{2}; \frac{5}{2}; 1-t\right)$$

Hence the general solution of the given differential equation about the point $x = 0$ is given by

$$y = c_1 F\left(1, -1; -\frac{1}{2}; 1-e^x\right) + c_2(1-e^x)^{3/2} F\left(\frac{5}{2}, \frac{1}{2}; \frac{5}{2}; 1-e^x\right)$$

Exercises 5.9

Find general solutions of Exercises 1–5 in terms of hypergeometric functions about the singular point $x = 0$.

1. $x(1 - x)y'' + \left(\dfrac{1}{2} - 4x\right)y' - 2y = 0$

2. $2x(1 - x)y'' + (1 - 6x)y' - 2y = 0$

3. $3x(1 - x)y'' + (1 - 27x)y' - 45y = 0$

4. $2x(1 - x)y'' + (3 - 10x)y' - 6y = 0$

5. $x(1 - x)y'' + \left(\dfrac{3}{2} - 2x\right)y' + 2y = 0$

Find general solutions of Exercises 6–10 by using hypergeometric functions in the powers of the indicated term.

6. $(2x^2 + 2x)y'' + (4 + 5x)y' + y = 0 \quad x + 1$

7. $(x^2 - 1)y'' + (5x + 4)y' + 4y = 0 \quad x - 1$

8. $(x + 2)(x - 3)y'' + (5 + 3x)y' + y = 0 \quad x - 3$

9. $(1 - x^2)y'' - xy' + y = 0 \quad 1 - x$

10. $(1 - x^2)y'' - xy' + 4y = 0 \quad 1 - x$

In Exercises 11 and 12, use one of the methods discussed in Chapter 3 to obtain two linearly independent solutions to the given hypergeometric equation.

11. $x(1 - x)y'' + (1 - 3x)y' - y = 0$

12. $x(1 - x)y'' + (2 - 2x)y' - \dfrac{1}{4}y = 0$

13. Prove all the relations listed in the property (5.8).

14. Show that

$$\dfrac{d^n}{dx^n} F(a, b; c; x) = \dfrac{\Gamma(a + n)\Gamma(b + n)\Gamma(c)}{\Gamma(a)\Gamma(b)\Gamma(c + n)} F(a + n, b + n; c + n; x)$$

15. (a) Obtain a power series solution about $x = 0$ to the *confluent hypergeometric* equation $xy'' + (c - x)y' - ay = 0$, $x > 0$, $c \neq 0$, $-1, -2, \ldots$ and show that it converges for all $x > 0$. (*Hint:* Put $t = bx$ in the hypergeometric equation (5.20) and then make $b \to \infty$. The equation will be reduced to the required equation and so the solution).

(b) Denoting the above series by $M(a, c; x)$, show that the general solution of the above equation, when c is not an integer, is $c_1 M(a, c; x) + c_2 x^{1-c} M(a - c + 1, 2 - c; x)$.

5.10 ORTHOGONAL POLYNOMIALS

In this section, we intend to discuss certain important properties of orthogonal polynomials in general before we study particular orthogonal polynomials like Legendre polynomials, Chebyshev polynomials and Laguerre polynomials. Orthogonality is an important property widely encountered in various branches of mathematics.

5.11 ORTHOGONAL FUNCTIONS

A set of functions $\{f_0(x), f_1(x), \ldots f_n(x), \ldots\}$ is said to be an *orthogonal set* with respect to the weight function $w(x)$ on the interval $[a, b]$ iff

$$\int_a^b w(x) f_m(x) f_n(x) \, dx = 0 \quad \text{for } m \neq n$$

$$\neq 0 \quad \text{for } m = n$$

where $w(x)$ is continuous, non-negative and has finitely many zeros in (a, b).

Example 5.8 The set of functions $\{1, \cos x, \sin x, \cos 2x, \sin 2x, \ldots\}$ is an orthogonal set with respect to the weight function $w(x) = 1$ on the interval $[-\pi, \pi]$. For

$$\int_{-\pi}^{+\pi} \sin mx \sin nx \, dx = 0 \quad \text{if } m \neq n$$

$$\int_{-\pi}^{+\pi} \sin mx \cos nx \, dx = 0$$

$$\int_{-\pi}^{+\pi} \cos mx \cos nx \, dx = 0 \quad \text{if } m \neq n$$

$$\int_{-\pi}^{+\pi} 1 \, dx = 2\pi \quad \text{and} \quad \int_{-\pi}^{+\pi} \sin^2 mx \, dx = \pi = \int_{-\pi}^{+\pi} \cos^2 mx \, dx$$

5.12 SIMPLE SET OF POLYNOMIALS

A set of polynomials $\{p_0(x), p_1(x), \ldots, p_n(x), \ldots\}$ is called a *simple set* if $p_n(x)$ is of degree precisely n. The set then contains one polynomial of each degree 0, 1, 2, 3,... . A simple set is also called a *standard set*.

Example 5.9 The set of polynomials $\{1, x, x^2, \ldots, x^n, \ldots\}$ is a simple set of polynomials.

5.13 PROPERTIES OF SIMPLE SET OF POLYNOMIALS

1. If $g_m(x)$ is a polynomial of degree m and $\{p_0(x), p_1(x), \ldots, p_n(x), \ldots\}$ is a simple set of polynomials, then there exists a set of constants $\{c_0, c_1, \ldots, c_m\}$ such that

$$g_m(x) = \sum_{k=0}^{m} c_k p_k(x) \tag{5.37}$$

Proof: To prove this, let the highest degree term in $g_m(x)$ be $a_m x^m$ and that in $p_m(x)$ be $b_m x^m$. We take $c_m = a_m/b_m$, noting that $b_m \neq 0$. Then clearly $g_m(x) - (a_m/b_m)p_m(x) = g_m(x) - c_m p_m(x)$ is a polynomial of degree at most $m-1$. Applying the same procedure to this polynomial, we can define c_{m-1}. Continuing this process through $m+1$ steps, we get all the required constants to obtain Eq. (5.37).

2. A simple set $\{p_0(x), p_1(x), \ldots\}$ of polynomials is orthogonal with respect to a weight function $w(x)$ on the interval $[a, b]$ iff

$$\int_a^b w(x) x^k p_n(x) \, dx = 0 \quad \text{for } k = 0, 1, \ldots, (n-1) \tag{5.38}$$
$$\neq 0 \quad \text{for } k = n$$

Proof: To prove this, first assume that the simple set of polynomials $\{p_0(x), p_1(x), \ldots\}$ is an orthogonal set with respect to the weight function $w(x)$ on the interval $[a, b]$.

We have

$$x^k = \sum_{m=0}^{k} b_m p_m(x) \quad \text{by the above Property 1.}$$

Then

$$\int_a^b w(x) x^k p_n(x) \, dx = \sum_{m=0}^{k} b_m \int_a^b w(x) p_m(x) p_n(x) \, dx = 0 \quad \text{if } k < n$$
$$\neq 0 \quad \text{if } k = n$$

This proves the condition (5.38).

Next, let us assume that the simple set of polynomials satisfy the condition (5.38).

Since $\{1, x, x^2, \ldots\}$ is a simple set of polynomials, we can have for any polynomial of the simple set $p_m(x) = \sum_{k=0}^{m} a_k x^k$. Then

$$\int_a^b w(x)p_m(x)p_n(x)\,dx = \sum_{k=0}^m a_k \int_a^b w(x)x^k p_n(x)\,dx = 0$$

if $m \neq n$, for, if $m < n$, then each k involved is less than n, and if $m > n$, then interchange the role of m and n and repeat the argument.

If $m = n$, then we have

$$\int_a^b w(x)p_n(x)p_n(x) = \sum_{k=0}^n a_n \int_a^b w(x)x^k p_n(x)\,dx \neq 0$$

since $a_n \neq 0$. Thus we proved that the simple set of polynomials is an orthogonal set.

3. If $\{p_0(x), p_1(x), \ldots\}$ is an orthogonal simple set of polynomials with respect to the weight function $w(x)$ on $[a, b]$, then $p_n(x)$ is orthogonal to any polynomial of degree less than n with respect to the weight function $w(x)$ on $[a, b]$.

Proof: If $g_m(x)$ is a polynomial of degree $< n$, then it can be written as

$$g_m(x) = \sum_{k=0}^m a_k x^k$$ by Property 1, since $\{1, x, x^2, \ldots\}$ is a simple set of polynomials.

Now $$\int_a^b w(x)g_m(x)p_n(x)\,dx = \sum_{k=0}^m a_k \int_a^b w(x)x^k p_n(x)\,dx = 0$$

since $m < n$ by Property 2.

Note: This property is also true even if $a = -\infty$, $b = +\infty$, provided all the integrals involved are convergent.

4. If $\{p_0(x), p_1(x), \ldots\}$ and $\{q_0(x), q_1(x), \ldots\}$ are two orthogonal simple sets of polynomials with respect to weight function $w(x)$ on $[a, b]$, then for each n, $p_n(x)$ and $q_n(x)$ are scalar multiples of one another.

Proof: Since $p_n(x)$ is any polynomial and $\{q_0(x), q_1(x), \ldots\}$ is a simple set of polynomials, we can have, by Property 1, that $p_n(x) = \sum_{k=0}^n a_k q_k(x)$.

In this presentation $a_0 = a_1 = \cdots = a_{n-1} = 0$, but $a_n \neq 0$, because

$$\int_a^b w(x)p_n(x)q_t(x)\,dx = \sum_{k=0}^n a_k \int_a^b w(x)q_t(x)q_k(x)\,dx$$

The left hand side is zero for $t < n$ by Property 3, and the right hand side gives $a_0, a_1, \ldots, a_{n-1}$ for values of $t = 0, 1, \ldots, n-1$, as $\{q_0(x), q_1(x), \ldots\}$ is an orthogonal set with respect to $w(x)$ on $[a, b]$. Further, when $t = n$, then we get

$$\int_a^b w(x)p_n(x)q_n(x)\,dx = a_n \int_a^b w(x)q_n(x)q_n(x)\,dx \tag{5.39}$$

and both the integrals are nonzeros. Hence we have $p_n(x) = a_n q_n(x)$, where a_n is a scalar given by the relation (5.39).

By interchanging the role of $p_n(x)$ and $q_n(x)$ in the above argument, we can show that $q_n(x)$ is a scalar multiple of $p_n(x)$. This completes the proof.

This last property suggests that up to scalar multiples, $\mathcal{PC}$ [−1, 1] contains only one orthogonal simple set of polynomials $\{p_0(x), p_1(x), p_2(x),\ldots\}$. This set will be unique if we impose an additional condition that each of $p_n(x)$ be such that its leading coefficient is 1. This observation leads to an important result in a piecewise continuous function space $\mathcal{PC}$ [−1, 1], which we give in the following proprty:

5. The only orthogonal simple set of polynomials $\{p_0(x), p_1(x), \ldots\}$ in $\mathcal{PC}$ [−1, 1] with the property that $p_n(x)$ has leading coefficient one for each n is the set obtained by orthogonalizing $\{1, x, x^2,\ldots\}$.

6. If $\{p_0(x), p_1(x),\ldots\}$ is an orthogonal simple set of polynomials with respect to the weight function $w(x)$ on $[a, b]$, then: (i) $p_{n+1}(x) = (a_n x + b_n)p_n(x) + c_n p_{n-1}(x)$ for $n = 1, 2, 3, \ldots$, where a_n, b_n, c_n are real numbers and (ii) if the set of polynomials satisfies the condition "$p_n(x)$ is even or odd according as n is even or odd", then $b_n = 0$ for $n = 1, 2, 3, \ldots$ in part (i).

Proof: (i) Since deg $\{xp_n(x)\} = n + 1$, $xp_n(x)$ can be written as

$$xp_n(x) = \sum_{k=0}^{n+1} \alpha_k p_k(x) \qquad \text{by Property 1}$$

Let j be a fixed index less than $n - 1$. Now, deg $\{xp_j(x)\} < n$, so $\int_a^b w(x)p_n(x)xp_j(x)\,dx = 0$, by Property 3.

Thus
$$\sum_{k=0}^{n+1} \alpha_k \int_a^b w(x)p_k(x)p_j(x)\,dx = 0$$

$$\Rightarrow \alpha_j \int_a^b w(x)p_j(x)p_j(x)\,dx = 0 \qquad \text{by orthogonality}$$

$$\Rightarrow \alpha_j = 0$$

Thus all $\alpha_k = 0$ for $k = 0, 1, 2,\ldots, n - 2$. So we are left with $xp_n(x) = \alpha_{n-1} p_{n-1}(x) + \alpha_n p_n(x) + \alpha_{n+1} p_{n+1}(x)$.

Comparing the coefficients of x^{n+1} from both the sides it is clear that $\alpha_{n+1} \neq 0$. Hence by dividing α_{n+1} both the sides and rearranging the result follows.

(ii) Since $p_n(x)$ contains only even or odd powers of x the coefficient of x^{n-1} in $p_n(x)$ is 0. Hence comparing coefficients of x^n on both the sides of $p_{n+1}(x) = (a_n x + b_n)p_n(x) + c_n p_{n-1}(x)$ gives $b_n = 0$.

7. If $\{p_0(x), p_1(x), \ldots\}$ is an orthogonal simple set of real polynomials with respect to weight function $w(x)$ on $[a, b]$ and if $w(x) > 0$ in (a, b), then the zeros of $p_n(x)$ are distinct and all lie in (a, b).

Proof: If $n = 0$, then $p_0(x)$ is a polynomial of degree zero and has zero root in (a, b).

Assume $n \geq 1$. Then $\int_a^b w(x) p_n(x) dx = 0$, taking $k = 0$ in (5.38). This implies the polynomial $p_n(x)$ must change sign at least once in the interval (a, b) as $w(x)$ does not change sign in (a, b).

Let $x_1, x_2, \ldots, x_m$ be the distinct zeros of $p_n(x)$ in (a, b). Since $p_n(x)$ is of degree n, it has n zeros. So $m \leq n$.

Consider
$$q(x) = (x - x_1)(x - x_2)\ldots(x - x_m) \quad (5.40)$$

which is a polynomial of degree m, has distinct m zeros and has the same zeros as those of $p_n(x)$ in (a, b). Thus $p_n(x)$, $q(x)$ change signs in (a, b) at the same points, which implies that $q(x)p_n(x) \geq 0$ in (a, b). This further implies that

$$\int_a^b w(x) q(x) p_n(x) dx \neq 0$$

But $\int_a^b w(x) q(x) p_n(x) dx = 0$ for $m < n$ by Property 3

Hence $m = n$ and this completes the proof of the assertion.

Exercises 5.13

1. A set of functions $\{f_0(x), f_1(x), \ldots, f_n(x), \ldots\}$ is said to be an *orthonormal set* with respect to the weight function $w(x)$ on the interval $[a, b]$ iff

$$\int_a^b w(x) f_n(x) f_m(x) dx = \begin{cases} 0 & \text{if } m \neq n \\ 1 & \text{if } m = n \end{cases}$$

Prove that the set $\left\{ \dfrac{1}{\sqrt{2\pi}}, \dfrac{\cos x}{\sqrt{\pi}}, \dfrac{\sin x}{\sqrt{\pi}}, \dfrac{\cos 2x}{\sqrt{\pi}}, \dfrac{\sin 2x}{\sqrt{\pi}}, \ldots \right\}$ is an orthonormal set with respect to the weight function $w(x) = 1$ on the interval $[-\pi, \pi]$.

2. Show that the set $\{\sin x, \sin 2x, \ldots, \sin nx, \ldots\}$ is an orthogonal set with respect to the weight function $w(x) = 1$ on the interval $[0, \pi]$. Do the same thing for the set $\{1, \cos x, \cos 2x, \ldots, \cos nx, \ldots\}$. Are they orthonormal with respect to the same weight function?

3. Let $P_n(x)$ be a polynomial of degree n and a solution of the differential equation
$$(1 - x^2)y'' - 2xy' + n(n + 1)y = 0$$
Show that the set $\{P_0(x), P_1(x), P_2(x), \ldots, P_n(x), \ldots\}$ is an orthogonal set with respect to the weight function $w(x) = 1$ on the interval $[-1, 1]$.

4. Let $L_n(x)$ be a polynomial of degree n and a solution of the differential equation
$$xy'' + (1 - x)y' + ny = 0$$
Show that the set $\{L_0(x), L_1(x), \ldots, L_n(x), \ldots\}$ is an orthogonal set of polynomials with respect to the weight function $w(x) = e^{-x}$ on the interval $[0, \infty)$. (*Hint:* Use the equation $[xe^{-x}L'_n(x)]' + ne^{-x}L_n(x) = 0$.)

5. Let $H_n(x)$ be a polynomial of degree n and a solution of the differential equation
$$y'' - 2xy' + 2ny = 0$$
Show that the set $\{H_0(x), H_1(x), \ldots, H_n(x), \ldots\}$ is an orthogonal set of polynomials with respect to the weight function $w(x) = e^{-x^2}$ over the interval $(-\infty, \infty)$. (*Hint:* Use the equation $[e^{-x^2}H'_n(x)]' + 2ne^{-x^2}H_n(x) = 0$.)

5.14 BESSEL FUNCTIONS

In this section we shall study another important second order homogeneous linear differential equation with variable coefficients, called the *Bessel Equation*. Its solutions are called *Bessel Functions*. Bessel functions have very diverse applications in physics and engineering, besides pure mathematics, in connection with the propagation of waves, elasticity, fluid motion, potential theory, etc. We shall study the solutions of Bessel equation and their important properties.

5.15 BESSEL EQUATION

The equation
$$x^2y'' + xy' + (x^2 - p^2)y = 0 \qquad (5.41)$$
where, since only p^2 enters the equation, $p \geq 0$, a fixed constant, is called, the *Bessel equation of order (or index) p*.

This equation has a regular singular point at $x = 0$ and no other singular points in the finite plane.

At $x = 0$ the roots of its indicial equation are p and $-p$.

Theorem 3.7 guarantees that (5.41) has a solution of the form
$$y_1 = x^p \sum_{n=0}^{\infty} a_n x^n \qquad a_0 \neq 0$$

Introduction to Differential Equations

Proceeding in the usual manner of Frobenius method, we obtain $a_1 = a_3 = a_5 = \cdots = 0$ and

$$a_n = \frac{-a_{n-2}}{n(2p+n)} \quad \text{for } n = 2, 4, 6, \ldots$$

which in turn gives

$$a_{2n} = \frac{(-1)^n}{2^{2n} n!(p+1)(p+2)\cdots(p+n)} a_0 \quad \text{for } n \geq 0$$

Since a_0 is an arbitrary constant, let us choose a particular value of a_0 as equal to $\dfrac{1}{2^p \Gamma(p+1)}$. Then

$$a_{2n} = \frac{(-1)^n}{2^{2n+p} n!(p+1)(p+2)\cdots(p+n)\Gamma(p+1)} = \frac{(-1)^n}{2^{2n+p} n!\Gamma(n+p+1)}$$

Thus a solution of (5.41) is given by

$$y_1 = \sum_{n=0}^{\infty} \frac{(-1)^n}{n!\Gamma(n+p+1)} \left(\frac{x}{2}\right)^{2n+p} \tag{5.42}$$

It is easy to show that this series converges absolutely for all $x \geq 0$.

This solution (5.42) of (5.41) is a function defined for $x \geq 0$ and called the *Bessel function of first kind of order* (or index) p. It is usually denoted by $J_p(x)$.

To get the general solution of (5.41), it is necessary to find a second solution linearly independent of $J_p(x)$. Again, by Theorem 3.7, we have different cases depending on the values of p.

Case 1 ($p > 0$, $2p$ is not an integer): Here the roots of the indicial equation associated with (5.41) do not differ by an integer, and a second linearly independent Frobenius series solution can be obtained by repeating the above argument with $-p$ in place of p. Since the gamma function is defined for the non-integral negative values, the coefficients of the series will have the same form as before, and the solution in question can be written as

$$J_{-p}(x) = \sum_{n=0}^{\infty} \frac{(-1)^n}{n!\Gamma(n-p+1)} \left(\frac{x}{2}\right)^{2n-p} \tag{5.43}$$

which converges absolutely for $x \geq 0$. (Check!)

Hence the general solution of the Bessel equation (5.41) is given by

$$y(x) = c_1 J_p(x) + c_2 J_{-p}(x)$$

where $p > 0$ and $2p$ is not an integer.

Case 2 ($p > 0$, $2p$ is an integer): Let $p = k + \dfrac{1}{2}$, where k an integer. We observe that (5.43) is defined for this value of p, and (5.43) yields a solution independent of $J_p(x)$. Hence the general solution in this case is also given by

$$y(x) = c_1 J_p(x) + c_2 J_{-p}(x)$$

Case 3 ($p = 0$): In this case the indicial equation associated with (5.41) has zero as a repeated root. Hence by Theorem 3.8 we can find a second linearly independent solution of the form

$$K_0(x) = J_0(x) \ln x + \sum_{n=1}^{\infty} b_n x^n \qquad (5.44)$$

Proceeding in the usual manner, we can calculate the coefficients b_n to obtain the solution (5.44) as

$$K_0(x) = J_0(x) \ln x + \sum_{n=1}^{\infty} \frac{(-1)^{n+1}}{(n!)^2}\left(1 + \frac{1}{2} + \cdots + \frac{1}{n}\right)\left(\frac{x}{2}\right)^{2n} \quad \text{for } x > 0 \quad (5.45)$$

Hence the general solution of (5.41) in this case is given by

$$y(x) = c_1 J_0(x) + c_2 K_0(x)$$

Case 4 (p is an integer): In this case the roots of the indicial equation associated with (5.41) differ by a positive integer. So the theorem asserts that the second linearly independent solution of (5.41) is given by

$$K_p(x) = cJ_p(x) \ln x + \sum_{n=0}^{\infty} b_n x^{n+p}$$

where c is a constant and $b_0 \neq 0$.

Proceeding in the usual manner, we can calculate the coefficients b_n and the constant c to obtain the second solution as

$$K_p(x) = J_p(x) \ln x - \frac{1}{2}\sum_{n=0}^{p-1} \frac{(p-n-1)!}{n!}\left(\frac{x}{2}\right)^{2n-p} - \frac{I_p}{2p!}\left(\frac{x}{2}\right)^p$$

$$- \frac{1}{2}\sum_{n=1}^{\infty} \frac{(-1)^n [I_n + I_{n+p}]}{n!(n+p)!}\left(\frac{x}{2}\right)^{2n+p}$$

where $I_p = 1 + \frac{1}{2} + \cdots + \frac{1}{p}$, for $x > 0$.

Hence the general solution of (5.41) in this case is given by

$$y(x) = c_1 J_p(x) + c_2 K_p(x)$$

Remark 5.1 In applications, many times the independent variable is time t, and one is interested in knowing the form of the solution when this variable tends to infinity. One can show, using rather advanced arguments, that $J_p(x)$ behaves like

$\sqrt{\dfrac{2}{\pi x}} \cos\left(x - \dfrac{\pi}{4} - \dfrac{p\pi}{2}\right)$ for large x. Now, of course, when p is not an integer, it should be expected that a linear combination of $J_p(x)$ and $J_{-p}(x)$ should behave like $\sqrt{\dfrac{2}{\pi x}} \sin\left(x - \dfrac{\pi}{4} - \dfrac{p\pi}{2}\right)$ for large x. Such a solution is defined by

$$Y_p(x) = \dfrac{J_p(x)\cos p\pi - J_{-p}(x)}{\sin p\pi}$$

and is called the *Bessel function of second kind and of order (or index) p*.

The problem of obtaining the second independent solution is rather complicated when $p = 0, 1, 2, \ldots$ as we have seen above; so we will not worry about getting the Bessel functions of second kind for integral order (or index) except for mentioning them.

$$Y_0(x) = \dfrac{2}{\pi} J_0(x)\left(\ln\dfrac{x}{2} + \gamma\right) - \dfrac{2}{\pi}\sum_{n=1}^{\infty} \dfrac{(-1)^n}{(n!)^2} I_n \left(\dfrac{x}{2}\right)^{2n} \quad \text{for } x > 0$$

$$Y_p(x) = \dfrac{2}{\pi} J_p(x)\left(\ln\dfrac{x}{2} + \gamma\right) - \dfrac{1}{\pi}\sum_{n=0}^{p-1} \dfrac{(p-n-1)!}{n!}\left(\dfrac{x}{2}\right)^{2n-p} - \dfrac{I_p}{\pi(n!)}\left(\dfrac{x}{2}\right)^p$$

$$- \dfrac{1}{\pi}\sum_{n=1}^{\infty} \dfrac{(-1)^n [I_n + I_{n+p}]}{n!(n+p)!}\left(\dfrac{x}{2}\right)^{2n+p}$$

where $p = 1, 2, 3, \ldots$, for $x > 0$.

5.16 PROPERTIES OF BESSEL FUNCTIONS

1. In general, $J_p(x)$ defined by (5.42) is a Frobenius series which becomes a power series for non-negative integral values of p. This power series is absolutely convergent for all values of x.

2. It is easy to observe that $J_0(x)$, $J_2(x)$, $J_4(x)$, ..., are all even functions, whereas $J_1(x)$, $J_3(x)$, $J_5(x)$, ..., are all odd functions.

3. It is easily seen from (5.42) that $J_0(0) = 1$ and $J_p(0) = 0$ for $p > 0$.

4. We have seen before that $J_p(x)$ and $J_{-p}(x)$ are linearly independent solutions when $p = k + \dfrac{1}{2}$, where k is a nonnegative integer. When $k = 0$, then

$$J_{\frac{1}{2}}(x) = \sqrt{\dfrac{2}{\pi x}} \sin x$$

$$J_{-\frac{1}{2}}(x) = \sqrt{\dfrac{2}{\pi x}} \cos x$$

Special Functions and Equations

Proof:
$$J_{-\frac{1}{2}}(x) = \sum_{n=0}^{\infty} \frac{(-1)^n}{n!\,\Gamma\left(n-\frac{1}{2}+1\right)}\left(\frac{x}{2}\right)^{2n-\frac{1}{2}}$$

$$= \sqrt{\frac{2}{x}} \sum_{n=0}^{\infty} \frac{(-1)^n}{n!\,\Gamma\left(n+\frac{1}{2}\right)}\left(\frac{x}{2}\right)^{2n}$$

$$= \sqrt{\frac{2}{x}} \sum_{n=0}^{\infty} \frac{(-1)^n}{n!\left(n-\frac{1}{2}\right)\left(n-\frac{3}{2}\right)\cdots\frac{1}{2}\Gamma\left(\frac{1}{2}\right)}\left(\frac{x^{2n}}{2^{2n}}\right)$$

$$= \sqrt{\frac{2}{\pi x}} \sum_{n=0}^{\infty} \frac{(-1)^n}{\{n!(2n-1)(2n-3)\cdots 1\}2^n}\, x^{2n}$$

$$= \sqrt{\frac{2}{\pi x}} \sum_{n=0}^{\infty} \frac{(-1)^n}{(2n)!}\, x^{2n} = \sqrt{\frac{2}{\pi x}}\cos x$$

A similar argument shows that $J_{\frac{1}{2}}(x) = \sqrt{\frac{2}{\pi x}}\sin x$. Later on, we shall show that every Bessel function of the first kind of $\left(k+\frac{1}{2}\right)$th order, where k is an integer, can be expressed in finite form in terms of elementary functions.

5. $\lim_{x\to 0^+} J_{-p}(x) = \pm\infty$ for non-integral values of p. This follows easily from (5.43).

6. $J_p(x)J'_{-p}(x) - J'_p(x)J_{-p}(x) = \frac{c}{x}$, where c is a constant and p is not an integer.

For non-integral values of p, $J_p(x)$ and $J_{-p}(x)$ are linearly independent solutions of (5.41), and then the result follows from Abel's formula for the Wronskian.

7. If p is a positive integer, then $J_{-p}(x) = (-1)^p J_p(x)$.

Proof:
$$J_{-p}(x) = \sum_{n=0}^{\infty} \frac{(-1)^n}{n!\,\Gamma(n-p+1)}\left(\frac{x}{2}\right)^{2n-p}$$

$$= \sum_{n=p}^{\infty} \frac{(-1)^n}{n!\,\Gamma(n-p+1)}\left(\frac{x}{2}\right)^{2n-p}$$

since $\Gamma(n-p+1) = \infty$

Introduction to Differential Equations

for $n = 0, 1, \ldots, p - 1$, as $n - p + 1 \leq 0$, so $\dfrac{1}{\Gamma(n - p + 1)} = 0$

$$= \sum_{n=0}^{\infty} \frac{(-1)^{n+p}}{(n+p)!\Gamma(n+1)} \left(\frac{x}{2}\right)^{2n+p} \quad \text{by shifting the index}$$

$$= (-1)^p \sum_{n=0}^{\infty} \frac{(-1)^n}{n!(n+1)(n+2)\cdots(n+p)\Gamma(n+1)} \left(\frac{x}{2}\right)^{2n+p}$$

$$= (-1)^p \sum_{n=0}^{\infty} \frac{(-1)^n}{n!\Gamma(n+p+1)} \left(\frac{x}{2}\right)^{2n+p} = (-1)^p J_p(x)$$

8. $\dfrac{d}{dx}\left(x^p J_p(x)\right) = x^p J_{p-1}(x).$

Proof: $\dfrac{d}{dx}\left(x^p J_p(x)\right) = \dfrac{d}{dx} \sum_{n=0}^{\infty} \frac{(-1)^n}{n!\Gamma(n+p+1)} \frac{x^{2n+2p}}{2^{2n+p}}$

$$= \sum_{n=0}^{\infty} \frac{(-1)^n (2n+2p)}{n!\Gamma(n+p+1)} \frac{x^{2n+2p-1}}{2^{2n+p}} \quad \text{by differentiating term by term}$$

$$= x^p \sum_{n=0}^{\infty} \frac{(-1)^n}{n!\Gamma(n+p)} \left(\frac{x}{2}\right)^{2n+p-1} = x^p J_{p-1}(x)$$

$\Rightarrow x^p J_p'(x) + p x^{p-1} J_p(x) = x^p J_{p-1}(x)$

$\Rightarrow x J_p'(x) + p J_p(x) = x J_{p-1}(x)$ \hfill (5.46)

9. $\dfrac{d}{dx}\left(x^{-p} J_p(x)\right) = -x^{-p} J_{p+1}(x).$

Proof: This can be proved by the same argument as above. Then this leads to

$$x^{-p} J_p'(x) - p x^{-p-1} J_p(x) = -x^{-p} J_{p+1}(x)$$

$\Rightarrow x J_p'(x) - p J_p(x) = -x J_{p+1}(x)$ \hfill (5.47)

10. Adding (5.46) and (5.47), we get

$$2 J_p'(x) = J_{p-1}(x) - J_{p+1}(x)$$

$\Rightarrow J_{p+1}(x) = J_{p-1}(x) - 2 J_p'(x)$ \hfill (5.48)

11. Subtracting (5.47) from (5.46), we get another relation

$$2 p J_p(x) = x J_{p-1}(x) + x J_{p+1}(x)$$

$\Rightarrow x J_{p+1}(x) = 2 p J_p(x) - x J_{p-1}(x)$ \hfill (5.49)

Relations (5.46) to (5.49) are recurrence relations. These are useful in finding Bessel functions of any order in terms of lower orders. For example, if we take $p = \frac{1}{2}$ in (5.49), we get

$$xJ_{\frac{3}{2}}(x) = J_{\frac{1}{2}}(x) - xJ_{-\frac{1}{2}}(x) = \sqrt{\frac{2}{\pi x}}\sin x - x\sqrt{\frac{2}{\pi x}}\cos x \text{ by Property 4}$$

$$\Rightarrow J_{\frac{3}{2}}(x) = \frac{1}{x}\sqrt{\frac{2}{\pi x}}\sin x - \sqrt{\frac{2}{\pi x}}\cos x = \sqrt{\frac{2}{\pi x}}\left(\frac{\sin x}{x} - \cos x\right)$$

12. Every solution of the Bessel equation (5.41) has infinitely many positive zeros. In particular, $J_p(x)$ and $J_{-p}(x)$ for non integral values of p, are oscillatory functions. This follows from Example 4.12.

In particular, $J_p(x)$ and $J_{-p}(x)$, for non-integral values of p, are oscillatory functions. This follows from Example 4.12.

13. The zeros of $J_p(x)$ and $J_{-p}(x)$, for non-integral values of p, are simple and they alternate.

Proof: Since $J_p(x)$ and $J_{-p}(x)$, for non-integral values of p, are linearly independent solutions of (5.41), the zeros of $J_p(x)$ and $J_{-p}(x)$ are simple by Theorem 4.1.

Further, by Theorem 4.4 the zeros of $J_p(x)$ and $J_{-p}(x)$ alternate.

14. The zeros of $J_p(x)$ and $J_{p+1}(x)$ are distinct and they alternate.

Proof: For, if $J_p(x_0) = J_{p+1}(x_0) = 0$ for some $x_0 > 0$, then

$$xJ'_p(x) = pJ_p(x) - xJ_{p+1}(x) \Rightarrow J'_p(x_0) = 0 \quad \text{as } x_0 > 0$$

and the uniqueness theorem for the initial value problem of the Bessel equation would imply that $J_p(x) = 0$, which is impossible. Hence the zeros of $J_p(x)$ and $J_{p+1}(x)$ are distinct.

Next, let $0 < x_1 < x_2$ be two consecutive zeros of $J_p(x)$.
Again
$$xJ'_p(x) = pJ_p(x) - xJ_{p+1}(x)$$
$$\Rightarrow J'_p(x_1) = -J_{p+1}(x_1)$$
and
$$J'_p(x_2) = -J_{p+1}(x_2)$$

Since x_1, x_2 are two consecutive zeros of $J_p(x)$, either $J_p(x)$ is increasing at x_1 and decreasing at x_2, or $J_p(x)$ is decreasing at x_1 and increasing at x_2. So $J'_p(x_1)$ and $J'_p(x_2)$ have opposite signs, implying that $J_{p+1}(x_1)$ and $J_{p+1}(x_2)$ have opposite signs. Hence there exists $x_3 \in (x_1, x_2)$ such that $J_{p+1}(x_3) = 0$.

A similar argument using $xJ'_p(x) = -pJ_p(x) + xJ_{p-1}(x)$ may be used to show that $J_p(x)$ has a zero between any two consecutive zeros of $J_{p+1}(x)$. This proves the assertion.

15. If α and β are two distinct positive zeros of $J_p(x)$, then $\int_0^1 xJ_p(\alpha x)J_p(\beta x)\,dx = 0$. On the other hand, if $\alpha = \beta$, then the integral becomes $\frac{1}{2}J_{p+1}^2(\alpha)$.

Proof: Take $u(x) = J_p(\alpha x)$ $\qquad u'(x) = \alpha J'_p(\alpha x)$ $\qquad u''(x) = \alpha^2 J''_p(\alpha x)$

Introduction to Differential Equations

and $\quad v(x) = J_p(\beta x) \qquad v'(x) = \beta J'_p(\beta x) \qquad v''(x) = \beta^2 J''_p(\beta x)$

We have $\quad x^2 J''_p(x) + x J'_p(x) + (x^2 - p^2) J_p(x) = 0$

which becomes $\quad \alpha^2 x^2 J''_p(\alpha x) + \alpha x J'_p(\alpha x) + (\alpha^2 x^2 - p^2) J_p(\alpha x) = 0$
(if we replace x by αx)

$$\Rightarrow u''(x) + \frac{1}{x} u'(x) + \alpha^2 u(x) = \frac{p^2}{x^2} u(x)$$

Similarly $\quad v''(x) + \frac{1}{x} v'(x) + \beta^2 v(x) = \frac{p^2}{x^2} v(x)$

Eliminating p^2, we have

$$[u''(x)v(x) - u(x)v''(x)] + \frac{1}{x}[u'(x)v(x) - u(x)v'(x)] + (\alpha^2 - \beta^2) u(x)v(x) = 0$$

$$\Rightarrow x \frac{d}{dx}[u'(x)v(x) - u(x)v'(x)] + [u'(x)v(x) - u(x)v'(x)] + x(\alpha^2 - \beta^2) u(x)v(x) = 0$$

Integrating from 0 to 1 gives

$$[x(u'(x)v(x) - u(x)v'(x))]_0^1 = (\beta^2 - \alpha^2) \int_0^1 x u(x) v(x) dx$$

$$\Rightarrow \alpha J'_p(\alpha) J_p(\beta) - \beta J_p(\alpha) J'_p(\beta) = (\beta^2 - \alpha^2) \int_0^1 x J_p(\alpha x) J_p(\beta x) dx$$

$$\Rightarrow \int_0^1 x J_p(\alpha x) J_p(\beta x) dx = 0$$

This completes the first part.

As for the second part, we know $u(x) = J_p(\alpha x)$ is a solution of

$$x^2 u''(x) + x u'(x) + \alpha^2 x^2 u(x) - p^2 u(x) = 0$$

$$\Rightarrow 2\left[x^2 u''(x) u'(x) + x\{u'(x)\}^2 + \alpha^2 x^2 u(x) u'(x) - p^2 u(x) u'(x) \right] = 0$$

$$\Rightarrow \frac{d}{dx}\left\{ x^2 (u'(x))^2 \right\} + 2\alpha^2 x^2 u(x) u'(x) - 2p^2 u(x) u'(x) = 0$$

Integrating from 0 to 1, this becomes

$$\left[x^2 (u'(x))^2 \right]_0^1 + 2\alpha^2 \int_0^1 x^2 u(x) u'(x) dx - 2p^2 \int_0^1 u(x) u'(x) dx = 0$$

$$\Rightarrow (u'(1))^2 + 2\alpha^2 \left[x^2 \frac{(u(x))^2}{2} \right]_0^1 - 2\alpha^2 \int_0^1 x(u(x))^2 \, dx - p^2 \left[(u(x))^2 \right]_0^1 = 0$$

$$\Rightarrow \alpha^2 \left(J_p'(\alpha) \right)^2 - 2\alpha^2 \int_0^1 x \left(J_p(\alpha x) \right)^2 dx + p^2 \left(J_p(0) \right)^2 = 0$$

$$\Rightarrow \int_0^1 x \left(J_p(\alpha x) \right)^2 dx = \frac{1}{2} \left(J_p'(\alpha) \right)^2$$

since $J_p(0) = 0$ for $p > 0$ and that term vanishes when $p = 0$.

Since $\quad xJ_p'(x) = pJ_p(x) - xJ_{p+1}(x) \quad$ by (5.47)

we have $\quad \alpha J_p'(\alpha) = -\alpha J_{p+1}(\alpha)$

$$\Rightarrow J_p'(\alpha) = -J_{p+1}(\alpha)$$

which brings the last integral to

$$\int_0^1 x \left(J_p(\alpha x) \right)^2 dx = \frac{1}{2} \left(J_{p+1}(\alpha) \right)^2$$

Property 15 is often referred to as the *orthogonality relation* of Bessel functions.

16. (The generating function) A remarkable property of Bessel functions is the generating function in the sense that the Bessel functions of integral order are linked together by the formula

$$e^{\frac{x}{2}\left(t - \frac{1}{t}\right)} = J_0(x) + \sum_{n=1}^{\infty} J_n(x) \{ t^n + (-1)^n t^{-n} \} \qquad (5.50)$$

Since $J_{-n}(x) = (-1)^n J_n(x)$, this is usually written as

$$e^{\frac{x}{2}\left(t - \frac{1}{t}\right)} = \sum_{n=-\infty}^{+\infty} J_n(x) t^n \qquad (5.51)$$

Unfortunately, the proof of either of (5.50) or (5.51) requires several results of advanced analysis and is beyond the scope of this book. Nevertheless, we can not go to the next topic without mentioning this important property.

Exercises 5.16

1. Show that $J_0'(x) = -J_1(x)$.

2. Show that $\frac{d}{dx}[xJ_1(x)] = xJ_0(x)$.

3. Express $J'_2(x)$ and $J_4(x)$ in terms of $J_0(x)$ and $J_1(x)$.

4. Express $J_5(x)$ in terms of $J_0(x)$ and $J'_0(x)$.

5. For $p > 0$, $p \neq 1, 2, 3, \ldots$, show that

$$\frac{d}{dx}\left(x^p J_{-p}(x)\right) = -x^p J_{1-p}(x)$$

6. For any α and $p \neq 1, 2, 3, \ldots$, show that

$$\alpha J_{p-1}(\alpha x) - \frac{p}{x} J_p(\alpha x) = \frac{d}{dx}(J_p(\alpha x))$$

$$= -\alpha J_{p+1}(\alpha x) + \frac{p}{x} J_p(\alpha x)$$

7. (a) Using the Frobenius series for $J_{\frac{1}{2}}(x)$, show that

$$J_{\frac{1}{2}}(x) = \sqrt{\frac{2}{\pi x}} \sin x$$

(b) Obtain $J_{\frac{5}{2}}(x)$ and $J_{-\frac{5}{2}}(x)$ in terms of elementary functions.

8. Show that $J_2(x) = J_0(x) + 2J''_0(x)$.

9. Use the relation $J_{n+1}(x) + J_{n-1}(x) = \frac{2n}{x} J_n(x)$ to show that

$$x J_1(x) = 4 \sum_{n=1}^{\infty} (-1)^{n+1} n J_{2n}(x)$$

10. Prove that $\left[J_{p-1}(x)\right]^2 - \left[J_{p+1}(x)\right]^2 = \frac{2p}{x} \frac{d}{dx}\left[J_p(x)\right]^2$.

11. Show that $\frac{d}{dx}\{x J_n(x) J_{n+1}(x)\} = x\{J_n^2(x) - J_{n+1}^2(x)\}$.

12. (a) Show that $J_0(x) = \frac{2}{\pi} \int_0^{\pi/2} \cos(x \sin t)\, dt$.

(b) Show that $|J_0(x)| \leq 1$ for $x \geq 0$.

13. Prove that $\int_0^{\infty} e^{-ax} J_0(bx)\, dx = \frac{1}{\sqrt{a^2 + b^2}}$, where $a > 0$.

(**Hint:** Use Exercise 12.)

14. (a) Prove that $\int_0^\infty J_{n+1}(x)\,dx = \int_0^\infty J_{n-1}(x)\,dx$ for all positive integers n.

   (**Hint:** Integrate $J_{n-1}(x) - J_{n+1}(x) = 2J'_n(x)$ and use $J_n(x) \to 0$ as $x \to \infty$)

   (b) Show that $\int_0^\infty J_n(x)\,dx = 1$ for all $n > 1$.

   (**Hint:** Use Exercise 13 and part (a).)

   (c) Show that $\int_0^\infty \dfrac{J_n(x)}{x}\,dx = \dfrac{1}{n}$ for all integers $n > 0$.

   (**Hint:** Use $\dfrac{2n}{x} J_n(x) = J_{n+1}(x) + J_{n-1}(x)$ and part (b).)

15. (a) If $x^2 y'' + (1 - 2a) xy + \left[b^2 c^2 x^{2c} + (a^2 - p^2 c^2) \right] y = 0$

   and $\quad v = \dfrac{y}{x^a}, \ u = bx^c$, then show that

   $$u^2 \frac{d^2 v}{du^2} + u \frac{dv}{du} + (u^2 - p^2) v = 0$$

   (b) Show that the general solution of Airy's equation $y'' + xy = 0$ is

   $$\sqrt{x} \left[c_1 J_{\frac{1}{3}}\left(\frac{2}{3} x^{\frac{3}{2}} \right) + c_2 J_{-\frac{1}{3}}\left(\frac{2}{3} x^{\frac{3}{2}} \right) \right].$$

   (c) Show that the general solution of $xy'' + \dfrac{1}{2} y' + \dfrac{1}{4} y = 0$

   is $x^{\frac{1}{4}} \left[c_1 J_{\frac{1}{2}}(\sqrt{x}) + c_2 J_{-\frac{1}{2}}(\sqrt{x}) \right].$

5.17 LEGENDRE POLYNOMIALS

An important differential equation that arises very frequently in various branches of applied mathematics is Legendre's differential equation. This equation occurs in the process of obtaining solutions of Laplace's equation in spherical coordinates and hence is of great importance in mathematical applications to physics and engineering. This section is devoted to the study of the solutions of Legendre's equation and to a discussion of their important properties.

5.18 LEGENDRE'S EQUATION

The equation $$(1 - x^2)y'' - 2xy' + p(p + 1)y = 0 \qquad (5.52)$$
where $-1 \leq x \leq 1$, for any real number p, is called *Legendre's equation of degree p*.

It is clear that this equation has regular singular points at $x = \pm 1$, and every other point is an ordinary point.

We have seen in Chapter 3 that the two linearly independent power series solutions of (5.52) at the ordinary point $x = 0$ are given by

$$y_1(x) = 1 + \sum_{n=1}^{\infty} (-1)^n \frac{p(p-2) \cdots (p-2n+2)(p+1)(p+3) \cdots (p+2n-1)}{(2n)!} x^{2n}$$

(5.53)

and

$$y_2(x) = x + \sum_{n=1}^{\infty} (-1)^n \frac{(p-1)(p-3) \cdots (p-2n+1)(p+2)(p+4) \cdots (p+2n)}{(2n+1)!} x^{2n+1}$$

(5.54)

If p is a non-negative integer n, then $y_1(x)$ is a polynomial of degree n for n even, and $y_2(x)$ is a polynomial of degree n for n odd. In such cases these solutions are, of course, bounded at ± 1. In applications, many times it becomes necessary to consider all solutions of (5.52) which are bounded at 1. Since the solutions of (5.52) at an ordinary point are going to be of the forms (5.53) and (5.54), let us find solutions at the regular singular points.

5.19 POLYNOMIAL SOLUTIONS

To find the solutions of (5.52) at the regular singular point 1, let us put $t = \frac{1}{2}(1 - x)$. Then Eq. (5.52) reduces to the equation

$$t(1-t)\frac{d^2y}{dt^2} + (1-2t)\frac{dy}{dt} + p(p+1)y = 0 \qquad (5.55)$$

This is a hypergeometric equation with parameters $-p, p + 1, 1$. Thus any solution of (5.55) bounded at $t = 0$ is a multiple of $F(-p, p + 1; 1; t)$. Since a solution of (5.55) at $t = 0$ is a solution of (5.52) at $x = 1$, we have solutions of (5.52) which are bounded at $x = 1$ and must be multiples of

$$F\left(-p, p+1; 1; \frac{1-x}{2}\right) = 1 + \frac{(-p)(p+1)}{(1!)^2}\left(\frac{1-x}{2}\right)$$
$$+ \frac{(-p)(-p+1)(p+1)(p+2)}{(2!)^2}\left(\frac{1-x}{2}\right)^2 + \cdots$$

Special Functions and Equations

This series terminates, for non-negative integral values n of p, at

$$\frac{(-n)(-n+1)\cdots(-n+n-1)(n+1)(n+2)\cdots(n+n)}{(n!)^2}\left(\frac{1-x}{2}\right)^2$$

i.e. it is a polynomial of degree n with the leading coefficient.

$$\frac{(-1)^n(-n)(-n+1)\cdots(-n+n-1)(n+1)\cdots(2n)}{(n!)^2\, 2^n} = \frac{(2n)!}{2^n(n!)^2}$$

Also at $x = 1$ the value of this polynomial is 1. Thus we have the following:

Definition: The polynomial solution $P_n(x)$ of the Legendre's equation

$$(1 - x^2)y'' - 2xy' + n(n + 1)y = 0 \tag{5.56}$$

of degree n, non-negative integer, which satisfies $P_n(1) = 1$ is called the *Legendre polynomial of degree n*.

Thus
$$P_n(x) = F\left(-n, n+1; 1; \frac{1-x}{2}\right) \tag{5.57}$$

in the Legendre polynomial of degree n. These polynomials have many interesting properties. Now we discuss their properties.

5.20 PROPERTIES OF LEGENDRE POLYNOMIALS

1. It is clear that $P_{2n}(x)$ is an even function and $P_{2n-1}(x)$ is an odd function. Further, $\{P_n(x)\}$ is a simple set of polynomials.

2. **(Rodrigues' formula)** $\quad P_n(x) = \dfrac{1}{2^n n!} \dfrac{d^n}{dx^n}\{(x^2 - 1)^n\} \tag{5.58}$

Proof: Let

$$r(x) = (x^2 - 1)^n \Rightarrow \ln r(x) = n \ln (x^2 - 1)$$

$\Rightarrow r'(x)(x^2 - 1) = 2nxr(x) \qquad$ differentiating and transposing

Differentiating both the sides $(n + 1)$ times using Leibnitz's rule, we get

$$r^{(n+2)}(x)(x^2 - 1) + (n+1)r^{(n+1)}(x)\, 2x + \frac{(n+1)n}{2}r^{(n)}(x)2$$

$$= 2nxr^{(n+1)}(x) + 2n(n+1)r^{(n)}(x)$$

$$\Rightarrow (1-x^2)\left[r^{(n)}(x)\right]'' - 2x\left[r^{(n)}(x)\right]' + n(n+1)\left[r^{(n)}(x)\right] = 0$$

This shows that $R_n(x)$ satisfies Eq. (5.56) if we denote $R_k(x) = r^{(k)}(x)$. It is clear that $R_n(x)$ is a polynomial of degree n. What is $R_n(1)$?

For that, let us consider $r(x) = (x^2 - 1)^n = (x - 1)^n(x + 1)^n$.
Differentiating both the sides n times using Leibnitz's rule, we get

$$r^{(n)}(x) = n!(x+1)^n + n^2(x-1)(x+1)^{n-1}n! + \cdots + n!(x-1)^n$$

$$\Rightarrow R_n(1) = 2^n \, n!$$

Thus, $\dfrac{1}{2^n n!} R_n(x)$ is a polynomial of degree n and satisfies the Legendre equation (5.56) of degree n. Its value at 1 is 1. So, by definition

$$P_n(x) = \frac{1}{2^n n!} R_n(x) = \frac{1}{2^n n!} \frac{d^n}{dx^n}\{(x^2-1)^n\}$$

which completes the proof of Rodrigues' formula.

3. **(Recurrence Relations)** To find recurrence relations, we consider

$$\{(x^2-1)^{n+1}\}' = 2(n+1)x(x^2-1)^n$$

Differentiating both the sides $(n+1)$ times

$$\left[\{(x^2-1)^{n+1}\}'\right]^{n+1} = 2(n+1)x\{(x^2-1)^n\}^{(n+1)} + 2(n+1)(n+1)\{(x^2-1)^n\}^{(n)}$$

$$\Rightarrow 2^{n+1}(n+1)! P'_{n+1}(x) = 2(n+1)\{x 2^n n! P'_n(x) + (n+1)2^n n! P_n(x)\}$$

by Rodrigues' formula

$$\Rightarrow P'_{n+1}(x) = xP'_n(x) + (n+1)P_n(x) \tag{5.59}$$

Consider again $\quad\{(x^2-1)^n\}' = 2nx(x^2-1)^{n-1}$

$$\Rightarrow x\{(x^2-1)^n\}' = 2nx^2(x^2-1)^{n-1} = 2n\{(x^2-1)(x^2-1)^{n-1} + (x^2-1)^{n-1}\}$$

$$\Rightarrow x\{(x^2-1)^n\}' = 2n\{(x^2-1)^n + (x^2-1)^{n-1}\}$$

Differentiating n times, we obtain

$$x\{(x^2-1)^n\}^{(n+1)} + n\{(x^2-1)^n\}^{(n)} = 2n\{(x^2-1)^n\}^{(n)} + 2n\{(x^2-1)^{n-1}\}^{(n)}$$

$$\Rightarrow x 2^n n! P'_n(x) = n 2^n n! P_n(x) + 2n 2^{n-1}(n-1)! P'_{n-1}(x) \qquad \text{by (5.58)}$$

$$\Rightarrow xP'_n(x) = nP_n(x) + P'_{n-1}(x) \tag{5.60}$$

Putting (5.60) in (5.59), we get

$$P'_{n+1}(x) = (2n+1)P_n(x) + P'_{n-1}(x) \tag{5.61}$$

Multiplying (5.59) by $n+1$ and (5.60) by n, and subtracting, we get

$$(n+1)P'_{n+1}(x) - nxP'_n(x) = (n+1) xP'_n(x) + (n+1)^2 P_n(x) - n^2 P_n(x) - nP'_{n-1}(x)$$

$\Rightarrow (n+1)P'_{n+1}(x) = (2n+1)\{xP'_n(x) + P_n(x)\} - nP'_{n-1}(x)$

$\Rightarrow (n+1)P'_{n+1}(x) = (2n+1)\dfrac{d}{dx}\{xP_n(x)\} - nP'_{n-1}(x)$

$\Rightarrow (n+1)\displaystyle\int_1^x P_{n+1}(x)\,dx = (2n+1)\int_1^x \dfrac{d}{dx}\{xP_n(x)\}\,dx - n\int_1^x P'_{n-1}(x)\,dx$

$\Rightarrow (n+1)P_{n+1}(x) - (n+1)P_{n+1}(1) = (2n+1)xP_n(x) - (2n+1)P_n(1) - nP_{n-1}(x) + nP_{n-1}(1)$

$\Rightarrow (n+1)P_{n+1}(x) = (2n+1)xP_n(x) - nP_{n-1}(x) \qquad (5.62)$

for $n = 1, 2, 3, \ldots$, since $P_n(1) = 1$ for all n.

4. $\{P_n(x)\}$ is a simple set of orthogonal polynomials with respect to the weight function 1 on the interval $[-1, 1]$, i.e.

$$\int_{-1}^{1} P_m(x)P_n(x)\,dx = \begin{cases} 0 & \text{if } m \neq n \\ \dfrac{2}{2n+1} & \text{if } m = n \end{cases}$$

Proof: Since $P_n(x)$ is a solution of (5.56), we have

$\Rightarrow (1-x^2)P''_n(x) - 2xP'_n(x) + n(n+1)P_n(x) = 0$

$\Rightarrow \dfrac{d}{dx}\{(1-x^2)P'_n(x)\} + n(n+1)P_n(x) = 0$

$\Rightarrow P_m(x)\dfrac{d}{dx}\{(1-x^2)P'_n(x)\} + n(n+1)P_m(x)P_n(x) = 0$

Similarly, $\quad P_n(x)\dfrac{d}{dx}\{(1-x^2)P'_m(x)\} + m(m+1)P_m(x)P_n(x) = 0$

Subtracting, $\quad P_m(x)\dfrac{d}{dx}\{(1-x^2)P'_n(x)\} - P_n(x)\dfrac{d}{dx}\{(1-x^2)P'_m(x)\}$

$= (m-n)(m+n+1)P_m(x)P_n(x)$

$\Rightarrow \dfrac{d}{dx}\{(1-x^2)P_m(x)P'_n(x)\} - \dfrac{d}{dx}\{(1-x^2)P'_m(x)P_n(x)\}$

$= (m-n)(m+n+1)P_m(x)P_n(x)$

Integrating from -1 to 1, we get

$(m-n)(m+n+1)\displaystyle\int_{-1}^{1} P_m(x)P_n(x)\,dx = \Big[(1-x^2)\{P_m(x)P'_n(x) - P'_m(x)P_n(x)\}\Big]_{-1}^{1} = 0$

Introduction to Differential Equations

If $m \neq n$,
$$\int_{-1}^{1} P_m(x) P_n(x) dx = 0 \tag{5.63}$$

If $m = n$
$$\int_{-1}^{1} R_n(x) R_n(x) dx = \left[R_n(x) \int R_n(x) dx \right]_{-1}^{1} - \int_{-1}^{1} \left(R_n'(x) \int R_n(x) dx \right) dx$$

integrating by parts once

$$= -\int_{-1}^{1} R_{n+1}(x) R_{n-1}(x) dx \qquad \because R_k(\pm 1) = 0 \text{ for } k < n$$

$$= (-1)^n \int_{-1}^{1} R_{2n}(x)(x^2 - 1)^n dx \quad \text{integrating by parts } n \text{ times}$$

$$= (-1)^n \int_{-1}^{1} (2n)!(x^2 - 1)^n dx \qquad R_{2n}(x) = (2n)!$$

Thus
$$\int_{-1}^{1} [R_n(x)]^2 dx = (-1)^n (2n)! \int_{-1}^{1} (x^2 - 1)^n dx = (-1)^n (2n)! \int_{-1}^{1} (x-1)^n (x+1)^n dx$$

$$= (2n)! \int_{-1}^{1} (1-x)^n (1+x)^n dx$$

$$= (2n)! \int_{-1}^{1} n(1-x)^{n-1} \frac{(1+x)^{n+1}}{n+1} dx \qquad \text{integrating by parts once}$$

$$= (2n)! \frac{(n!)^2}{(2n)!} \int_{-1}^{1} (1+x)^{2n} dx \qquad \text{integrating by parts } n \text{ times}$$

$$= \frac{(n!)^2 2^{2n+1} (2n)!}{(2n+1)!} = \frac{(n!)^2 2^{2n+1}}{2n+1}$$

Thus we have
$$\int_{-1}^{1} \{P_n(x)\}^2 dx = \frac{1}{2^{2n}(n!)^2} \int_{-1}^{1} \{R_n(x)\}^2 dx = \frac{(n!)^2 2^{2n+1}}{2^{2n}(n!)^2 (2n+1)} = \frac{2}{2n+1} \tag{5.64}$$

5. $P_n(x)$ has n distinct (real) zeros in $(-1, 1)$.
This follows from Property 7 of Section 5.13.

6. (**Generating Function**) $\dfrac{1}{\sqrt{1-2xt+t^2}}$ is the generating function for the Legendre polynomials $P_n(x)$ i.e. if we expand $\dfrac{1}{\sqrt{1-2xt+t^2}}$ in a Taylor series about $t = 0$, treating x as a fixed parameter then the coefficients of t^n are Legendre polynomials. Thus

$$\frac{1}{\sqrt{1-2xt+t^2}} = \sum_{n=0}^{\infty} P_n(x) t^n \qquad (5.65)$$

We shall not prove this relation (5.65) even if it is not difficult, though it is lengthy and involved.

5.21 LEGENDRE SERIES

One of the simplest physical applications of Legendre polynomials occurs in potential theory in terms of expanding a given function in a series of Legendre polynomials. It is easy to see that this can be always done when the given function is itself a polynomial. Clearly, from (5.57) we find

$$P_0(x) = 1, \; P_1(x) = x, \; P_2(x) = \frac{1}{2}(3x^2 - 1), \; P_3(x) = \frac{1}{2}(5x^3 - 3x), \dots$$

and $\;1 = P_0(x), \; x = P_1(x), \; x^2 = \dfrac{1}{3}P_0(x) + \dfrac{2}{3}P_2(x), \; x^3 = \dfrac{3}{5}P_1(x) + \dfrac{2}{5}P_3(x), \dots$

Thus, any third degree polynomial $p(x) = b_0 + b_1 x + b_2 x^2 + b_3 x^3$ can be written as

$$p(x) = b_0 P_0(x) + b_1 P_1(x) + b_2 \left(\frac{1}{3} P_0(x) + \frac{2}{3} P_2(x) \right) + b_3 \left(\frac{3}{5} P_1(x) + \frac{2}{5} P_3(x) \right)$$

$$= \left(b_0 + \frac{b_2}{3} \right) P_0(x) + \left(b_1 + \frac{3b_3}{5} \right) P_1(x) + \frac{2b_2}{3} P_2(x) + \frac{2b_3}{5} P_3(x)$$

$$= \sum_{n=0}^{3} a_n P_n(x)$$

where a_0, a_1, a_2, a_3 are the coefficients in terms of the b's as seen from the preceding line.

The next question is about the possibility of expanding a given function in a series of Legendre polynomials. We know that many poblems of potential theory depend on this. The following theorem answers this question in affirmative for the "nice" functions. We shall state the theorem without proof for obvious reasons.

Theorem 5.1 If f is a nice function defined in the interval $[-1, 1]$, this $f(x)$ can be expressed as

$$f(x) = \sum_{n=0}^{\infty} a_n P_n(x) \qquad (5.66)$$

where the coefficients a_n are given by the formula

$$a_n = \left(n + \frac{1}{2}\right) \int_{-1}^{1} f(x) P_n(x)\, dx \qquad (5.67)$$

The series (5.66) is called the *Legendre Series* of the function f.

Exercises 5.21

1. If $f(x)$ is a polynomial of degree less than n, then show that

$$\int_{-1}^{1} f(x) P_n(x)\, dx = 0,$$ and in particular show that

$$\int_{-1}^{1} P_n(x)\, dx = 0 \quad \text{for} \quad n > 0$$

2. Show that $\int_{-1}^{1} x P_n(x) P_{n-1}(x)\, dx = \dfrac{2n}{4n^2 - 1}$.

3. Using the relation (5.65), show that
   (i) $P_n(1) = 1$;
   (ii) $P_n(-1) = (-1)^n$;
   (iii) $P_{2n+1}(0) = 0$;
   (iv) $P_{2n}(0) = (-1)^n \dfrac{1 \cdot 3 \cdots (2n-1)}{2^n\, n!}$;
   (v) $P'_n(1) = \dfrac{1}{2} n(n+1)$;
   (vi) $P'_n(-1) = (-1)^{n-1} \dfrac{n(n+1)}{2}$;
   (vii) $P_n(-x) = (-1)^n P_n(x)$.

4. Using the relations (5.63) and (5.65), prove the relation (5.64).

5. If $x > 1$, prove that $P_n(x) < P_{n+1}(x)$.

6. Show that $\left\{P_n(x) - \left(\dfrac{2n-1}{n}\right) x P_{n-1}(x)\right\}$ is a polynomial of degree $< n$.

7. Prove that $\displaystyle\int_{-1}^{1} P_n(x) P'_{n+1}(x)\, dx = 2$.

8. Prove that $\displaystyle\int_{-1}^{1} x P'_n(x) P_n(x)\, dx = \dfrac{2n}{2n+1}$.

9. Prove that $(1 - x^2) P'_n(x) = n P_{n-1}(x) - n x P_n(x)$.

10. Prove that $P'_n(0) = n P_{n-1}(0),\ n > 0$.

11. Prove that $\displaystyle\int_{0}^{1} P_n(x)\, dx = \dfrac{1}{n+1} P_{n-1}(0),\quad n > 0$.

12. Prove that $(n, -n;\ \tfrac{1}{2};\ x)$ is orthogonal to $P_{n+1}(x)$ with respect to the weight function $w(x) = 1$ in the interval $[-1, 1]$.

13. Using the relation (5.65), prove that
$$(x - t) \sum_{n=0}^{\infty} P_n(x) t^n = (1 - 2xt + t^2) \sum_{n=1}^{\infty} n P_n(x) t^{n-1},$$
and hence show the relation (5.62).

14. Show that $\dfrac{1 - t^2}{(1 - 2xt + t^2)^{3/2}} = \displaystyle\sum_{n=0}^{\infty} (2n + 1) t^n P_n(x)$.

15. Find the Legendre series expansion of the function
$$f(x) = \begin{cases} 0 & -1 \le x < 0 \\ 1 & 0 < x \le 1 \end{cases}$$

16. Find the Legendre series expansion of $f(x) = |x|,\ -1 \le x \le 1$.

17. Find the first three terms of the Legendre series of $f(x) = e^x$.

18. Prove that $P'_n(x) = (2n - 1) P_{n-1}(x) + (2n - 5) P_{n-3}(x) + (2n - 9) P_{n-5}(x) + \ldots$ for all $n \ge 1$, by expanding $P'_n(x)$ in a Legendre series.

19. If $g(x) = x^n + h(x)$, where degree of $h(x) < n$, and c is a constant such that $c P_n(x) = x^n + q(x)$, where degree of $q(x) < n$, then show that
$$\int_{-1}^{1} \{g(x)\}^2\, dx \ge \int_{-1}^{1} c^2 (P_n(x))^2\, dx$$

(This result suggests that $cP_n(x)$ is the best approximation to zero in $(-1, 1)$ among the monic polynomials of nth degree in the sense of least squares.)

20. Using the following figure with $r_2 > 0$, show that

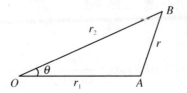

Figure 5.1 Exercise 20.

$$\frac{1}{r} = \frac{1}{r_2}\left\{P_0 + \left(\frac{r_1}{r_2}\right)P_1(\cos\theta) + \left(\frac{r_1}{r_2}\right)^2 P_2(\cos\theta) + \left(\frac{r_1}{r_2}\right)^3 P_3(\cos\theta) + \cdots\right\}$$

5.22 CHEBYSHEV POLYNOMIALS

The equation

$$(1 - x^2)y'' - xy' + p^2 y = 0 \tag{5.68}$$

is called the *Chebyshev equation* of degree p, where p is a real number.

As in the case of Legendre equation, the origin is an ordinary point and $x = \pm 1$ are regular singular points for Eq. (5.68).

The power series solutions near $x = 0$ are convergent in $-1 < x < 1$. The recurrence relation that the coefficients of power series must satisfy is given by

$$a_{n+2} = \frac{(n^2 - p^2)}{(n+1)(n+2)} a_n \qquad n = 0, 1, 2, \ldots \tag{5.69}$$

Thus the two linearly independent solutions of (5.68), valid for $|x| < 1$, are given by

$$y_1(x) = 1 - \frac{p \cdot p}{2!} x^2 + \frac{p(p-2)p(p+2)}{4!} x^4 - \cdots$$

and

$$y_2(x) = x - \frac{(p-1)(p+2)}{3} x^3 + \frac{(p-1)(p-3)(p+1)(p+3)}{5!} x^5 - \cdots$$

If p is a non-negative integer n, then one of these solutions is a polynomial of degree n. In the case of the Chebyshev equation, many times the interest lies in these solutions, the derivatives of which are bounded at ± 1. The polynomial solutions, of course, satisfy this condition.

To get the solution of (5.68) near $x = 1$, let us change the independent variable from x to t by using $t = \dfrac{1-x}{2}$. Then Eq. (5.68) transforms to

Special Functions and Equations

$$t(1-t)\frac{d^2y}{dt^2} + \left(\frac{1}{2}-t\right)\frac{dy}{dt} + p^2 y = 0 \tag{5.70}$$

which is a hypergeometric equation with parameters $p, -p, \frac{1}{2}$. Thus a fundamental set of solutions near the origin for Eq. (5.70) consists of $F(p, -p; \frac{1}{2}; t)$ and $\sqrt{t}\, F\left(p+\frac{1}{2}, -p+\frac{1}{2}; \frac{3}{2}; t\right)$.

Thus the general solution of (5.68) near $x = 1$ is given by

$$y(x) = AF\left(p, -p; \frac{1}{2}, \frac{1-x}{2}\right) + B\left(\frac{1-x}{2}\right)^{\frac{1}{2}} F\left(p+\frac{1}{2}, -p+\frac{1}{2}; \frac{3}{2}, \frac{1-x}{2}\right)$$

Differentiating this, we get

$$y'(x) = Ap^2 F\left(p+1, -p+1; \frac{3}{2}, \frac{1-x}{2}\right) - \frac{B}{4}\left(\frac{1-x}{2}\right)^{-\frac{1}{2}} F\left(p+\frac{1}{2}, -p+\frac{1}{2}; \frac{3}{2}, \frac{1-x}{2}\right)$$

$$+ \frac{B\left(p+\frac{1}{2}\right)\left(p-\frac{1}{2}\right)}{3}\left(\frac{1-x}{2}\right)^{\frac{1}{2}} F\left(p+\frac{3}{2}, -p+\frac{3}{2}; \frac{5}{2}, \frac{1-x}{2}\right)$$

$y'(x)$ is bounded near $x = 1$, provided $B = 0$. Thus only those solutions of (5.68) whose derivatives are bounded near $x = 1$ are given by $y(x) = AF\left(p, -p; \frac{1}{2}; \frac{1-x}{2}\right)$.

This is a polynomial for non-negative integral values of p. Hence the only solutions of

$$(1 - x^2)y'' - xy' + n^2 y = 0 \tag{5.71}$$

where $n = 0, 1, 2, \ldots$, whose derivatives near $x = \pm 1$ are bounded must be multiples of

$$F\left(n, -n; \frac{1}{2}; \frac{1-x}{2}\right) \tag{5.72}$$

The polynomial (5.72) is called the *Chebyshev polynomial* and is usually denoted by $T_n(x)$. It is a polynomial of degree n and the leading coefficient is 2^{n-1}. These polynomials have many interesting applications to the theory of approximations, which cannot be discussed in this book. Nevertheless, to know more about this polynomials, let us discuss some of its properties.

5.23 PROPERTIES OF CHEBYSHEV POLYNOMIALS

1. $T_n(1) = 1$ for $n = 0, 1, 2, \ldots$
This follows directly from the definition of $T_n(x)$ in (5.72).
2. $T_{2n}(x)$ is an even function and $T_{2n+1}(x)$ is an odd function i.e.

$$T_n(-x) = (-1)^n T_n(x).$$

This follows from Definition 5.72.

3. $T_n(x) = \cos(n \cos^{-1} x)$
Equation (5.71) reduces, with the substitution $x = \cos\theta$, to

$$\frac{d^2 y}{d\theta^2} + n^2 y = 0 \tag{5.73}$$

$$\Rightarrow y = c_1 \cos n\theta + c_2 \sin n\theta$$

The solutions of Eq. (5.71), whose derivatives are bounded near $x = 1$ are the solutions of Eq. (5.73) whose derivatives are bounded at $\theta = 0$.
Thus $T_n(x) = \cos n\theta = \cos(n \cos^{-1} x)$, taking $c_1 = 1$, $c_2 = 0$. This is due to the fact that $T_n(x) \to 1$ as $x \to 1$ and $\cos n\theta \to 1$ as $\theta \to 0$.

$$T_n(x) = \cos(n \cos^{-1} x) \tag{5.74}$$

is another definition of Chebyshev polynomials.

4. Another expression for Chebyshev polynomials is given by

$$T_n(x) = \frac{1}{2}\left\{\left(x + i\sqrt{1-x^2}\right)^n + \left(x - i\sqrt{1-x^2}\right)^n\right\}$$

We know $\cos n\theta = \dfrac{1}{2}\{(\cos n\theta + i \sin n\theta) + (\cos n\theta - i \sin n\theta)\}$

$$= \frac{1}{2}\left\{(\cos\theta + i\sin\theta)^n + (\cos\theta - i\sin\theta)^n\right\}$$

by de Moivre's formula

$$= \frac{1}{2}\left\{\left(x + i\sqrt{1-x^2}\right)^n + \left(x - i\sqrt{1-x^2}\right)^n\right\}$$

5. (Recurrence Relation)

$\cos n\theta + \cos(n-2)\theta$
$= \cos\{(n-1)+1\}\theta + \cos\{(n-1)-1\}\theta$
$= 2\cos\theta \cos(n-1)\theta$

$\Rightarrow \cos n\theta = 2\cos\theta \cos(n-1)\theta - \cos(n-2)\theta$

$$\Rightarrow T_n(x) = 2x T_{n-1}(x) - T_{n-2}(x) \qquad n \geq 2 \tag{5.75}$$

6. (Orthogonality Relation) $\{T_0(x), T_1(x), T_2(x), \ldots\}$ is an orthogonal simple set of polynomials with respect to the weight function $(1 - x^2)^{-1/2}$ on the interval $-1 \leq x \leq 1$.

$$\int_{-1}^{1} \frac{T_m(x)T_n(x)}{\sqrt{1-x^2}} dx = \int_{\pi}^{0} \frac{\cos m\theta \cos n\theta}{\sin \theta}(-\sin \theta) d\theta \qquad \text{due to (5.74)}$$

$$= \int_{0}^{\pi} \cos m\theta \cos n\theta \, d\theta \qquad (5.76)$$

$$= \frac{1}{2}\int_{0}^{\pi} \{\cos(m+n)\theta + \cos(m-n)\theta\} d\theta \qquad \text{if } m \ne n$$

$$= \frac{1}{2}\left[\frac{\sin(m+n)\theta}{m+n} + \frac{\sin(m-n)\theta}{m-n}\right]_{0}^{\pi} = 0$$

If $m = n \ne 0$, then Eq. (5.76) becomes

$$\int_{-1}^{1} \frac{T_n^2(x)}{\sqrt{1-x^2}} dx = \int_{0}^{\pi} \cos^2 n\theta \, d\theta = \frac{1}{2}\int_{0}^{\pi} (1+\cos 2\theta) d\theta = \frac{\pi}{2} \qquad (5.77)$$

If $m = n = 0$, then Eq (5.76) becomes

$$\int_{-1}^{1} \frac{T_0^2(x)}{\sqrt{1-x^2}} dx = \int_{0}^{\pi} d\theta = \pi$$

Thus, we have the orthogonality relations for the Chebyshev polynomials as

$$\int_{-1}^{1} \frac{T_m(x)T_n(x)}{\sqrt{1-x^2}} dx = \begin{cases} 0 & \text{if } m \ne n \\ \frac{\pi}{2} & \text{if } m = n \ne 0 \\ \pi & \text{if } m = n = 0 \end{cases} \qquad (5.78)$$

7. (Generating function) $\dfrac{1-t^2}{1-2xt+t^2}$ is the generating function for the Chebyshev polynomials $T_n(x)$, i.e., if we expand $\dfrac{1-t^2}{1-2xt+t^2}$ in a Taylor series about $t = 0$, treating x as a fixed parameter, then the coefficients of t^n are Chebyshev polynomials.
Thus

$$\frac{1-t^2}{1-2xt+t^2} = T_0(x) + 2\sum_{n=1}^{\infty} T_n(x)t^n \qquad (5.79)$$

We shall not prove this result (5.79), as the treatment is beyond the scope of this book.

8. (Chebyshev Series) Just like the Legendre series, any "nice" function f defined in $[-1, 1]$ can be expressed as the Chebyshev series

$$f(x) = \sum_{n=0}^{\infty} a_n T_n(x) \tag{5.80}$$

where the coefficients a_n are given by the formula

$$a_0 = \frac{1}{\pi} \int_{-1}^{1} \frac{f(x)}{\sqrt{1-x^2}} dx$$

$$a_n = \frac{2}{\pi} \int_{-1}^{1} \frac{T_n(x) f(x)}{\sqrt{1-x^2}} dx \quad n > 0$$

due to the orthogonality property (5.78).

Example 5.10 If $f(x)$ is a polynomial of degree less than n, then show that

$$\int_{-1}^{1} \frac{f(x) T_n(x)}{\sqrt{1-x^2}} dx = 0$$

Since $\{T_0(x), T_1(x), T_2(x), \ldots\}$ is an orthogonal simple set of polynomials on the interval $-1 \leq x \leq 1$ with respect to the weight function $(1-x^2)^{-1/2}$, the result follows directly from Property 3 of Section 5.13. However, the following proof is obtained with the help of the Chebyshev series and orthogonality relations.

We know that $f(x)$ is a polynomial of degree less than n

$$\Rightarrow f(x) = \sum_{k=0}^{n-1} a_k T_k(x) \quad \text{by (5.80)}$$

Then $\quad \int_{-1}^{1} \frac{f(x) T_n(x)}{\sqrt{1-x^2}} dx = \sum_{k=1}^{n-1} \int_{-1}^{1} \frac{a_k T_k(x) T_n(x)}{\sqrt{1-x^2}} dx = 0 \quad \text{by (5.78)}$

Exercises 5.23

1. Prove that $F\left(n, -n; \frac{1}{2}; x\right)$ and $P_n(x)$ are both orthogonal to $T_{n+1}(x)$ on the interval $-1 \leq x \leq 1$ with respect to the weight function $(1-x^2)^{-1/2}$.

2. Prove that $\int_{-1}^{1} \dfrac{x T_{n-1}(x) T_n(x)}{\sqrt{1-x^2}}\, dx = \dfrac{\pi}{4}$.

3. Prove that $\int_{-1}^{1} \dfrac{x T_{n-1}^2(x)}{\sqrt{1-x^2}}\, dx = 0$.

4. Show that $T_n(x) = 2^{n-1} x^n + g(x)$, where $g(x)$ is a polynomial of degree less than n, and $|T_n(x)| \le 1$.

5. Find the Chebyshev series expansion of the function
$$f(x) = \begin{cases} 0 & -1 \le x < 0 \\ x & 0 \le x \le 1 \end{cases}$$

6. Show that the roots of $T_7(x) = 0$ are $\cos \dfrac{\pi}{14},\ \cos \dfrac{3\pi}{14},\ \ldots,\ \cos \dfrac{13\pi}{14}$.

7. Show that $T_n(x)$ has n distinct zeros in $(-1, 1)$.
   (**Hint:** Use Property 7 of Section 5.13.)

5.24 HERMITE POLYNOMIALS

The equation $y'' - 2xy' + 2py = 0$, $-\infty < x < +\infty$, p a real number, is called *Hermite equation of degree p*. (5.81)

Since this equation has no singular points, we have two LI series solutions that converge for all x. By power series method, we obtain two LI solutions in powers of x as

$$y_1(x) = 1 + \sum_{n=1}^{\infty} \dfrac{2^n(-p)(-p+2)\cdots(-p+2n-2)}{(2n)!} x^{2n}$$

and

$$y_2(x) = x + \sum_{n=1}^{\infty} \dfrac{2^n(1-p)(1-p+2)\cdots(1-p+2n-2)}{(2n+1)!} x^{2n+1} \qquad (5.82)$$

If $p \ge 0$ and an integer, then: (i) for p even, $y_1(x)$ terminates, each term for $n \ge \dfrac{1}{2}(p+2)$ being zero, and $y_1(x)$ is a polynomial of degree p; (ii) for p odd, $y_2(x)$ terminates, each term for $n \ge \dfrac{1}{2}(p+1)$ being zero, and $y_2(x)$ is a polynomial of degree p.

For $p = 0, 1, 2, 3, 4, \ldots$, these polynomials are $1,\ x,\ 1 - 2x^2,\ x - \dfrac{2}{3}x^3,\ 1 - 4x^2 + \dfrac{4}{3}x^4, \ldots$

Introduction to Differential Equations

Thus the Hermite equation of degree n always has a polynomial solution $H_n(x)$ of degree n, for non-negative integers. It is elementary but tedious to obtain a single expression for this polynomial solution from the above two LI solutions, i.e.

$$H_n(x) = \sum_{k=0}^{\left[\frac{n}{2}\right]} \frac{(-1)^k n!}{k!(n-2k)!} (2x)^{n-2k} \tag{5.83}$$

where $\left[\frac{n}{2}\right]$ stands for the greatest integer not greater than $\frac{n}{2}$. $H_n(x)$ is a constant multiple of the polynomials described in (5.82) for $p = n$ and the term containing, the highest power of x looks like $2^n x^n$. The polynomial $H_n(x)$ is called *Hermite polynomial* which is a solution of the Hermite equation

$$y'' - 2xy' + 2ny = 0 \tag{5.84}$$

of degree $n = 0, 1, 2, 3, \ldots$.

5.25 PROPERTIES OF HERMITE POLYNOMIALS

1. $H_n(x)$ is an even function for n even and an odd function for n odd, i.e. $H_n(-x) = (-1)^n H_n(x)$.
    2. (Recurrence relation) $H_{n+1}(x) = 2xH_n(x) - 2nH_{n-1}(x)$.
    The derivation of recurrence relation is by method of induction.
    From (5.83), we clearly get $H_0(x) = 1$, $H_1(x) = 2x$, $H_2(x) = 2^2 x^2 - 2$, ...
    Let $f_{n+1}(x) = 2xH_n(x) - 2nH_{n-1}(x)$ \hfill (5.85)
We prove by method of induction that $f_{n+1}(x) = H_{n+1}(x)$.

$$f_2(x) = 2xH_1(x) - 2H_0(x) = 2x \cdot 2x - 2 = 2^2 x^2 - 2 = H_2(x)$$

Let us assume $f_n(x) = H_n(x)$; then

$$f_{n+1}(x) = 2xH_n(x) - 2nH_{n-1}(x)$$

$$= 2x \sum_{k=0}^{\left[\frac{n}{2}\right]} \frac{(-1)^k n!}{k!(n-2k)!} (2x)^{n-2k} - 2n \sum_{k=0}^{\left[\frac{n-1}{2}\right]} \frac{(-1)^k (n-1)!}{k!(n-1-2k)!} (2x)^{n-1-2k}$$

$$= 2x \sum_{k=0}^{\left[\frac{n}{2}\right]} \frac{(-1)^k n!}{k!(n-2k)!} (2x)^{n-2k} - 2n \sum_{k=1}^{\left[\frac{n+1}{2}\right]} \frac{(-1)^{k-1}(n-1)!}{(k-1)!(n+1-2k)!} (2x)^{n+1-2k} \tag{5.86}$$

by replacing k by $k - 1$ in the second summation.

If $n = 2m$, an even integer, then we have from (5.86)

$$f_{2m+1}(x) = 2x \sum_{k=0}^{m} \frac{(-1)^k (2m)!}{k!(2m-2k)!} (2x)^{2m-2k}$$

$$- 4m \sum_{k=1}^{m} \frac{(-1)^{k-1}(2m-1)!}{(k-1)!(2m+1-2k)!} (2x)^{2m+1-2k}$$

$$= (2x)^{2m+1} + \sum_{k=1}^{m} \frac{(-1)^k (2m)!}{k!(2m-2k)!} (2x)^{2m+1-2k}$$

$$+ 2 \sum_{k=1}^{m} \frac{(-1)^k (2m)!}{(k-1)!(2m+1-2k)!} (2x)^{2m+1-2k}$$

$$= (2x)^{2m+1} + \sum_{k=1}^{m} \frac{(-1)^k (2m)! (2x)^{2m+1-2k}}{(k-1)!(2m-2k)!} \left\{ \frac{1}{k} + \frac{2}{2m+1-2k} \right\}$$

$$= (2x)^{2m+1} + \sum_{k=1}^{m} \frac{(-1)^k (2m+1)!}{k!(2m+1-2k)!} (2x)^{2m+1-2k}$$

$$= \sum_{k=0}^{m} \frac{(-1)^k (2m+1)!}{k!(2m+1-2k)!} (2x)^{2m+1-2k} = H_{2m+1}(x)$$

On the other hand, if $n = 2m + 1$, an odd integer, then (5.86) becomes

$$f_{2m+2}(x) = 2x \sum_{k=0}^{m} \frac{(-1)^k (2m+1)!}{k!(2m+1-2k)!} (2x)^{2m+1-2k}$$

$$-2(2m+1) \sum_{k=1}^{m+1} \frac{(-1)^{k-1}(2m)!}{(k-1)!(2m+2-2k)!} (2x)^{2m+2-2k}$$

$$= (2x)^{2m+2} + \sum_{k=1}^{m} \frac{(-1)^k (2m+1)!}{k!(2m+1-2k)!} (2x)^{2m+2-2k}$$

$$-2(2m+1) \sum_{k=1}^{m+1} \frac{(-1)^{k-1}(2m)!}{(k-1)!(2m-2k+2)!} (2x)^{2m-2k+2}$$

$$= (2x)^{2m+2} + \sum_{k=1}^{m} \frac{(-1)^k (2m+1)!}{k!(2m+1-2k)!} (2x)^{2m+2-2k}$$

Introduction to Differential Equations

$$+ 2\sum_{k=1}^{m+1} \frac{(-1)^k (2m+1)!}{(k-1)!(2m-2k+2)!}(2x)^{2m-2k+2}$$

$$= (2x)^{2m+2} + \sum_{k=1}^{m} \frac{(-1)^k (2m+1)!}{(k-1)!(2m-2k+1)!}(2x)^{2m+2-2k}\left\{\frac{1}{k} + \frac{2}{2m-2k+2}\right\}$$

$$+ \frac{2(-1)^{m+1}(2m+1)!}{m!}$$

$$= (2x)^{2m+2} + \sum_{k=1}^{m} \frac{(-1)^k (2m+2)!}{k!(2m+2-2k)!}(2x)^{2m+2-2k}$$

$$+ \frac{2(-1)^{m+1}(2m+1)!(m+1)}{(m+1)!}$$

$$= \sum_{k=0}^{m} \frac{(-1)^k (2m+2)!}{k!(2m+2-2k)!}(2x)^{2m+2-2k} + \frac{(-1)^{m+1}(2m+2)!}{(m+1)!}$$

$$= \sum_{k=0}^{m+1} \frac{(-1)^k (2m+2)!}{k!(2m+2-2k)!}(2x)^{2m+2-2k} = H_{2m+2}(x)$$

Hence the recurrence relation

$$H_{n+1}(x) = 2xH_n(x) - 2nH_{n-1}(x) \tag{5.87}$$

3. (Generating Function) The function e^{2xt-t^2} is called the *generating function* of the Hermite polynomials $H_n(x)$, i.e., the Taylor series expansion of e^{2xt-t^2} about $t = 0$, treating x as a fixed parameter, is given by

$$e^{2xt-t^2} = \sum_{n=0}^{\infty} \frac{H_n(x)}{n!} t^n \tag{5.88}$$

We shall not prove this result (5.88) as the treatment is beyond the scope of this book.

4. (Rodrigues' Formula) $H_n(x) = (-1)^n e^{x^2} \dfrac{d^n}{dx^n} e^{-x^2}$.

We know by Mclaurin's Theorem

$$e^{2xt-t^2} = \sum_{n=0}^{\infty} \frac{1}{n!}\left(\frac{\partial^n}{\partial t^n} e^{2xt-t^2}\right)_{t=0} t^n \tag{5.89}$$

Also, from generating function (5.88) we have

$$e^{2xt-t^2} = \sum_{n=0}^{\infty} \frac{H_n(x)}{n!} t^n \tag{5.90}$$

Special Functions and Equations

From (5.89) and (5.90), we have

$$H_n(x) = \left(\frac{\partial^n}{\partial t^n} e^{2xt-t^2}\right)_{t=0} = e^{x^2}\left(\frac{\partial^n}{\partial t^n} e^{-(x-t)^2}\right)_{t=0}$$

$$= (-1)^n e^{x^2} \left(\frac{d^n}{dz^n} e^{-z^2}\right)_{z=x} \qquad \text{putting } z = x - t$$

$$= (-1)^n e^{x^2} \frac{d^n}{dx^n} e^{-x^2} \tag{5.91}$$

5. (Orthogonality Relation) $\displaystyle\int_{-\infty}^{+\infty} e^{-x^2} H_m(x) H_n(x) dx = \begin{cases} 0 & \text{if } m \neq n \\ 2^n n! \sqrt{\pi} & \text{if } m = n \end{cases}$

We have
$$H_m''(x) - 2x H_m'(x) + 2m H_m(x) = 0$$

$$H_n''(n) - 2x H_n'(x) + 2n H_n(x) = 0$$

since $H_m(x)$ and $H_n(x)$ are solutions of the Hermite equation (5.84). Multiply the first equation by $H_n(x)$ and the second equation by $H_m(x)$, then subtract the resulting second equation from the resulting first equation to obtain

$$\{H_m''(x) H_n(x) - H_m(x) H_n''(x)\} - 2x\{H_m'(x) H_n(x) - H_m(x) H_n'(x)\}$$
$$+ 2(m-n) H_m(x) H_n(x) = 0$$

Multiplying both the sides by e^{-x^2}

$$2(n-m) H_m(x) H_n(x) e^{-x^2} = e^{-x^2}\{H_m''(x) H_n(x) - H_m(x) H_n''(x)\}$$
$$- 2x e^{-x^2}\{H_m'(x) H_n(x) - H_m(x) H_n'(x)\}$$
$$= \frac{d}{dx}\left\{e^{-x^2}\left(H_m'(x) H_n(x) - H_m(x) H_n'(x)\right)\right\}$$

Integrating from $-\infty$ to $+\infty$, we get

$$2(n-m)\int_{-\infty}^{+\infty} e^{-x^2} H_m(x) H_n(x) dx = \left[e^{-x^2}\left(H_m'(x) H_n(x) - H_m(x) H_n'(x)\right)\right]_{-\infty}^{+\infty} = 0$$

since the product of e^{-x^2} and a polynomial vanishes as $x \to +\infty$ or $x \to -\infty$.

Hence $\displaystyle\int_{-\infty}^{+\infty} e^{-x^2} H_m(x) H_n(x) dx = 0 \qquad \text{if } m \neq n \tag{5.92}$

If $m = n$, then

$$\int_{-\infty}^{+\infty} e^{-x^2} H_n(x) H_n(x) dx$$

$$= (-1)^n \int_{-\infty}^{+\infty} e^{-x^2} H_n(x) e^{x^2} \frac{d^n}{dx^n} e^{-x^2} dx \quad \text{due to (5.91)}$$

$$= (-1)^n \int_{-\infty}^{+\infty} H_n(x) \frac{d^n}{dx^n} e^{-x^2} dx$$

$$= (-1)^n \left[H_n(x) \frac{d^{n-1}}{dx^{n-1}} e^{-x^2} \right]_{-\infty}^{+\infty} + (-1)^{n+1} \int_{-\infty}^{+\infty} H_n'(x) \frac{d^{n-1}}{dx^{n-1}} e^{-x^2} dx$$

$$\text{integrating by parts}$$

$$= (-1)^{n+1} \int_{-\infty}^{+\infty} H_n'(x) \frac{d^{n-1}}{dx^{n-1}} e^{-x^2} dx$$

$\therefore e^{-x^2}(H_n(x)p(x))$, where $p(x)$ is a polynomial, is the term inside the bracket and the product of e^{-x^2} and a polynomial is zero when $x \to -\infty$ and $x \to +\infty$

$$= (-1)^{n+2} \int_{-\infty}^{+\infty} H_n''(x) \frac{d^{n-2}}{dx^{n-2}} e^{-x^2} dx \quad \text{integrating by parts}$$

$$= \cdots = (-1)^{n+n} \int_{-\infty}^{+\infty} H_n^{(n)}(x) \frac{d^{n-n}}{dx^{n-n}} e^{-x^2} dx \quad \text{by repeated integration by parts}$$

$$= \int_{-\infty}^{+\infty} 2^n n! \, e^{-x^2} dx$$

since $H_n(x)$ is a polynomial of degree n with the coefficient of x^n as 2^n

$$= 2^n n! \int_{-\infty}^{+\infty} e^{-x^2} dx = 2^n n! \sqrt{\pi} \quad (5.93)$$

6. Every nontrivial solutions of Hermite equation $y'' - 2xy' + 2py = 0$ of degree p has at most finitely many zeros in $(-\infty, +\infty)$. See Example 4.11.

7. (Hermite Series) Just like the Legendre series, any 'nice' function f defined in $-\infty < x < +\infty$ can be expressed in terms of Hermite polynomials in a Hermite Series as

$$f(x) = \sum_{n=0}^{\infty} a_n H_n(x) \quad (5.94)$$

Special Functions and Equations

where the coefficients are given due to orthogonality relations (5.92) and (5.93) by

$$a_n = \frac{1}{2^n n! \sqrt{\pi}} \int_{-\infty}^{+\infty} e^{-x^2} H_n(x) f(x) dx \qquad (5.95)$$

Example 5.11 Evaluate, by using recurrence relation, the integral

$$\int_{-\infty}^{+\infty} e^{-x^2} x H_{n-1}(x) H_{n-2}(x) dx$$

Solution: By recurrence relation (5.87) we have

$H_{n+1}(x) = 2xH_n(x) - 2nH_{n-1}(x)$

$\Rightarrow H_n(x) = 2xH_{n-1}(x) - 2(n-1)H_{n-2}(x)$ replacing n by $n-1$

$\Rightarrow e^{-x^2} H_n(x) H_{n-2}(x) = 2xe^{-x^2} H_{n-1}(x) H_{n-2}(x) - 2(n-1)e^{-x^2} H_{n-2}^2(x)$

multiplying both the sides by $e^{-x^2} H_{n-2}(x)$

$\Rightarrow \int_{-\infty}^{+\infty} e^{-x^2} H_n(x) H_{n-2}(x) dx = \int_{-\infty}^{+\infty} 2xe^{-x^2} H_{n-1}(x) H_{n-2}(x) dx - \int_{-\infty}^{+\infty} 2(n-1)e^{-x^2} H_{n-2}^2(x) dx$

integrating from $-\infty$ to $+\infty$

$\Rightarrow 0 = \int_{-\infty}^{+\infty} 2xe^{-x^2} H_{n-1}(x) H_{n-2}(x) dx - 2(n-1)2^{n-2}(n-2)!\sqrt{\pi}$ by (5.92) and (5.93)

$\Rightarrow \int_{-\infty}^{+\infty} xe^{-x^2} H_{n-1}(x) H_{n-2}(x) dx = 2^{n-2}(n-1)!\sqrt{\pi}$

Now try the following problems in the exercises.

Exercises 5.25

1. Differentiating the relation (5.88) with respect to t, prove the recurrence relation (5.87).
2. Derive the recurrence relation (5.87) by using Rodrigue's formula (5.91).
3. Using the recurrence relation (5.87), prove the following:

   (i) $\int_{-\infty}^{+\infty} xe^{-x^2} H_n^2(x) dx = 0$

   (ii) $\int_{-\infty}^{+\infty} xe^{-x^2} H_n(x) H_{n+1}(x) dx = 2^n (n+1)!\sqrt{\pi}$

4. Prove that $\int_{-1}^{1} P_3(x)H_8(x)T_6(x)dx = 0$.

5. If $f(x) = \begin{cases} 0 & -\infty < x < 0 \\ x^2 & 0 < x < +\infty \end{cases}$

then find the Hermite series

$$f(x) = a_0 H_0(x) + a_1 H_1(x)t + a_2 H_2(x)t^2 + \cdots$$

Determine the first three coefficients a_0, a_1, a_2.

6. If $f(x)$ is a polynomial of degree $< n$, then prove that

$$\int_{-\infty}^{+\infty} e^{-x^2} f(x) H_n(x) dx = 0$$

5.26 LAGUERRE POLYNOMIALS

The equation

$$xy'' + (1-x)y' + py = 0 \qquad (5.96)$$

where $x > 0$ and p is a constant, is called the *Laguerre equation* of degree p.

The origin is a regular singular point. But zero is a repeated root of the indicial equation. A power series solution exists. Proceeding in the usual manner, it is easy to obtain a solution as

$$\sum_{k=0}^{\infty} a_k x^k \quad \text{where} \quad a_{k+1} = \frac{-p+k}{(k+1)^2} a_k$$

Thus the series is

$$a_0 \left\{ 1 - \frac{p}{(1!)^2} x + \frac{p(p-1)}{(2!)^2} x^2 - \frac{p(p-1)(p-2)}{(3!)^2} x^3 + \cdots \right\} \qquad (5.97)$$

We can obtain the general solution of (5.96) by applying Theorem 3.8. The series (5.97) converges for all values of x, and the only solutions bounded near the origin are constant multiples of (5.97). Thus a solution of (5.96) which is independent of (5.97) is unbounded at the origin.

Further, if p is a nonnegative integer n, then the series (5.97) reduces to a polynomial of degree n with the leading coefficient $\dfrac{(-1)^n}{n!}$. As usual, this polynomial is of our interest.

Thus the polynomial solution of the equation

$$xy'' + (1-x)y' + ny = 0 \qquad n = 0, 1, 2, \ldots \qquad (5.98)$$

Special Functions and Equations

is called the *Laguerre Polynomial* and is denoted by $L_n(x)$.

$$\therefore \quad L_n(x) = \sum_{k=0}^{n} (-1)^k \binom{n}{k} \frac{x^k}{k!} \quad (5.99)$$

Laguerre polynomials have important applications in the quantum mechanics of the hydrogen atom. Let us discuss some of the properties of these polynomials.

5.27 PROPERTIES OF LAGUERRE POLYNOMIALS

1. (Rodrigues' Formula) Laguerre polynomials satisfy the following relation

$$L_n(x) = \frac{e^x}{n!} \frac{d^n}{dx^n} \left(e^{-x} x^n \right) \quad (5.100)$$

To prove this relation, let us differentiate $(e^{-x} x^n)$ n times by using the Leibnitz rule. Then we have

$$\frac{d^n}{dx^n}\left(e^{-x} x^n\right) = (-1)^n e^{-x} x^n + \binom{n}{1}(-1)^{n-1} e^{-x} n x^{n-1}$$

$$+ \binom{n}{2}(-1)^{n-2} e^{-x} n(n-1) x^{n-2} + \cdots + e^{-x} n!$$

$$= e^{-x} n! \left\{ \frac{(-1)^n}{n!} x^n + (-1)^{n-1} \binom{n}{1} \frac{x^{n-1}}{(n-1)!} + (-1)^{n-2} \binom{n}{2} \frac{x^{n-2}}{(n-2)!} + \cdots + 1 \right\}$$

$$= e^{-x} n! \left\{ 1 - \binom{n}{1} \frac{x}{1!} + \binom{n}{2} \frac{x^2}{2!} - \cdots + (-1)^n \frac{x^n}{n!} \right\}$$

writing in the reverse order from (5.99)

$$= e^{-x} n! L_n(x)$$

$$\Rightarrow L_n(x) = \frac{e^x}{n!} \frac{d^n}{dx^n} (e^{-x} x^n)$$

2. (Recurrence Relation) The recurrence relation satisfied by the sequence $\{L_n(x)\}$ of Laguerre polynomials is

$$(n + 1)L_{n+1}(x) = (2n + 1 - x)L_n(x) - nL_{n-1}(x) \quad (5.101)$$

To prove this relation, we shall assume that the sequence $\{L_n(x)\}$ is an orthogonal simple set of polynomials. We shall prove this in the next property. Then by Property 6(i) of Section 5.13, we have

$$L_{n+1}(x) = (a_n x + b_n)L_n(x) + c_n L_{n-1}(x) \quad \text{for } n = 1, 2, 3, \ldots$$

where a_n, b_n, c_n are real numbers.

Now, comparing the coefficients of x^0, x, and x^{n+1} from both the sides, we get

140 Introduction to Differential Equations

$$1 = b_n + c_n \qquad -(n+1) = a_n - nb_n - (n-1)c_n$$

and $$\frac{(-1)^{n+1}}{(n+1)!} = a_n \frac{(-1)^n}{n!}$$

$$\Rightarrow a_n = -\frac{1}{n+1} \qquad b_n = \frac{2n+1}{n+1} \qquad c_n = -\frac{n}{n+1}$$

Putting the values of a_n, b_n, c_n in the above equation, we get

$$L_{n+1}(x) = \left(-\frac{x}{n+1} + \frac{2n+1}{n+1}\right)L_n(x) - \frac{n}{n+1}L_{n-1}(x)$$

$$\Rightarrow (n+1)L_{n+1}(x) = (2n+1-x)L_n(x) - nL_{n-1}(x)$$

which is the required recurrence relation for $n = 1, 2, 3, \ldots$.

3. (Orthogonality Relation) The sequence $\{L_n(x)\}$ of Laguerre polynomials is an orthonormal sequence of polynomials with respect to the weight function e^{-x} on the interval $[0, \infty)$, i.e.

$$\int_0^\infty e^{-x} L_m(x) L_n(x) dx = \begin{cases} 0 & \text{if } m \neq n \\ 1 & \text{if } m = n \end{cases} \qquad (5.102)$$

Proof: Since $L_m(x)$ and $L_n(x)$ are solutions of Laguerre equations of degree m and n respectively, we have

$$xL_m''(x) + (1-x)L_m'(x) + mL_m(x) = 0$$

and $$xL_n''(x) + (1-x)L_n'(x) + nL_n(x) = 0$$

Multiplying the first equation by $L_n(x)$ and the second equation by $L_m(x)$ and subtracting, we get

$$x(L_m''(x)L_n(x) - L_m(x)L_n''(x)) + (1-x)(L_m'(x)L_n(x) - L_m(x)L_n'(x)) + (m-n)(L_m(x)L_n(x)) = 0$$

$$\Rightarrow xe^{-x}\frac{d}{dx}\{L_m'(x)L_n(x) - L_m(x)L_n'(x)\} + \left(e^{-x} - xe^{-x}\right)$$

$$\{L_m'(x)L_n(x) - L_m(x)L_n'(x)\} + (m-n)e^{-x}L_m(x)L_n(x) = 0 \quad \text{multiplying by } e^{-x}$$

$$\Rightarrow \frac{d}{dx}\{xe^{-x}(L_m'(x)L_n(x) - L_m(x)L_n'(x))\} + (m-n)e^{-x}L_m(x)L_n(x) = 0$$

$$\Rightarrow (n-m)\int_0^\infty e^{-x}L_m(x)L_n(x)\,dx = xe^{-x}(L_m'(x)L_n(x) - L_m(x)L_n'(x))\big|_0^\infty$$

integrating from 0 to ∞

$$\Rightarrow \int_0^\infty e^{-x}L_m(x)L_n(x)\,dx = 0$$

Special Functions and Equations

∴ $x(L'_m(x)L_n(x) - L_m(x)L'_n(x))$ is a polynomial; the product of this polynomial with e^{-x} tends to zero as $x \to \infty$, also $m \neq n$.

Thus
$$\int_0^\infty e^{-x} L_m(x) L_n(x)\, dx = 0 \qquad m \neq n$$

Next, if $m = n$, $\int_0^\infty e^{-x} L_n(x) L_n(x)\, dx$

$$= \int_0^\infty e^{-x} L_n(x) \frac{e^x}{n!} \frac{d^n}{dx^n}(e^{-x} x^n)\, dx \qquad \text{using (5.100)}$$

$$= \frac{1}{n!} \int_0^\infty L_n(x) \frac{d^n}{dx^n}(e^{-x} x^n)\, dx$$

$$= \frac{1}{n!}\left[L_n(x) \frac{d^{n-1}}{dx^{n-1}}\left(e^{-x} x^n\right) \right]_0^\infty - \frac{1}{n!} \int_0^\infty L'_n(x) \frac{d^{n-1}}{dx^{n-1}}(e^{-x} x^n)\, dx$$

integrating by rts

$$= \frac{-1}{n!} \int_0^\infty L'_n(x) \frac{d^{n-1}}{dx^{n-1}}(e^{-x} x^n)\, dx$$

Here the first term, being the product of a polynomial and e^{-x}, vanishes at 0 and ∞. Therefore the integral becomes

$$\frac{(-1)^n}{n!} \int_0^\infty L_n^{(n)}(x) e^{-x} x^n\, dx \qquad \text{integrating by parts } n \text{ times}$$

$$= \frac{(-1)^n}{n!} \int_0^\infty \frac{d^n}{dx^n}\left\{ \frac{(-1)^n}{n!} x^n + \text{terms involving lower powers of } x \right\} e^{-x} x^n\, dx$$

$$= \frac{1}{n!} \int_0^\infty e^{-x} x^n\, dx = \frac{\Gamma(n+1)}{n!} = 1 \qquad \text{by (5.2)}$$

Thus $\int_0^\infty e^{-x} L_n^2(x)\, dx = 1$.

This completes the proof.

With this, we conclude this chapter and also the discussion of a single differential equation in one unknown function. In the next chapter, we shall discuss systems of equations involving more than one unknown function.

Exercises 5.27

1. Find the series solution of the generalized Laguerre equation $xy'' + (\alpha + 1 - x)y' + py = 0$ of order α and degree p.

2. Show that the generalized Laguerre polynomial

$$L_n^{(\alpha)}(x) = \sum_{k=0}^{\infty} (-1)^n \binom{n}{k} \frac{(\alpha+n)(\alpha+n-1)(\alpha+n-2)\cdots(\alpha+n-k+1)}{k!} x^k$$

satisfies the generalized Laguerre equation of degree n.

3. Show that $L_n^{(\alpha)}(x) = \dfrac{e^x x^{-\alpha}}{n!} \dfrac{d^n}{dx^n}\left(e^{-x} x^{n+\alpha}\right)$.

4. Show that $\dfrac{d}{dx}\left[L_n^{(\alpha)}(x)\right] = -L_{n-1}^{(\alpha+1)}(x)$.

5. Show that, if $m \neq n$, then

$$\int_0^\infty x^{-\alpha} e^{-x} L_m^{(\alpha)}(x) L_n^{(\alpha)}(x)\,dx = 0$$

6. Prove the recurrence relation

$$(n+1)L_{n+1}^{(\alpha)}(x) = (\alpha + 2n + 1)L_n^{(\alpha)}(x) - (\alpha + n)L_{n-1}^{(\alpha)}(x)$$

CHAPTER 6

Systems of First Order Linear Equations

6.1 INTRODUCTION

In the previous chapters, we have been concerned with one differential equation in one unknown function. In this chapter, we shall consider systems of two differential equations in two unknown functions, and more generally, systems of n differential equations in n unknown functions. Biological, physical and engineering sciences often give rise to problems involving systems of equations in more than one unknown function. These systems may be linear systems or nonlinear systems. We shall restrict our attention to linear systems only, and that too to first order linear systems. We will denote the independent variable by t and the dependent variables by x and y.

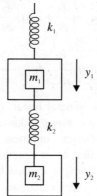

Figure 6.1 Spring–mass system.

For example, consider a system consisting of a spring, a mass, another spring and another mass, all suspended one below the other, oscillating up and down under gravity in such a way that the tensions in the springs will satisfy the following equations:

$$m_1 y_1''(t) = -k_1 y_1(t) + k_2(y_2(t) - y_1(t))$$
$$m_2 y_2''(t) = -k_2(y_2(t) - y_1(t))$$

where m_1 and m_2 are the masses, k_1 and k_2 are the spring constants and $y_1(t)$ and $y_2(t)$ are the displacements (from the equilibrium position) of the two masses respectively (see Fig. 6.1).

6.2 LINEAR SYSTEMS

The general linear system of two first order differential equations in two unknown functions $x(t)$ and $y(t)$ is of the form

$$a_1(t)\frac{dx}{dt} + a_2(t)\frac{dy}{dt} + a_3(t)x + a_4(t)y = F_1(t)$$

$$b_1(t)\frac{dx}{dt} + b_2(t)\frac{dy}{dt} + b_3(t)x + b_4(t)y = F_2(t)$$

(6.1)

where $a_1(t)$, $a_2(t)$, $b_1(t)$, $b_2(t)$ are not zero in $a \leq x \leq b$.

A solution of system (6.1) is an ordered pair of real valued functions (f, g) such that $x = f(t)$, $y = g(t)$ simultaneously satisfy both the equations of the system (6.1) on some interval $a \leq t \leq b$.

The general linear system of three or more first order differential equations in three or more unknown functions can similarly be written.

The general linear system of two second order differential equations in two unknown functions $x(t)$ and $y(t)$ is of the form

$$a_1(t)\frac{d^2x}{dt^2} + a_2(t)\frac{d^2y}{dt^2} + a_3(t)\frac{dx}{dt} + a_4(t)\frac{dy}{dt} + a_5(t)x + a_6(t)y = F_1(t)$$

$$b_1(t)\frac{d^2x}{dt^2} + b_2(t)\frac{d^2y}{dt^2} + b_3(t)\frac{dx}{dt} + b_4(t)\frac{dy}{dt} + b_5(t)x + b_6(t)y = F_2(t)$$

(6.2)

Similarly, we can write the general linear system of two or more nth order differential equations in two or more unknown functions. For the sake of convenience and clarity, we restrict our attention, throughout the rest of this chapter, to system of only two first order equations in two unknown functions of the form (6.1). All that we are going to discuss for this system can be very easily extended to a system of n first order equations in n unknown functions.

The system of equations in (6.1) can be written, after calculation, in the form

$$\frac{dx}{dt} = a_{11}(t)x + a_{12}(t)y + G_1(t)$$

$$\frac{dy}{dt} = a_{21}(t)x + a_{22}(t)y + G_2(t)$$

(6.3)

This is called the *normal form* of the system of equations.

This system of equations can be written in the matrix form as

Systems of First Order Linear Equations

$$\begin{pmatrix} \dfrac{dx}{dt} \\ \dfrac{dy}{dt} \end{pmatrix} = \begin{pmatrix} a_{11}(t) & a_{12}(t) \\ a_{21}(t) & a_{22}(t) \end{pmatrix} \begin{pmatrix} x \\ y \end{pmatrix} + \begin{pmatrix} G_1(t) \\ G_2(t) \end{pmatrix}$$

i.e.
$$\begin{pmatrix} x' \\ y' \end{pmatrix} = \begin{pmatrix} a_{11}(t) & a_{12}(t) \\ a_{21}(t) & a_{22}(t) \end{pmatrix} \begin{pmatrix} x \\ y \end{pmatrix} + \begin{pmatrix} G_1(t) \\ G_2(t) \end{pmatrix}$$

i.e.
$$\begin{pmatrix} x \\ y \end{pmatrix}' = \begin{pmatrix} a_{11}(t) & a_{12}(t) \\ a_{21}(t) & a_{22}(t) \end{pmatrix} \begin{pmatrix} x \\ y \end{pmatrix} + \begin{pmatrix} G_1(t) \\ G_2(t) \end{pmatrix} \qquad (6.4)$$

i.e. $X' = A(t)X + B(t)$, where $X = (x \;\; y)^T$, $B(t) = (G_1(t), G_2(t))^T$, and $A(t)$
$= \begin{pmatrix} a_{11}(t) & a_{12}(t) \\ a_{21}(t) & a_{22}(t) \end{pmatrix}$, called the *coefficient matrix*.

The system (6.3) is called a *Homogeneous System* if $B(t) = 0$, otherwise it is called a *Nonhomogeneous System*.

We shall be concerned with systems in which the entries in the coefficient matrix are all constants. Then the system becomes

$$X' = AX + B(t) \qquad (6.5)$$

Example 6.1 The linear system of equations

$$\dfrac{dx}{dt} = t^2 x - e^t y + 7t$$

$$\dfrac{dy}{dt} = (2t+1)x + 3t^3 y + 8t - 5$$

is an example of a linear system which is the normal form. Its matrix representation is

$$\begin{pmatrix} x \\ y \end{pmatrix}' = \begin{pmatrix} t^2 & -e^t \\ 2t+1 & 3t^3 \end{pmatrix} \begin{pmatrix} x \\ y \end{pmatrix} + \begin{pmatrix} 7t \\ 8t - 5 \end{pmatrix}$$

Example 6.2 The linear system

$$\dfrac{dx}{dt} = 4x + y$$

$$\dfrac{dy}{dt} = 2x - y$$

is homogeneous with constant coefficients.

Example 6.3 The linear system

$$\frac{dx}{dt} = 4x + y + 5t - 1$$

$$\frac{dy}{dt} = 2x - y - 8t + 3$$

is nonhomogeneous with constant coefficients.

6.3 RELATIONSHIP

An important fundamental property of a normal linear system (6.3) is its relationship to a single second order linear differential equation in one unknown function. Specifically, consider the normal second order linear differential equation

$$\frac{d^2y}{dt^2} + a_1(t)\frac{dy}{dt} + a_0(t)y = F(t) \qquad (6.6)$$

Let $\qquad x_1 = y \quad \text{and} \quad x_2 = \frac{dy}{dt}$

then

$$\frac{dx_1}{dt} = x_2$$

$$\frac{dx_2}{dt} = F(t) - a_0(t)x_1 - a_1(t)x_2 \qquad (6.7)$$

which is a normal linear system of two equations in two unknown functions $x_1(t)$ and $x_2(t)$. Thus we see that a single second order linear differential equation of the form (6.6) in one unknown function is indeed intimately related to a normal linear system (6.3) of first order differential equations in two unknown functions.

Example 6.4 Transform the differential equation

$$y'' - x^2 y' - xy = 0$$

into a system of first order linear system of equations.

Let $\qquad y = x_1 \quad \text{and} \quad y' = x_2$

then

$$\frac{dx_1}{dx} = x_2$$

and the above equation becomes

$$\frac{dx_2}{dx} = \frac{d^2x_1}{dx^2} = \frac{d^2y}{dx^2} = x^2 x_2 + x x_1$$

Thus the equivalent linear system is

$$\frac{dx_1}{dx} = x_2$$

$$\frac{dx_2}{dx} = x x_1 + x^2 x_2$$

Example 6.5 Find an equivalent linear system of the following linear differential equation:

$$y''' = y'' - x^2(y')^2$$

Let $y = x_1$, $y' = x_2$ and $y'' = x_3$. Then the above equation becomes

$$\frac{dx_3}{dx} = x_3 - x^2 x_2^2$$

Thus the equivalent linear system is given by

$$\frac{dx_1}{dx} = x_2$$

$$\frac{dx_2}{dx} = x_3$$

$$\frac{dx_3}{dx} = -x^2 x_2^2 + x_3$$

6.4 APPLICATION TO ELECTRICAL CIRCUITS

Systems of first order equations often arise in electrical circuits. We shall now consider electrical networks that consist of two or more loops as shown in Fig. 6.2. (A closed path in an electrical network is called a loop.)

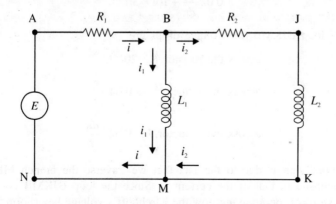

Figure 6.2 Electrical network with three loops.

Introduction to Differential Equations

This network consists of three loops ABMNA, BJKMB, and ABJKMNA. Points such as B and M at which two or more circuits join are called *junction points* or *branch points*. The direction of current flow has been arbitrarily assigned and indicated by arrows.

In order to solve problems involving multiple loop networks, we shall need the following fundamental laws:

Kirchhoff's Voltage Law (Form 1): The algebraic sum of the instantaneous voltage drops around a closed circuit in a specific direction is zero.

Kirchhoff's Voltage Law (Form 2): The sum of the voltage drops across resistors, inductors, and capacitors is equal to the total electromotive force in a closed circuit.

Kirchhoff's Current Law: In an eletrical network the total current flowing into a junction point is equal to the total current flowing away from the junction point.

We need to find the currents in the electrical network of the above figure. Given that E is an electromotive force of 30 V, R_1 is a resistor of 10 Ω, R_2 is a resistor of 20 Ω, L_1 is an inductor of 0.02 H, L_2 is an inductor of 0.04 H and the currents are initially zero. Suppose that current flowing in the branch MNAB is denoted by i, that on BM by i_1 and that on BJKM by i_2.

Applying Kirchhoff's voltage law to each of the above three loops, we have:
For the loop ABMNA, the voltage drops are

across the resistor R_1: $10i$

across the inductor L_1: $0.02\dfrac{di_1}{dt}$

Thus applying the Kirchhoff's voltage law (Form 2) to the loop ABMNA, we have the equation

$$0.02\frac{di_1}{dt} + 10i = 30 \qquad (6.8)$$

For the loop BJKMB, the voltage drops are

across the resistor R_2: $20i_2$

across the inductor L_2: $0.04\dfrac{di_2}{dt}$

across the inductor L_1: $-0.02\dfrac{di_1}{dt}$

The minus sign is due to the fact that we traverse the branch MB in the direction opposite to that of the current i_1. Since the loop BJKMB contains no electromotive force, on applying now the Kirchhoff's voltage law (Form 2) to the loop BJKMB, we have the equation

$$-0.02\frac{di_1}{dt} + 0.04\frac{di_2}{dt} + 20i_2 = 0 \qquad (6.9)$$

For the loop ABJKMNA, the voltage drops are

across the resistor R_1: $10i$

across the resistor R_2: $20i_2$

across the inductor L_2: $0.04\dfrac{di_2}{dt}$

Applying Kirchhoff's voltage law (Form 2) to the loop ABJKMNA, we have the equation

$$10i + 0.04\frac{di_2}{dt} + 20i_2 = 30 \qquad (6.10)$$

We observe that the three equations (6.8), (6.9) and (6.10) thus obtained are not independent, because Eq. (6.9) can be obtained by subtracting (6.8) from (6.10). Thus it is enough to consider Eqs. (6.8) and (6.10).

Now applying Kirchhoff's current law to the junction point B, we get

$$i = i_1 + i_2$$

which when replaced in both Eqs. (6.8) and (6.10), we get the linear system of equations with two unknown functions i_1 & i_2 as

$$0.02\frac{di_1}{dt} + 10i_1 + 10i_2 = 30$$

$$10i_1 + 0.04\frac{di_2}{dt} + 30i_2 = 30 \qquad (6.11)$$

Since the currents are initially zero, we have the initial conditions

$$i_1(0) = 0 \quad \text{and} \quad i_2(0) = 0$$

6.5 SOLUTIONS

A solution of the system (6.3) is an ordered pair of real-valued functions $x = f(t)$ and $y = g(t)$, defined and having continuous derivatives on the interval $[a, b]$, such that

$$\frac{df(t)}{dt} = a_{11}(t)f(t) + a_{12}g(t) + G_1(t)$$

$$\frac{dg(t)}{dt} = a_{21}(t)f(t) + a_{22}g(t) + G_2(t)$$

for all t such that $a \leq t \leq b$.

Theorem 6.1 Let the functions $a_{11}(t)$, $a_{12}(t)$, $a_{21}(t)$, $a_{22}(t)$, $G_1(t)$ and $G_2(t)$ in the system (6.3) be all continuous on $[a, b]$. Then the general solution of the system

(6.3) is the sum of the general solution of the associated homogeneous linear system $X' = A(t)X$ and a particular solution of the system (6.3).

Theorem 6.2 The associated homogeneous system

$$\frac{dx}{dt} = a_{11}(t)x + a_{12}(t)y$$

$$\frac{dy}{dt} = a_{21}(t)x + a_{22}(t)y \tag{6.12}$$

of the system (6.3) has two linearly independent solutions. The general solution of the linear system (6.12) is a linear combination of any two linearly independent solutions of (6.12).

Thus, if $x = f_0(t)$ and $y = g_0(t)$ is a particular solution of the linear system (6.3), and

$$\begin{array}{cc} x = f_1(t) & x = f_2(t) \\ y = g_1(t) & \text{and} \quad y = g_2(t) \end{array}$$

are two linearly independent solutions of the linear system (6.12), then Theorems 6.1 and 6.2 guarantee that the general solution of the system (6.3) is given by

$$x = c_1 f_1(t) + c_2 f_2(t) + f_0(t)$$

$$y = c_1 g_1(t) + c_2 g_2(t) + g_0(t)$$

We shall not prove these theorems. They will be used in the next section.

6.6 METHOD OF OBTAINING SOLUTIONS

For obtaining the solutions of the linear system (6.3), we assume the coefficients to be constants. So the linear system (6.3) becomes

$$\frac{dx}{dt} = a_{11}x + a_{12}y + G_1(t)$$

$$\frac{dy}{dt} = a_{21}x + a_{22}y + G_2(t) \tag{6.13}$$

and the associated homogeneous linear system is

$$\frac{dx}{dt} = a_{11}x + a_{12}y$$

$$\frac{dy}{dt} = a_{21}x + a_{22}y \tag{6.14}$$

Systems of First Order Linear Equations **151**

6.7 HOMOGENEOUS LINEAR SYSTEMS

First, we shall discuss the method of finding the general solution of the linear system (6.14). How shall we proceed?

If we take $a_{12} = 0$ and $a_{21} = 0$, then $x = Ae^{mt}$ and $y = Be^{mt}$ may form a solution. So let us assume that

$$x = Ae^{mt}$$
$$y = Be^{mt}$$

is a solution of the system (6.14).

Putting this solution in the system (6.14), we get

$$\left. \begin{array}{l} mA = a_{11}A + a_{12}B \\ mB = a_{21}A + a_{22}B \end{array} \right\} \Rightarrow \begin{array}{l} (a_{11} - m)A + a_{12}B = 0 \\ a_{21}A + (a_{22} - m)B = 0 \end{array}$$

A and B must satisfy these two equations. If $A = B = 0$, then $x = 0$ and $y = 0$ is a trivial solution.

So, for a nontrivial solution we must have

$$\begin{vmatrix} a_{11} - m & a_{12} \\ a_{21} & a_{22} - m \end{vmatrix} = 0$$

This equation is called the *auxiliary equation* of the system (6.14).

Let m_1 and m_2 be two roots of this equation. Then 3 cases arise.

(i) m_1 and m_2 are distinct real roots: The two LI solutions of the system (6.14) are given by

$$x = A_1 e^{m_1 t} \qquad \text{and} \qquad x = A_2 e^{m_2 t}$$
$$y = B_1 e^{m_1 t} \qquad\qquad\qquad y = B_2 e^{m_2 t}$$

where A_1, A_2, B_1, B_2 are real constants.

(ii) m_1 and m_2 are distinct and complex roots, say $m_1 = a + ib$ and $m_2 = a - ib$: The two linearly independent solutions are

$$x = A_1^* e^{(a+ib)t} \qquad \text{and} \qquad x = A_2^* e^{(a-ib)t}$$
$$y = B_1^* e^{(a+ib)t} \qquad\qquad\qquad y = B_2^* e^{(a-ib)t}$$

where A_1^*, A_2^*, B_1^* and B_2^* are complex constants. We have to find two linearly independent real solutions.

(iii) $m_1 = m_2$ and real: The two linearly independent solutions are

$$x = Ae^{m_1 t} \qquad \text{and} \qquad x = (A_1 + tA_2)e^{m_1 t}$$
$$y = Be^{m_1 t} \qquad\qquad\qquad y = (B_1 + tB_2)e^{m_1 t}$$

The methods of obtaining these solutions will be explained through the following examples. This method is called *Euler's Method*.

Example 6.6 Consider the system of equations

$$\frac{dx}{dt} = -3x + 4y$$
$$\frac{dy}{dt} = -2x + 3y$$
(6.15)

The auxiliary equation is

$$\begin{vmatrix} -3-m & 4 \\ -2 & 3-m \end{vmatrix} = 0 \Rightarrow (3+m)(m-3) + 8 = 0$$

$$\Rightarrow m^2 - 1 = 0 \Rightarrow m_1 = 1, m_2 = -1$$

So the two LI solutions are

$$X_1 = \begin{cases} x = A_1 e^t \\ y = B_1 e^t \end{cases} \quad \text{and} \quad X_2 = \begin{cases} x = A_2 e^{-t} \\ y = B_2 e^{-t} \end{cases}$$

Putting X_1 in the given system, we have

$$\begin{aligned} A_1 e^t &= -3A_1 e^t + 4B_1 e^t \\ B_1 e^t &= -2A_1 e^t + 3B_1 e^t \end{aligned} \Rightarrow \begin{aligned} 4A_1 - 4B_1 &= 0 \\ 2A_1 - 2B_1 &= 0 \end{aligned} \Rightarrow A_1 = B_1$$

We can choose $A_1 = B_1 = 1$ to get a solution of the system as

$$X_1 = \begin{cases} x = e^t \\ y = e^t \end{cases}$$

Similarly putting X_2 in the system, we get

$$\begin{aligned} -A_2 e^{-t} &= -3A_2 e^{-t} + 4B_2 e^{-t} \\ -B_2 e^{-t} &= -2A_2 e^{-t} + 3B_2 e^{-t} \end{aligned} \Rightarrow \begin{aligned} 2A_2 - 4B_2 &= 0 \\ 2A_2 - 4B_2 &= 0 \end{aligned} \Rightarrow A_2 = 2B_2$$

We can choose $B_2 = 1$ and then $A_2 = 2$ to get another solution, which is independent of the first one, as

$$x = 2e^{-t}$$
$$y = e^{-t}$$

Thus the general solution of the system is given by

$$x = c_1 e^t + 2c_2 e^{-t}$$
$$y = c_1 e^t + c_2 e^{-t}$$

where c_1 and c_2 are arbitrary constants.

Systems of First Order Linear Equations

Example 6.7 In case of the linear system

$$\frac{dx}{dt} = 4x - 2y$$

$$\frac{dy}{dt} = 5x + 2y$$

(6.16)

The auxiliary equation is

$$\begin{vmatrix} 4-m & -2 \\ 5 & 2-m \end{vmatrix} = 0 \Rightarrow (4-m)(2-m) + 10$$

$$\Rightarrow m^2 - 6m + 18 = 0 \Rightarrow m = 3 \pm 3i$$

Thus one solution of the system is given by

$$x = A_1 e^{(3+3i)t}$$
$$y = B_1 e^{(3+3i)t}$$

Putting this in the given system, we have

$$(3+3i)A_1 e^{(3+3i)t} = 4A_1 e^{(3+3i)t} - 2B_1 e^{(3+3i)t}$$

$$(3+3i)B_1 e^{(3+3i)t} = 5A_1 e^{(3+3i)t} + 2B_1 e^{(3+3i)t}$$

$$\Rightarrow \begin{array}{l} (1-3i)A_1 - 2B_1 = 0 \\ 5A_1 - (1+3i)B_1 = 0 \end{array}$$

A simple nontrivial solution of this system is $A_1 = 2$ and $B_1 = 1 - 3i$
So

$$x = 2e^{(3+3i)t} = 2e^{3t}(\cos 3t + i\sin 3t) = 2e^{3t}\cos 3t + i2e^{3t}\sin 3t$$

$$y = (1-3i)e^{(3+3i)t} = e^{3t}(1-3i)(\cos 3t + i\sin 3t) = e^{3t}(\cos 3t + 3\sin 3t)$$

$$+ ie^{3t}(\sin 3t - 3\cos 3t)$$

Since this is a solution of the linear system, their real and imaginary parts are also solutions of the given linear system, and they are linearly independent.
Thus two linearly independent real solutions of the given linear system are

$$x = 2e^{3t}\cos 3t \qquad \qquad x = 2e^{3t}\sin 3t$$
$$y = e^{3t}(\cos 3t + 3\sin 3t) \quad \text{and} \quad y = e^{3t}(\sin 3t - 3\cos 3t)$$

Hence the general solution of the given linear system is given by

$$x = c_1 2e^{3t}\cos 3t + c_2 2e^{3t}\sin 3t$$
$$y = c_1 e^{3t}(\cos 3t + 3\sin 3t) + c_2 e^{3t}(\sin 3t - 3\cos 3t)$$

where c_1 and c_2 are arbitrary constants.

Example 6.8 Now we consider a linear system

$$\frac{dx}{dt} = -4x - y$$
$$\frac{dy}{dt} = x - 2y$$
(6.17)

whose auxiliary equation $\begin{vmatrix} -4-m & -1 \\ 1 & -2-m \end{vmatrix} = 0$ has two equal real roots $m_1 = m_2 = -3$.

This gives one solution of the form

$$x = Ae^{-3t}$$
$$B = Be^{-3t}$$

which when put in the linear system (6.17) gives an algebraic system

$$A + B = 0$$
$$A + B = 0$$

A simple nontrivial solution of this system being $A = -B = 1$, we obtain a nontrivial solution

$$x = e^{-3t}$$
$$y = -e^{-3t}$$
(6.18)

of the given system (6.17).

Since the roots of the auxiliary equation are both equal to -3, we must seek a second solution of the form

$$x = (A_1 + tA_2)e^{-3t}$$
$$y = (B_1 + tB_2)e^{-3t}$$
(6.19)

of the system (6.17).

Substituting (6.18) in (6.17), we obtain

$$(-3A_1 - 3tA_2 + A_2)e^{-3t} = (-4A_1 - 4tA_2)e^{-3t} - (B_1 + tB_2)e^{-3t}$$
$$(-3B_1 - 3tB_2 + B_2)e^{-3t} = (A_1 + tA_2)e^{-3t} - (2B_1 + 2tB_2)e^{-3t}$$

These equations reduce to

$$(A_1 + B_1 + A_2) + t(A_2 + B_2) = 0$$
$$(A_1 + B_1 - B_2) + t(A_2 + B_2) = 0$$

In order for these equations to be identities, we must have

$$A_1 + B_1 + A_2 = 0 \qquad A_2 + B_2 = 0$$
$$A_1 + B_1 - B_2 = 0 \qquad A_2 + B_2 = 0$$

These equations lead to $A_1 + B_1 = -A_2 = B_2$.

Systems of First Order Linear Equations

We may choose any convenient nonzero values for A_2 and B_2. So we choose $A_2 = 1$ and then $B_2 = -1$. We can choose $A_1 = 1$ and $B_1 = 0$. We are thus led to the second solution

$$x = (1+t)e^{-3t}$$
$$y = -te^{-3t}$$
(6.20)

Thus we may get two linearly independent solutions (6.18) and (6.20) of the system (6.17). Hence the general solution of the system (6.17) is given by

$$x = c_1 e^{-3t} + c_2(1+t)e^{-3t}$$
$$y = -c_1 e^{-3t} - c_2 t e^{-3t}$$

where c_1 and c_2 are arbitrary constants.

6.8 NONHOMOGENEOUS LINEAR SYSTEMS

Now we turn our attention to nonhomogeneous linear systems with constant coefficients. We know from Theorem 6.1 that the general solution of the system of equation (6.13) is the sum of the general solution $Y_c(t)$ of the associated homogeneous system of equations (6.14) and a particular solution $Y_p(t)$ of the system of equations (6.13). Thus the general solution $Y(t)$ of the system of equation (6.13) is given by

$$Y(t) = Y_c(t) + Y_p(t)$$

To find $Y_p(t)$, we use the method of *variation of parameters*, which worked in the case of ordinary nonhomogeneous differential equations. We shall illustrate these ideas through the following example.

Example 6.9 Solve the following system of linear differential equations:

$$\frac{dx}{dt} = 5x + 4y + t + 1$$
$$\frac{dy}{dt} = -x + y + 2t$$
(6.21)

The auxiliary equation is

$$\begin{vmatrix} 5-m & 4 \\ -1 & 1-m \end{vmatrix} = 0 \Rightarrow (5-m)(1-m) + 4 = 0$$

$$\Rightarrow m_1 = m_2 = 3$$

One solution is given by $X_1 = \begin{cases} x = Ae^{3t} \\ y = Be^{3t} \end{cases}$ by case (iii) of Section 6.7.

Putting this X_1 in the associated homogeneous system of equations

$$\frac{dx}{dt} = 5x + 4y$$
$$\frac{dy}{dt} = -x + y$$
(6.22)

of the nonhomogeneous system of equations (6.21), we obtain

$$3Ae^{3t} = 5Ae^{3t} + 4Be^{3t} \quad \Rightarrow \quad 2A + 4B = 0 \quad \Rightarrow \quad A = -2B$$
$$3Be^{3t} = -Ae^{3t} + Be^{3t} \quad \quad -A - 2B = 0$$

Taking $A = 2$ and $B = -1$, we obtain a solution of the system (6.22) as

$$X_1 = \begin{cases} x = 2e^{3t} \\ y = -e^{3t} \end{cases}$$

Next to find the second solution X_2 such that X_1, X_2 are linearly independent, we assume, by case (iii) of Section 6.7 that

$$X_2 = \begin{cases} x = (A_1 + tA_2)e^{3t} \\ y = (B_1 + tB_2)e^{3t} \end{cases}$$

Putting this in the system of equations (6.22), we obtain

$$(3A_1 + 3tA_2 + A_2)e^{3t} = (5A_1 + 5tA_2)e^{3t} + (4B_1 + 4tB_2)e^{3t}$$

$$(3B_1 + 3B_2 t + B_2)e^{3t} = (-A_1 - tA_2)e^{3t} + (B_1 + tB_2)e^{3t}$$

$$\Rightarrow 2A_1 - A_2 + 4B_1 + 2tA_2 + 4tB_2 = 0$$

$$-A_1 - 2B_1 - B_2 - tA_2 - 2tB_2 = 0$$

$$\Rightarrow (2A_1 - A_2 + 4B_1) + t(2A_2 + 4B_2) = 0$$

$$(A_1 + 2B_1 + B_2) + t(A_2 + 2B_2) = 0$$

In order for these equations to be identities, we must have

$$2A_1 - A_2 + 4B_1 = 0 \quad \quad A_2 + 2B_2 = 0$$
$$A_1 + 2B_1 + B_2 = 0 \quad \quad A_2 + 2B_2 = 0$$

This constants A_1, A_2, B_1 and B_2 must be chosen to satisfy the above identities so that X_2 is a solution of the system (6.22).

We have $A_2 = -2B_2$, and other two equations suggest that A_1 and B_1 must satisfy $A_1 + 2B_1 = -B_2 = \frac{1}{2}A_2$.

Thus we may choose any convenient nonzero values for A_2 and B_2. We choose $A_2 = 2$ and $B_2 = -1$, so that $A_1 + 2B_1 = 1$ and we can choose any convenient values for A_1 and B_1. We choose $A_1 = 1$ and $B_1 = 0$.

Thus
$$X_2 = \begin{cases} x = (1+2t)e^{3t} \\ y = -te^{3t} \end{cases}$$

and

$$X_c = c_1 X_1 + c_2 X_2 = \begin{cases} x = c_1 2e^{3t} + c_2(1+2t)e^{3t} \\ y = c_1(-1)e^{3t} - c_2 t e^{3t} \end{cases}$$

is the general solution of the system of Eqs. (6.22).

Let
$$Y_p(t) = \begin{cases} x = v_1(t)2e^{3t} + v_2(t)(1+2t)e^{3t} \\ y = -v_1(t)e^{3t} - v_2(t)te^{3t} \end{cases}$$

be a particular solution of the system of Eqs. (6.21).

Putting $Y_p(t)$ in the system (6.21), we obtain

$$v_1'(t)2e^{3t} + v_2'(t)(1+2t)e^{3t} + 6v_1(t)e^{3t} + v_2(t)3(1+2t)e^{3t} + 2v_2(t)e^{3t}$$

$$= 10v_1(t)e^{3t} + v_2(t)5(1+2t)e^{3t} - 4v_1(t)e^{3t} - 4v_2(t)te^{3t} + t + 1$$

$$-v_1'(t)e^{3t} - v_2'(t)te^{3t} - 3v_1(t)e^{3t} - 3v_2(t)te^{3t} - v_2(t)e^{3t}$$

$$= -v_1(t)2e^{3t} - v_2(t)(1+2t)e^{3t} - v_1(t)e^{3t} - v_2(t)te^{3t} + 2t$$

$$\Rightarrow 2v_1'(t)e^{3t} + (1+2t)v_2'(t)e^{3t} = t+1$$

$$v_1'(t)e^{3t} + v_2'(t)te^{3t} = -2t$$

By using Cramer's rule, we obtain

$$v_1'(t)e^{3t} = \frac{\begin{vmatrix} t+1 & 1+2t \\ -2t & t \end{vmatrix}}{\begin{vmatrix} 2 & 1+2t \\ 1 & t \end{vmatrix}} = \frac{t^2+t+2t+4t^2}{2t-1-2t} = 5t^2+3t$$

$$\Rightarrow v_1(t) = \int (5t^2+3t)e^{-3t}\,dt = -\frac{5}{3}t^2 e^{-3t} - \frac{19}{9}te^{-3t} - \frac{19}{27}e^{-3t}$$

and

$$v_2'(t)e^{3t} = \frac{\begin{vmatrix} 2 & t+1 \\ 1 & -2t \end{vmatrix}}{1} = -5t-1$$

$$\Rightarrow v_2(t) = -\int (5t+1)e^{-3t}\,dt = \frac{5}{3}te^{-3t} + \frac{8}{9}e^{-3t}$$

Hence the general solution to the nonhomogeneous linear system (6.21) is given by

$$x = 2c_1 e^{3t} + c_2(1+2t)e^{3t} - 2\left(\frac{5}{3}t^2 + \frac{19}{9}t + \frac{19}{27}\right) + (1+2t)\left(\frac{5}{3}t + \frac{8}{9}\right)$$

$$y = -c_1 e^{3t} - c_2 t e^{3t} + \frac{5}{3}t^2 + \frac{19}{9}t + \frac{19}{27} - \frac{5}{3}t^2 - \frac{8}{9}t$$

We conclude this chapter here. Now try the following problems in the Exercises.

Exercises 6.8

Find the general solution of each of the following systems of equations:

1. $x' = 4x - 3y$
   $y' = 8x - 6y$

2. $x' = 2x$
   $y' = 3y$

3. $x' = -4x - y$
   $y' = x - 2y$

4. $x' = x - 2y$
   $y' = 4x + 5y$

5. $x' = y$
   $y' = z$
   $z' = 8x - 14y + 7z$

6. $x' = y$
   $y' = z$
   $z' = 12x - 16y + 7z$

7. $x' = 4x + 5y$
   $y' = -4x - 4y$

8. $x' = x + 2y - z$
   $y' = 2x + y + z$
   $z' = -x + y$

9. $x' = -2x - 5y$
   $y' = x + 2y$

10. $x' = x + 2y - z$
    $y' = y + z$
    $z' = 2z$

11. $\begin{pmatrix} x \\ y \\ z \end{pmatrix}' = \begin{pmatrix} 1 & 0 & 0 \\ 2 & 1 & -2 \\ 3 & 2 & 1 \end{pmatrix} \begin{pmatrix} x \\ y \\ z \end{pmatrix}$

12. $\begin{pmatrix} x \\ y \end{pmatrix}' = \begin{pmatrix} -4 & -3 \\ -2 & -5 \end{pmatrix} \begin{pmatrix} x \\ y \end{pmatrix} + \begin{pmatrix} t \\ e^t \end{pmatrix}$

13. $x' = 2x + y + 3e^{2t}$
    $y' = -4x + 2y + te^{2t}$

14. $x' = 4x + y + te^{6t}$
    $y' = -4x + 8y + 2te^{6t}$

15. $\begin{pmatrix} x \\ y \end{pmatrix}' = \begin{pmatrix} 0 & 1 \\ -2 & 3 \end{pmatrix} \begin{pmatrix} x \\ y \end{pmatrix} + \begin{pmatrix} 0 \\ 3 \end{pmatrix} e^t$

16. $\begin{pmatrix} x \\ y \end{pmatrix}' = \begin{pmatrix} 2 & 1 \\ 0 & 2 \end{pmatrix} \begin{pmatrix} x \\ y \end{pmatrix} + \begin{pmatrix} \sin t \\ \cos t \end{pmatrix}$

Prove Exercises 17 and 18 for the homogeneous system

$$\begin{pmatrix} x \\ y \end{pmatrix}' = \begin{pmatrix} a & b \\ c & d \end{pmatrix} \begin{pmatrix} x \\ y \end{pmatrix} \quad (*)$$

17. Find the roots of the auxiliary equation of the system (*). Show that the roots of the auxiliary equation are complex only if $(a - d)^2 + 4bc < 0$. In particular, note that complex roots occur in conjugate pairs and that they occur only if b and c are not zeros.

18. Suppose the roots of the auxiliary equation of the system (*) are complex numbers $\alpha + i\beta$ and $\alpha - i\beta$ with $\beta \neq 0$. Show that the corresponding solutions of the system (*) are conjugate pairs.

CHAPTER 7

Total Differential Equations

7.1 INTRODUCTION

We continue discussing in this chapter the differential equations involving more than two variables. Out of these variables, one is independent and the rest are dependent. First we shall consider a single differential equation and then we shall consider a system of simultaneous differential equations. These equations arise frequently in mathematical physics, electrical engineering, etc.

An ordinary differential equation of the form

$$P\,dx + Q\,dy + R\,dz = 0 \tag{7.1}$$

$$\Rightarrow P + Q\,\frac{dy}{dx} + R\,\frac{dz}{dx} = 0 \tag{7.2}$$

where P, Q, R are functions of x, y and z, is called a *total differential equation*.

If the left hand side expression of Eq. (7.1) is integrable, then we can easily get the solution of Eq. (7.1). Therefore, our first task is to find the condition under which the expression is integrable. If integrable, then there must exist a function $f(x, y, z)$ such that

$$df = P\,dx + Q\,dy + R\,dz$$

and Eq. (7.1) is called *exact* (refer to Section 2.4). For example,

$$yz\,dx + zx\,dy + xy\,dz = 0$$

is an exact equation because

$$d(xyz) = yz\,dx + zx\,dy + xy\,dz$$

But the equation

$$y\,dx + (z - y)\,dy + x\,dz = 0$$

is not exact. How do we know this? The following section answers this question.

7.2 CONDITION FOR INTEGRABILITY OF $P\,dx + Q\,dy + R\,dz = 0$

Theorem 7.1 provides the condition for integrability of the equation.

Theorem 7.1 A necessary and sufficient condition for the integrability of Eq. (7.1) is

$$P\left(\frac{\partial Q}{\partial z} - \frac{\partial R}{\partial y}\right) + Q\left(\frac{\partial R}{\partial x} - \frac{\partial P}{\partial z}\right) + R\left(\frac{\partial P}{\partial y} - \frac{\partial Q}{\partial x}\right) = 0 \quad (7.3)$$

Proof: First let us assume there exists a function $u(x, y, z)$ such that

$$du = P\,dx + Q\,dy + R\,dz$$

Then $u(x, y, z) = c$ is an integral of (7.1).

But we know that $du = \frac{\partial u}{\partial x}dx + \frac{\partial u}{\partial y}dy + \frac{\partial u}{\partial z}dz$, which means

$$\frac{\frac{\partial u}{\partial x}}{P} = \frac{\frac{\partial u}{\partial y}}{Q} = \frac{\frac{\partial u}{\partial z}}{R} = \lambda(x,y,z) \quad \text{say}$$

$$\Rightarrow \frac{\partial u}{\partial x} = \lambda P \qquad \frac{\partial u}{\partial y} = \lambda Q \qquad \frac{\partial u}{\partial z} = R\lambda$$

From the first two relations we have

$$\frac{\partial(\lambda P)}{\partial y} = \frac{\partial^2 u}{\partial y \partial x} = \frac{\partial^2 u}{\partial x \partial y} = \frac{\partial(\lambda Q)}{\partial x}$$

$$\Rightarrow \frac{\partial \lambda}{\partial y}P + \lambda\frac{\partial P}{\partial y} = \frac{\partial \lambda}{\partial x}Q + \lambda\frac{\partial Q}{\partial x}$$

$$\Rightarrow \lambda\left(\frac{\partial P}{\partial y} - \frac{\partial Q}{\partial x}\right) = Q\frac{\partial \lambda}{\partial x} - P\frac{\partial \lambda}{\partial y} \quad (7.4)$$

Similarly we have

$$\lambda\left(\frac{\partial Q}{\partial z} - \frac{\partial R}{\partial y}\right) = R\frac{\partial \lambda}{\partial y} - Q\frac{\partial \lambda}{\partial z} \quad (7.5)$$

and

$$\lambda\left(\frac{\partial R}{\partial x} - \frac{\partial P}{\partial z}\right) = P\frac{\partial \lambda}{\partial z} - R\frac{\partial \lambda}{\partial x} \quad (7.6)$$

Multiplying (7.4), (7.5) and (7.6) by R, P and Q respectively and then adding, we get

$$P\left(\frac{\partial Q}{\partial z} - \frac{\partial R}{\partial y}\right) + Q\left(\frac{\partial R}{\partial x} - \frac{\partial P}{\partial z}\right) + R\left(\frac{\partial P}{\partial y} - \frac{\partial Q}{\partial x}\right) = 0$$

This proves that the condition (7.3) is necessary.

Next let us suppose that the condition (7.3) is valid. Then the same condition is satisfied by $P_1 = \mu P$, $Q_1 = \mu Q$, $R_1 = \mu R$, where μ is a function of x, y, z and we have

Introduction to Differential Equations

$$\frac{\partial Q_1}{\partial z} - \frac{\partial R_1}{\partial y} = \mu \frac{\partial Q}{\partial z} + Q \frac{\partial \mu}{\partial z} - \mu \frac{\partial R}{\partial y} - R \frac{\partial \mu}{\partial y}$$

$$\Rightarrow \frac{\partial Q_1}{\partial z} - \frac{\partial R_1}{\partial y} = \mu \left(\frac{\partial Q}{\partial z} - \frac{\partial R}{\partial y} \right) + Q \frac{\partial \mu}{\partial z} - R \frac{\partial \mu}{\partial y}$$

Similarly, we have

$$\frac{\partial R_1}{\partial x} - \frac{\partial P_1}{\partial z} = \mu \left(\frac{\partial R}{\partial x} - \frac{\partial P}{\partial z} \right) + R \frac{\partial \mu}{\partial x} - P \frac{\partial \mu}{\partial z}$$

$$\frac{\partial P_1}{\partial y} - \frac{\partial Q_1}{\partial x} = \mu \left(\frac{\partial P}{\partial y} - \frac{\partial Q}{\partial x} \right) + P \frac{\partial \mu}{\partial y} - Q \frac{\partial \mu}{\partial x}$$

All these suggest that P_1, Q_1 and R_1 also satisfy the condition (7.3). In other words, the coefficients of

$$\mu P dx + \mu Q dy + \mu R dz = 0 \tag{7.7}$$

satisfy the condition (7.3).

If $P dx + Q dy$ is not an exact differential with respect to x and y, an integrating factor μ can be found for it, and (7.7) can then be taken as the equation to be considered. Hence there is no loss of generality in regarding $Pdx + Qdy$ as an exact differential.

Then we have
$$\frac{\partial P}{\partial y} = \frac{\partial Q}{\partial x}$$

and let
$$v = \int (P\,dx + Q\,dy)$$

It follows then that
$$P = \frac{\partial v}{\partial x} \quad \text{and} \quad Q = \frac{\partial v}{\partial y}$$

$$\Rightarrow \frac{\partial P}{\partial z} = \frac{\partial^2 v}{\partial z \partial x} \quad \text{and} \quad \frac{\partial Q}{\partial z} = \frac{\partial^2 v}{\partial z \partial y}$$

Hence from (7.3) we get

$$\frac{\partial v}{\partial x}\left(\frac{\partial^2 v}{\partial z \partial y} - \frac{\partial R}{\partial y} \right) + \frac{\partial v}{\partial y}\left(\frac{\partial R}{\partial x} - \frac{\partial^2 v}{\partial z \partial x} \right) = 0$$

$$\Rightarrow \frac{\partial v}{\partial x} \frac{\partial}{\partial y}\left(\frac{\partial v}{\partial z} - R \right) - \frac{\partial v}{\partial y} \cdot \frac{\partial}{\partial x}\left(\frac{\partial v}{\partial z} - R \right) = 0$$

$$\Rightarrow \begin{vmatrix} \dfrac{\partial v}{\partial x} & \dfrac{\partial}{\partial x}\left(\dfrac{\partial v}{\partial z} - R \right) \\[6pt] \dfrac{\partial v}{\partial y} & \dfrac{\partial}{\partial y}\left(\dfrac{\partial v}{\partial z} - R \right) \end{vmatrix} = 0$$

This shows that a relation independent of x and y exists between v and $\frac{\partial v}{\partial z} - R$. Therefore $\frac{\partial v}{\partial z} - R$ can be expressed as a function of z and v alone.

Let
$$\frac{\partial v}{\partial z} - R = \phi(z, v) \tag{7.8}$$

Since
$$P\,dx + Q\,dy + R\,dz = \frac{\partial v}{\partial x}dx + \frac{\partial v}{\partial y}dy + \frac{\partial v}{\partial z}dz + \left(R - \frac{\partial v}{\partial z}\right)dz$$

Equation (7.1) may be written, taking account of (7.8), as
$$dv - \phi(z, v)\,dz = 0$$

This is an equation in two variables. Its integration will lead to an equation of the form $F(v, z) = 0$.

Hence the condition is sufficient. This completes the proof.

The necessary and sufficient condition (7.3) for the integrability of Eq. (7.1) can be easily remembered in the form

$$\begin{vmatrix} P & Q & R \\ \dfrac{\partial}{\partial x} & \dfrac{\partial}{\partial y} & \dfrac{\partial}{\partial z} \\ P & Q & R \end{vmatrix} = 0 \tag{7.9}$$

$$\Rightarrow (iP + jQ + kR) \cdot \left\{ i\left(\frac{\partial R}{\partial y} - \frac{\partial Q}{\partial z}\right) + j\left(\frac{\partial P}{\partial z} - \frac{\partial R}{\partial x}\right) + k\left(\frac{\partial R}{\partial x} - \frac{\partial P}{\partial y}\right) \right\} = 0$$

$$\Rightarrow X \cdot \text{Curl } X = 0$$

where $X = iP + jQ + kR$.

Thus the condition for Eq. (7.1) to be exact is
$$\frac{\partial P}{\partial y} = \frac{\partial Q}{\partial x}, \quad \frac{\partial Q}{\partial z} = \frac{\partial R}{\partial y}, \quad \frac{\partial R}{\partial x} = \frac{\partial P}{\partial z}$$

which clearly shows that every exact equation satisfies the condition of integrability (7.3).

Example 7.1 Show that the equation
$$(y^2 + yz)dx + (xz + z^2)dy + (y^2 - xy)dz = 0$$
is integrable.

Here $\quad P = y^2 + yz \quad Q = xz + z^2 \quad R = y^2 - xy$

So the condition (7.3) becomes
$$(y^2 + yz)(x + 2z - 2y + x) + (xz + z^2)(-y - y) + (y^2 - xy)(2y + z - z)$$
$$= 2xy^2 + 2xyz + 2y^2z + 2yz^2 - 2y^3 - 2y^2z - 2xyz - 2yz^2 + 2y^3 - 2xy^2 = 0$$

Thus the condition of integrability for the given equation is satisfied.

7.3 GEOMETRICAL INTERPRETATION OF $P\,dx + Q\,dy + R\,dz = 0$

Suppose that Eq. (7.1) satisfies the condition of integrability (7.3), and that its solution is

$$u(x, y, z) = c \qquad (7.10)$$

Equation (7.10) represents a single infinite system of surfaces, as there is one arbitrary constant. This constant can be determined, so that (7.10) will represent the surface which passes through any given point of space.

We know that $\dfrac{\partial u}{\partial x}, \dfrac{\partial u}{\partial y}, \dfrac{\partial u}{\partial z}$ are proportional to the direction cosines of the normal at (x, y, z) on the surface.

From Eq. (7.10) we have

$$\frac{\partial u}{\partial x}dx + \frac{\partial u}{\partial y}dy + \frac{\partial u}{\partial z}dz = 0$$

Comparing this with Eq. (7.1), we find that $\dfrac{\partial u}{\partial x}, \dfrac{\partial u}{\partial y}$ and $\dfrac{\partial u}{\partial z}$ are respectively proportional to P, Q and R.

So (7.1) represents a single infinite system of surfaces so that the direction cosines of the normal at any point (x, y, z) on a surface belonging to this system are proportional to the values of P Q and R at that point. Thus it follows that a point moving in such a way that its coordinates and the direction cosines of its path always satisfy (7.1) and can pass through any point in an infinity of directions. But when passing through any point, it must remain on the particular surface represented by the integral (7.10) which pass through the point, hence all the possible curves, infinite in number, which it can describe through that point must lie on that surface.

Next, we discuss various methods of obtaining solutions of the total differential equation (7.1).

7.4 METHODS OF SOLVING TOTAL DIFFERENTIAL EQUATIONS

We consider the total differential equation (7.1) for which the integrability condition (7.3) is satisfied. Various methods can be used for obtaining the solutions of such equations.

7.5 SOLUTION BY INSPECTION

Once the condition of integrability has been verified, it is often possible to derive the solution by inspection.

In particular, if for a given equation we have curl $X = 0$, then $iP + jQ + kR = X$ must be of the form

$$i\frac{\partial u}{\partial x} + j\frac{\partial u}{\partial y} + k\frac{\partial u}{\partial z}$$

Then the given equation is

$$\frac{\partial u}{\partial x}dx + \frac{\partial u}{\partial y}dy + \frac{\partial u}{\partial z}dz = 0$$

$$\Rightarrow du = 0$$

$$\Rightarrow u(x, y, z) = c$$

which is the required solution.

On the other hand, if Curl $X \neq 0$, but $X \cdot$ Curl $X = 0$, then it is possible to arrange the terms or divide by a suitable function of x, y, z or both, so that the solution can be found by inspection as illustrated in the following examples.

Example 7.2 Solve $yzdx + zxdy + xydz = 0$.

Here $\qquad P = yz \qquad Q = zx$ and $R = xy$

so that $\qquad X = iyz + jzx + kxy$

and $\qquad$ curl $X = \begin{vmatrix} i & j & k \\ \frac{\partial}{\partial x} & \frac{\partial}{\partial y} & \frac{\partial}{\partial z} \\ yz & zx & xy \end{vmatrix} = i(x - x) + j(y - y) + k(z - z) = 0$

Hence $X \cdot$ curl $X = 0$ and the given equation is integrable. By inspection of the given equation we can immediately write it as

$$d(xyz) = 0 \Rightarrow xyz = c$$

which is the required solution.

Example 7.3 Solve $(yz + xyz)dx + (zx + xyz)dy + (xy + xyz)dz = 0$.

Here $\qquad P = yz + xyz \quad Q = zx + xyz \quad R = xy + xyz$

so that $\qquad X = i(yz + xyz) + j(zx + xyz) + k(xy + xyz)$

It can be easily checked that curl $X \neq 0$, but the condition of integrability $X \cdot$ curl $X = 0$ is satisfied.

Dividing the given equation by xyz, we find that

$$\left(\frac{1}{x} + 1\right)dx + \left(\frac{1}{y} + 1\right)dy + \left(\frac{1}{z} + 1\right)dz = 0$$

$\Rightarrow \ln x + x + \ln y + y + \ln z + z = c \qquad$ on integration

$\Rightarrow \ln(x\, y\, z) + x + y + z = c$

is the required solution.

Example 7.4 Solve $(y - z)(y + z - 2x)dx + (z - x)(z + x - 2y)dy + (x - y)(x + y - 2z)dx = 0$.

Check that the condition of integrability is satisfied. Then by rearranging the terms the equation becomes

$$(y^2dx + 2xydy) - (z^2dx + 2zxdz) + (z^2dy + 2zydz) - (x^2dy + 2xydx)$$
$$+ (x^2dz + 2xzdx) - (y^2dz + 2yzdy) = 0$$
$$\Rightarrow xy^2 - z^2x + z^2y - x^2y + x^2z - y^2z = c \quad \text{on integration}$$
$$\Rightarrow x(y^2 - z^2) + y(z^2 - x^2) + z(x^2 - y^2) = c$$

is the required solution.

Example 7.5 Solve $(x^2z - y^3)dx + 3xy^2dy + x^3dz = 0$.

Here
$P = x^2z - y^3 \qquad Q = 3xy^2 \qquad R = x^3 \quad \text{and} \quad x = i(x^2z - y^3) + j3xy^2 + kx^3$

$$\text{Curl } X = \begin{vmatrix} i & j & k \\ \dfrac{\partial}{\partial x} & \dfrac{\partial}{\partial y} & \dfrac{\partial}{\partial z} \\ x^2z - y^3 & 3xy^2 & x^3 \end{vmatrix}$$

$$= i(0) + j(x^2 - 3x^2) + k(3y^2 + 3y^2)$$
$$= -2x^2 j + 6y^2 k$$

and
$X \cdot \text{Curl } X = (i(x^2z - y^3) + j\, 3xy^2 + kx^3) \cdot (-2x^2j + 6y^2k) = -6x^3y + 6x^3y = 0$
Thus the given equation is integrable.
We can rewrite the given equation by rearranging the terms as

$$x^2(zdx + xdz) - y^3dx + 3xy^2dy = 0$$

$$\Rightarrow zdx + xdz + \frac{3xy^2dy - y^3dx}{x^2} = 0 \qquad \text{dividing by } x^2$$

$$\Rightarrow d(zx) + d\left(\frac{y^3}{x}\right) = 0$$

$$\Rightarrow zx + \frac{y^3}{x} = c$$

is the required solution.
And now a few exercises for you.

Exercises 7.5

Solve the following equations:

1. $yzdx - zxdy - y^2dz = 0$
2. $(y^2 + z^2)dx + xydy + xzdz = 0$

3. $(y + z)dx + dy + dz = 0$
4. $y^2 dx - zdy + ydz = 0$
5. $(x^2y - y^3 - y^2z)dx + (xy^2 - x^2z - x^3)dy + (xy^2 + x^2y)dz = 0$

7.6 ONE VARIABLE ASSUMED TO BE CONSTANT

If the condition of integrability is satisfied, then in order to solve the equation we consider any one of the variables to be constant. If any two terms, like $P\ dx + Q\ dy = 0$ can be solved easily, then we assume one variable z as a constant, so that $dz = 0$. Then we solve $P\ dx + Q\ dy = 0$.

Let $u(x, y) = f(z)$ be the solution, where $f(z)$ is constant with respect to x and y. (7.11)

Now differentiate (7.11) with respect to x, y, and z, and compare with the given Eq. (7.1). Then we shall get a relation, independent of dx and dy, between $\dfrac{df}{dz}$ and z. By integrating this we can obtain the complete solution of the given equation.

This method is illustrated by means of the following examples.

Example 7.6 Solve $(y^2 + yz)dx + (xz + z^2)dy + (y^2 - xy)dz = 0$.

Check that the condition of integrability is satisfied.
Let us suppose that z is constant, so that $dz = 0$. Then the given equation reduces to

$$(y^2 + yz)dx + (xz + z^2)dy = 0$$
$$\Rightarrow y(y + z)dx + z(x + z)dy = 0$$
$$\Rightarrow \frac{dx}{x + z} + \frac{zdy}{y(y + z)} = 0$$
$$\Rightarrow \frac{dx}{x + z} + \left(\frac{1}{y} - \frac{1}{y + z}\right)dy = 0$$

Integrating, we get

$$\ln(x + z) + \ln y - \ln(y + z) = \text{constant}$$
$$\Rightarrow \frac{y(x + z)}{y + z} = \text{constant} = f(z) \qquad (7.12)$$

Differentiating with respect to x, y and z, we get

$$\frac{y(y+z)dx + \{(x+z)(y+z) - y(x+z)\}dy + \{y(y+z) - y(x+z)\}dz}{(y+z)^2} = f'(z)dz$$

$$\Rightarrow \frac{(y^2 + yz)dx + (xz + z^2)dy + (y^2 - xy)dz}{(y+z)^2} = f'(z)dz$$

Comparing this with the given equation, we found the numerator of the above equation is zero, so that we have $df = 0$

$\therefore$ $\qquad\qquad\qquad f = c \qquad$ a constant

Hence the required solution is

$$\frac{y(x+z)}{y+z} = c$$

$$\Rightarrow y(x+z) = c(y+z)$$

by putting the value of f in (7.12).

Example 7.7 Solve $xy\,dx + (x^2y - zx)dy + (x^2z - xy)dz = 0$.

Check that the integrability condition is satisfied.
Here we treat x as a constant, so that $dx = 0$. Then the given equation becomes

$$(x^2y - zx)dy + (x^2z - xy)dz = 0$$
$$\Rightarrow x^2(y\,dy + z\,dz) - x(y\,dz + z\,dy) = 0$$

On integration, we get

$$x^2\left(\frac{y^2}{2} + \frac{z^2}{2}\right) - x(yz) = f(x) \qquad \text{a constant} \qquad (7.13)$$

Now differentiating

$$x(y^2 + z^2)dx - yz\,dx + x^2y\,dy - xz\,dy + x^2z\,dz - xy\,dz = f'(x)dx$$

$$\Rightarrow (xy^2 + xz^2 - yz)dx + (x^2y - xz)dy + (x^2z - xy)dz = f'(x)dx$$

Comparing this equation with the given equation, we have

$$xy = xy^2 + xz^2 - yz - f'(x)$$
$$\Rightarrow f'(x) = xy^2 + xz^2 - yz - xy$$
$$\Rightarrow f(x) = \frac{1}{2}x^2y^2 + \frac{1}{2}x^2z^2 - xyz - \frac{1}{2}x^2y + c'$$

where c' is a constant.

Putting this value of f in (7.13), we obtain the required solution as

$$\frac{1}{2}x^2(y^2 + z^2) - xyz = \frac{1}{2}x^2(y^2 + z^2) - xyz - \frac{1}{2}x^2y + c'$$

$$\Rightarrow x^2y = 2c' = c$$

where c is a constant.

You may now try the following exercises.

Exercises 7.6

Solve the following equations:

1. $2yz\,dx + zx\,dy - xy(1+z)\,dz = 0$
2. $(e^x y + e^z)\,dx + (e^y z + e^x)\,dy + (e^y - e^x y - e^y z)\,dz = 0$
3. $(y+z)\,dx + dy + dz = 0$
4. $(2x^2 + 2xy + 2xz^2 + 1)\,dx + dy + 2z\,dz = 0$
5. $z^2\,dx + (z^2 - 2yz)\,dy + (2y^2 - yz - zx)\,dz = 0$

7.7 METHOD OF AUXILIARY EQUATIONS

Let Eq. (7.1), viz. $P\,dx + Q\,dy + R\,dz = 0$, be the given equation, which satisfies the condition for integrability (7.3), viz.

$$P\left(\frac{\partial R}{\partial y} - \frac{\partial Q}{\partial z}\right) + Q\left(\frac{\partial P}{\partial z} - \frac{\partial R}{\partial x}\right) + R\left(\frac{\partial Q}{\partial x} - \frac{\partial P}{\partial y}\right) = 0$$

Comparing these two equations, we get

$$\frac{dx}{\dfrac{\partial R}{\partial y} - \dfrac{\partial Q}{\partial z}} = \frac{dy}{\dfrac{\partial P}{\partial z} - \dfrac{\partial R}{\partial x}} = \frac{dz}{\dfrac{\partial Q}{\partial x} - \dfrac{\partial P}{\partial y}} \qquad (7.14)$$

Let this be

$$\frac{dx}{U} = \frac{dy}{V} = \frac{dz}{W}$$

$$U = \frac{\partial R}{\partial y} - \frac{\partial Q}{\partial z} \qquad V = \frac{\partial P}{\partial z} - \frac{\partial R}{\partial x} \qquad W = \frac{\partial Q}{\partial x} - \frac{\partial P}{\partial y}$$

We know from elementary algebra that

$$\frac{dx}{U} = \frac{dy}{V} = \frac{dz}{W} = \frac{l\,dx + m\,dy + n\,dz}{lU + mV + nW} = \frac{l'dx + m'dy + n'dz}{l'U + m'V + n'W} \qquad (7.15)$$

It may be possible by a proper choice of multipliers l, m, n, l', m' and n' to obtain equations which are easily solved, and whose solutions are the solutions of (7.14). In particular, l, m, n may be found such that

$$lU + mV + nW = 0$$

and consequently

$$l\,dx + m\,dy + n\,dz = 0$$

If $l\,dx + m\,dy + n\,dz$ is an exact differential, say du, then

$$u = a$$

is one equation of the complete solution.

170 Introduction to Differential Equations

If l', m' and n' can be chosen in such a way so that
$$l'U + m'V + n'W = 0$$
and consequently
$$l'dx + m'dy + n'dz = 0$$
If $l'dx + m'dy + n'dz$ be an exact differential, say dv, then
$$v = b$$
is the second equation of the complete solution. These two component solutions are linearly independent.

Then we find the value of $Adu + Bdv = 0$ and compare with the given equation to find the values of A and B in terms of u and v. Putting these values in $Adu + Bdv = 0$ and integrating, we shall get the required solution.

We shall illustrate this method through the following examples:

Example 7.8 Solve $(y^2 + yz) dx + (xz + z^2) dy + (y^2 - xy) dz = 0$ (7.16)

The condition of integrability is

$$P\left(\frac{\partial Q}{\partial z} - \frac{\partial R}{\partial y}\right) + Q\left(\frac{\partial R}{\partial x} - \frac{\partial P}{\partial z}\right) + R\left(\frac{\partial P}{\partial y} - \frac{\partial Q}{\partial x}\right) = 0$$

$\Rightarrow (y^2 + yz)\{x + 2z - 2y + x\} + (xz + z^2)(-y - y) + (y^2 - xy)(2y + z - z) = 0$ (7.17)
$\Rightarrow 2xy^2 + 2xyz + 2y^2z + 2yz^2 - 2y^3 - 2y^2z - 2xyz - 2yz^2 + 2y^3 - 2xy^2 = 0$

Thus the condition of integrability is satisfied.

Comparing (7.16) and (7.17), we get

$$\frac{dx}{2x - 2y + 2z} = \frac{dy}{-2y} = \frac{dz}{2y}$$

From the last two expressions, we get
$$dy + dz = 0 \Rightarrow y + z = u$$
Again, from the first and last expressions we get
$$\frac{dx + dz}{x + z} = \frac{dy}{-y} \Rightarrow \ln(x + z) = -\ln y + \ln v$$
$$\Rightarrow y(x + z) = v$$
where u, v are two integrals obtained by solving the auxiliary equations.

Consider $Adu + Bdv = 0$ where A and B are to be determined
$$A(dy + dz) + B(xdy + zdy + ydx + ydz) = 0$$
$$\Rightarrow Bydx + (A + Bx + Bz)dy + (A + By)dz = 0$$
Comparing this with (7.16), we get
$$By = y^2 + yz$$
$$A + Bx + Bz = xz + z^2$$
$$A + By = y^2 - xy$$
$$\Rightarrow B = y + z = u$$

$$A = -y(x + z) = -v$$

Thus $A\,du + B\,dv = 0$ becomes $-v\,du + u\,dv = 0$.

On integration, we get

$$\frac{v}{u} = c \Rightarrow y(x + z) = c\,(y + z)$$

which is the required solution.

Now try the following exercises.

Exercises 7.7

Check if the following equations are integrable. If they are, then solve them.

1. $(yz + 2x)dx + (xz + 2y)dy + (xy + 2z)dz = 0$
2. $(1 + yz)dx + x(z - x)dy - (1 + xy)dz = 0$
3. $(2xz - yz)dx + (2yz - xz)dy - (x^2 - xy + y^2)dz = 0$
4. $(y^2 + yz + z^2)dx + (x^2 + zx + z^2)dy + (x^2 + xy + y^2)dz = 0$
5. $(x^2y - y^3 - y^2z)dx + (xy^2 - x^2z - x^3)dy + (xy^2 + x^2y)dz = 0$

7.8 HOMOGENEOUS EQUATIONS

If the functions P, Q and R in (7.1) are homogeneous functions of x, y, and z, then Eq. (7.1) is called a homogeneous equation. In such equations, one variable may be separated from the other two by substituting $x = zu$ and $y = zv$.

This implies that $dx = z\,du + u\,dz$ and $dy = z\,dv + v\,dz$.

Then the given equation will be reduced to a form in which the coefficient of dz is either zero or nonzero. In either case the new equation can be solved easily. We shall illustrate this through the following examples.

Example 7.9 Solve $(x^2y - y^3 - y^2z)dx + (xy^2 - x^2z - x^3)dy + (xy^2 + x^2y)dz = 0$.

This is a homogeneous equation. So let us put $x = uz$ and $y = vz$ in the given equation.

$$(vu^2z^3 - v^3z^3 - v^2z^3)(u\,dz + z\,du) + (uv^2z^3 - u^2z^3 - u^3z^3)(z\,dv + v\,dz)$$
$$+ (uv^2z^3 + u^2vz^3)dz = 0$$

$$\Rightarrow z(u^2v - v^3 - v^2)du + z(uv^2 - u^2 - u^3)dv$$
$$+ (vu^3 - v^3u - v^2u + uv^3 - u^2v - u^3v + uv^2 + u^2v)dz = 0$$

$$\Rightarrow v(u^2 - v^2 - v)du + u(v^2 - u^2 - u)dv + 0\,dz = 0$$

$$\Rightarrow (u^2 - v^2)(v\,du - u\,dv) - v^2\,du - u^2\,dv = 0$$

$$\Rightarrow \frac{v\,du - u\,dv}{v^2} - \frac{v\,du - u\,dv}{u^2} - \frac{du}{u^2} - \frac{dv}{v^2} = 0 \quad \text{dividing by } u^2v^2$$

172 Introduction to Differential Equations

$$\Rightarrow \frac{u}{v} + \frac{v}{u} + \frac{1}{u} + \frac{1}{v} = c \quad \text{integrating}$$

$$\Rightarrow \frac{x}{y} + \frac{y}{x} + \frac{z}{x} + \frac{z}{y} = c \quad \text{putting the values of } u \text{ and } v$$

$$\Rightarrow x^2 + y^2 + z(x + y) = cxy$$

which is the required solution.

Example 7.10 Solve $(y^2 + yz + z^2)dx + (z^2 + zx + x^2)dy + (x^2 + xy + y^2)dz = 0$.

Let us put $x = uz$ and $y = vz$, since this is a homogeneous equation. Then the given equation reduces to

$(z^2v^2 + z^2v + z^2)(udz + zdu) + (z^2 + uz^2 + u^2z^2)(vdz + zdv)$
$+ (z^2u^2 + z^2uv + z^2v^2)dz = 0$

$\Rightarrow z^3(v^2 + v + 1)du + z^3(u^2 + u + 1)dv + z^2(uv^2 + uv + u + v + uv + u^2v + u^2 + uv + v^2)dz = 0$

$\Rightarrow z\{(v^2 + v + 1)du + (u^2 + u + 1)dv\} + \{(u + v) + (u + v)^2 + uv(u + v) + uv\}dz = 0$

$\Rightarrow z\{(v^2 + v + 1)du + (u^2 + u + 1)dv\} + \{(u + v)(1 + u + v) + uv(u + v + 1)\}dz = 0$

$\Rightarrow z\{(v^2 + v + 1)du + (u^2 + u + 1)dv\} + \{(1 + u + v)(u + v + uv)\}dz = 0$

$\Rightarrow \dfrac{(v^2 + v + 1)du + (u^2 + u + 1)dv}{(1 + u + v)(u + v + uv)} + \dfrac{dz}{z} = 0$

$\Rightarrow \dfrac{(u^2dv + v^2du) + (udv + vdu) + (du + dv)}{(1 + u + v)(u + v + uv)} + \dfrac{dz}{z} = 0$

$\Rightarrow \dfrac{(1 + u + v)d(u + v + uv) - (u + v + uv)d(1 + u + v)}{(1 + u + v)(u + v + uv)} + \dfrac{dz}{z} = 0$

This step is obtained by taking the denominator and simplifying the numerator.

$\Rightarrow \dfrac{d(u + v + uv)}{u + v + uv} - \dfrac{d(1 + u + v)}{1 + u + v} + \dfrac{dz}{z} = 0$

$\Rightarrow \dfrac{z(u + v + uv)}{1 + u + v} = c \quad \text{putting the values of } u \text{ and } v$

$\Rightarrow z(x + y) + xy = c(x + y + z)$

which is the required solution.
Now attempt the following exercises.

Exercises 7.8

Check the following equations for integrability. If integrable, then solve them

1. $(x^2y - y^3 - y^2z)dx + (xy^2 - x^2z - x^3)dy + (x^2y + xy^2)\,dz = 0$
2. $(x^2 - y^2 - z^2 + 2xz + 2xy)dx + (y^2 - z^2 - x^2 + 2yz + 2yx)dy$
   $+ (z^2 - x^2 - y^2 + 2zx + 2zy)dz = 0$
3. $(y^2 + z^2 - x^2)dx - 2xy\,dy - 2xz\,dz = 0$
4. $(z^2 + 2xy)dx + (x^2 + 2yz)dy + (y^2 + 2zx)dz = 0$
5. $zy\,dx = zx\,dy + y^2\,dz$

7.9 NON-INTEGRABLE EQUATIONS

We discussed different methods of obtaining solutions to Eq. (7.1) which satisfies the integrability condition (7.3). Now suppose that Eq. (7.1) does not satisfy the integrability conditions (7.3). Then it is not possible to find a solution to Eq. (7.1). However, we can find an infinite number of solutions which satisfy (7.1) and a given condition.

Geometrically this means that the solutions to the non-integrable equation (7.1) are curves lying on a given system of surfaces. We shall illustrate this in the following examples.

Example 7.11 Obtain the system of curves lying on the system of surfaces $zx = c$ and satisfying the differential equation

$$yz\,dx + z^2\,dy + y(z + x)dz = 0 \qquad (7.18)$$

It can be easily checked that Eq. (7.18) does not satisfy the condition of integrability.

Differentiating the given equation of surfaces $zx = c$, we get

$$z\,dx + x\,dz = 0 \qquad (7.19)$$

Solving by method of cross-multiplication from Eqs. (7.18) and (7.19), we get

$$\frac{dx}{z^2x} = \frac{dy}{yz^2} = \frac{dz}{-z^3}$$

From the first two fractions we get

$$\frac{dx}{x} = \frac{dy}{y} \Rightarrow x = c_1 y$$

From the first and last fractions we get

$$\frac{dx}{x} = \frac{dz}{-z} \Rightarrow zx = c_2$$

which is the same as the system of given surfaces.

Hence the system of curves is given by $x = c_1 y$, $zx = c_2$.

Example 7.12 Find the curves which satisfy the differential equation
$$ydx + zdy - ydy + xdz = 0$$
and lie on the plane $2x - y - z = 1$.

The given differential equation is
$$ydx + (z - y)dy + xdz = 0 \qquad (7.20)$$

Differentiating the equation of the plane, we get
$$2dx - dy - dz = 0 \qquad (7.21)$$

From (7.20) and (7.21) by the method of cross-multiplication, we get
$$\frac{dx}{x+y-z} = \frac{dy}{2x+y} = \frac{dz}{y-2z}$$

Each of these is equal to
$$\frac{2dx - dy - dz}{2(x+y-z) - (2x+y) - (y-2z)}$$
$$\Rightarrow 2dx - dy - dz = 0$$
$$\Rightarrow 2x - y - z = c_1$$

For $c_1 = 1$, we get the given surface.

From the first two fractions above we have
$$\frac{dx}{x+y-z} = \frac{dy}{2x+y}$$
$$\Rightarrow \frac{dx}{2y-x+1} = \frac{dy}{2x+y}$$

putting the value of z from the given plane
$$\Rightarrow (2x+y)dx + (x-2y-1)dy = 0$$
$$\Rightarrow 2xdx - 2ydy + (ydx + xdy) - dy = 0$$
$$\Rightarrow x^2 - y^2 + xy - y = c_2 \qquad \text{on integration}$$

Hence the system of curves is $2x - y - z = 1$ and $x^2 - y^2 + xy - y = c_2$.

Exercises 7.9

1. Find the curves lying in the plane $z = x + y$ and satisfying the differential equation $dz = 2ydx + xdy$.
2. Find the system of curves lying on the surface
$$\frac{x^2}{a^2} + \frac{y^2}{b^2} + \frac{z^2}{c^2} = 1$$
and satisfying the differential equation
$$xdx + ydy + c\sqrt{1 - \frac{x^2}{a^2} - \frac{y^2}{b^2}}\, dz = 0$$

3. Obtain the system of curves lying on the system of surfaces $zx = c$ and satisfying the differential equation
$$yz\,dx + z^2\,dy + y(z + x)\,dz = 0$$
4. Find the curves represented by the solutions of the differential equation
$$(y^2 + xz)\,dx + (x^2 + yz)\,dy + z^2\,dz = 0$$
which lie on the sphere $x^2 + y^2 + z^2 = a^2$.
5. Find all the curves lying on the surface $zx = a^2$ and satisfying the differential equation
$$z\,dx + (x + y)\,dy + x\,dz = 0$$

7.10 SIMULTANEOUS EQUATIONS

So far we were discussing a single total differential equation. In this section we shall discuss a system of total differential equations. Let us consider a system of two simultaneous total differential equations of first order and first degree involving three variables. The method, however, is general and can be applied to equations having any number of variables.

The general type of a set of simultaneous equations of the first order with three variables is

$$\left.\begin{array}{l} P_1\,dx + Q_1\,dy + R_1\,dz = 0 \\ P_2\,dx + Q_2\,dy + R_2\,dz = 0 \end{array}\right\} \qquad (7.22)$$

where the coefficients are functions of x, y and z.

Solving (7.22) by the method of cross-multiplications, we get

$$\frac{dx}{Q_1 R_2 - Q_2 R_1} = \frac{dy}{R_1 P_2 - R_2 P_1} = \frac{dz}{P_1 Q_2 - P_2 Q_1}$$

$$\Rightarrow \frac{dx}{P} = \frac{dy}{Q} = \frac{dz}{R} \qquad (7.23)$$

where $P = Q_1 R_2 - Q_2 R_1$, $Q = R_1 P_2 - R_2 P_1$ and $R = P_1 Q_2 - P_2 Q_1$.

Equations (7.23) will be taken as the type of a set of simultaneous equations of the first order with three variables. What does Eq. (7.23) mean geometrically? First, let us discuss the geometrical meaning of Eq. (7.23) and then the different methods of solving the equation.

7.11 GEOMETRICAL INTERPRETATION

We know that the direction cosines of the tangents to a curve at any point (x, y, z) are $\dfrac{dx}{ds}, \dfrac{dy}{ds}$ and $\dfrac{dz}{ds}$. In other words, the direction ratios are dx, dy and dz.

Therefore, geometrically, Eq. (7.23) represents a system of curves in space such that the direction ratios of the tangent to it at any point (x, y, z) are proportional to P, Q and R.

Next, let us discuss different methods of obtaining solutions to Eq. (7.23).

7.12 SOLUTION OF SIMULTANEOUS EQUATIONS

There are two methods of finding solutions for the simultaneous equations (7.22) or (7.23).

First Method: By taking any two members of (7.23) and solving, we may get another equation which will be one of the equations of the general solution. Next we may take another set of two members of (7.23) and solve to get another equation of the general solution.

We can also proceed as follows. Having found one equation of the general solution, we can find the value of one variable in terms of the other two. Then we can get a differential equation in two variables, which when solved will give us the other equation of the general solution. This method is illustrated in the following examples.

Example 7.13 Solve $\dfrac{dx}{z^2 y} = \dfrac{dy}{z^2 x} = \dfrac{dz}{y^2 x}$

Taking the first two fractions, we get

$$\frac{dx}{y} = \frac{dy}{x}$$

$$\Rightarrow xdx - ydy = 0 \qquad (7.24)$$

$$\Rightarrow x^2 - y^2 = c_1$$

Again, taking the second and third fraction of the given equation, we get

$$y^2 dy = z^2 dz$$

$$\Rightarrow y^3 - z^3 = c_2 \qquad (7.25)$$

Equations (7.24) and (7.25) together give the required solution.

Example 7.14 Solve $\dfrac{dx}{x+z} = \dfrac{dy}{y} = \dfrac{dz}{z+y^2}$

Taking the last two ratios, we have

$$\frac{dz}{dy} = \frac{z+y^2}{y}$$

$$\Rightarrow \frac{dz}{dy} - \frac{z}{y} = y$$

which is a linear equation, and its solution is given by

$$z - y^2 = c_1 y \qquad (7.26)$$

Total Differential Equations 177

Putting the value of z from (7.26) in the first ratio and then taking the first two ratios, we have

$$\frac{dx}{dy} = \frac{x + c_1 y + y^2}{y}$$

$$\Rightarrow \frac{dx}{dy} - \frac{x}{y} = c_1 + y$$

which is a linear equation, and its solution is given by

$$x = c_1 y \ln y + y^2 + c_2 y = (z - y^2) \ln y + y^2 + c_2 y \qquad (7.27)$$

putting the value of c_1 from (7.26).
Equations (7.26) and (7.27) together give the required solution.

Second Method: We know from elementary algebra that each ratio of (7.23) can be equal to

$$\frac{dx}{P} = \frac{dy}{Q} = \frac{dz}{R} = \frac{l\,dx + m\,dy + n\,dz}{lP + mQ + nR} = \frac{l'dx + m'dy + n'dz}{l'P + m'Q + n'R} \qquad (7.28)$$

It may be possible by proper choice of multipliers l, m, n, l', m', and n', to obtain equation which are easily solved, and whose solutions are the solutions of (7.23). In particular, l, m and n may be found such that $lP + mQ + nR = 0$ and consequently $l\,dx + m\,dy + n\,dz = 0$.

If $l\,dx + m\,dy + n\,dz$ is an exact differential, say du, then $du = 0 \Rightarrow u = a$ is one equation of the complete solution.

Similarly, if l', m', n' can be so chosen so that $l'P + m'Q + n'R = 0$, and $l'dx + m'dy + n'dz$ is an exact differential, say dv, then

$$dv = 0$$
$$\Rightarrow v = b$$

is the second equation of the complete solution.

Two equations $u = a$ and $v = b$ together constitute the complete solution. They must be linearly independent.

We illustrate this method with the help of the following examples.

Example 7.15 Solve $\dfrac{dx}{x(y-z)} = \dfrac{dy}{y(z-x)} = \dfrac{dz}{z(x-y)}$

We choose one set of multipliers as 1, 1 and 1. Then we have

$$x(y - z) + y(z - x) + z(x - y) = 0$$

and $$dx + dy + dz = 0 \qquad (7.29)$$

$$\Rightarrow x + y + z = a$$

Again, we may choose another set of multipliers as $\dfrac{1}{x}, \dfrac{1}{y}$ and $\dfrac{1}{z}$; then we have

$$\frac{x(y-z)}{x} + \frac{y(z-x)}{y} + \frac{z(x-y)}{z} = 0$$

and

$$\frac{dx}{x} + \frac{dy}{y} + \frac{dz}{z} = 0$$

$$\Rightarrow \ln x + \ln y + \ln z = \ln b \quad (7.30)$$

$$\Rightarrow xyz = b$$

Equations (7.29) and (7.30) together give the required solution.

Example 7.16 Solve $\dfrac{dx}{z(x+y)} = \dfrac{dy}{z(x-y)} = \dfrac{dz}{x^2 + y^2}$

Using multipliers x, $-y$ and $-z$, we get each of the fractions equal to

$$\frac{xdx - ydy - zdz}{0}$$

∴ $xdx - ydy - zdz = 0$

$$\Rightarrow x^2 - y^2 - z^2 = a \quad (7.31)$$

Again using y, x and $-z$ as multipliers, we get

$$ydx + xdy - zdz = 0$$

$$\Rightarrow d(xy) = d\left(\frac{z^2}{2}\right)$$

$$\Rightarrow xy = \frac{z^2}{2} + b'$$

$$\Rightarrow 2xy - z^2 = b \quad (7.32)$$

Equations (7.31) and (7.32) constitute the required solution.

Example 7.17 Solve $\dfrac{dx}{y^3 x - 2x^4} = \dfrac{dy}{2y^4 - x^3 y} = \dfrac{dz}{9z(x^3 - y^3)}$

Using $\dfrac{1}{x}, \dfrac{1}{y}$ and $\dfrac{1}{3z}$ as multipliers, we get

$$\frac{dx}{x} + \frac{dy}{y} + \frac{dz}{3z} = 0$$

$$\Rightarrow \ln x + \ln y + \frac{1}{3}\ln z = \ln a$$

$$\Rightarrow xyz^{\frac{1}{3}} = a \quad (7.33)$$

Again, from the first two fractions in the question, we get

$$2y^4\, dx - x^3 y\, dx - y^3 x\, dy + 2x^4\, dy = 0$$

$$\Rightarrow \frac{2y}{x^3}\, dx - \frac{dx}{y^2} - \frac{dy}{x^2} + \frac{2x}{y^3}\, dy = 0 \qquad \text{dividing by } x^3 y^3$$

$$\Rightarrow \left(\frac{1}{x^2}\, dy - \frac{2y}{x^3}\, dx\right) + \left(\frac{1}{y^2}\, dx - \frac{2x}{y^3}\, dy\right) = 0 \qquad \text{rearranging}$$

$$\Rightarrow d\left(\frac{y}{x^2}\right) + d\left(\frac{x}{y^2}\right) = 0$$

$$\Rightarrow \frac{y}{x^2} + \frac{x}{y^2} = b \tag{7.34}$$

Equations (7.33) and (7.34) together form a solution of the given equation.

Example 7.18 Solve $\dfrac{dx}{x^2 - yz} = \dfrac{dy}{y^2 - zx} = \dfrac{dz}{z^2 - xy}$

The given fractions are equal to

$$\frac{dx - dy}{(x - y)(x + y + z)} = \frac{dy - dz}{(y - z)(x + y + z)} = \frac{dz - dx}{(z - x)(x + y + z)}$$

From the first two fractions we get

$$\ln(x - y) = \ln(y - z) + \ln a$$

$$\Rightarrow \frac{x - y}{y - z} = a \tag{7.35}$$

From the last two fractions in the question, we get

$$\ln(y - z) = \ln(z - x) + \ln b$$

$$\Rightarrow \frac{y - z}{z - x} = b \tag{7.36}$$

Equations (7.35) and (7.36) together form the required solution to the given equation.

Now try the following exercises.

Exercises 7.12

Solve the following pairs of equations:

1. $\dfrac{x\, dx}{y^3 z} = \dfrac{dy}{x^2 z} = \dfrac{dz}{y^3}$

2. $\dfrac{dx}{z} = \dfrac{dy}{-z} = \dfrac{dz}{z^2 + (x + y)^2}$

3. $\dfrac{dx}{mz-ny} = \dfrac{dy}{nx-lz} = \dfrac{dz}{ly-mx}$

4. $\dfrac{adx}{(b-c)yz} = \dfrac{bdy}{(c-a)zx} = \dfrac{cdz}{(a-b)xy}$

5. $\dfrac{dx}{x(y^2-z^2)} = \dfrac{dy}{y(z^2-x^2)} = \dfrac{dz}{z(x^2-y^2)}$

6. $\dfrac{dx}{xz(z^2+xy)} = \dfrac{dy}{-yz(z^2+xy)} = \dfrac{dz}{z^4}$

7. $\dfrac{l\,dx}{mn(y-z)} = \dfrac{m\,dy}{nl(z-x)} = \dfrac{n\,dz}{lm(x-y)}$

8. $\dfrac{dx}{x^2+y^2} = \dfrac{dy}{2xy} = \dfrac{dz}{(x+y)z}$

9. $\dfrac{dx}{x^2-y^2-yz} = \dfrac{dy}{x^2-y^2-zx} = \dfrac{dz}{z(x-y)}$

10. $\dfrac{-dx}{x(x+y)} = \dfrac{dy}{y(x+y)} = \dfrac{dz}{(x-y)(2x+2y+z)}$

CHAPTER 8

The Laplace Transform

8.1 INTRODUCTION

We are familiar with some operators that transform functions into functions. A familiar example is the differential operator D.

We have already found that the operator D is useful in the treatment of linear differential equations with constant coefficients. In this chapter we shall introduce another transformation that is especially useful in the solution of initial value problems. When this transform is applied in connection with an initial value problem involving a linear differential equation in an unknown function of t, it transforms the given initial value problem into an algebraic problem involving the variable s. In this chapter we shall illustrate how the differential equations can be solved easily with the help of this new transform, called the *Laplace* Transform*.

8.2 DEFINITION AND EXISTENCE OF LAPLACE TRANSFORM

The Laplace transform of a function f of a real variable t, denoted by $L(f)$, is a complex-valued function F of a complex variable p, defined by

$$L(f) = \int_0^\infty e^{-pt} f(t)\, dt = F(p) \qquad (8.1)$$

whenever the integral is convergent. Here $f(t)$ may be either real-valued or complex-valued.

The integral (8.1) does not converge for every function f nor for every complex number p. We shall illustrate this in the following example.

Example 8.1 If $f(t) = \cos at$, then

$$L(\cos at) = \int_0^\infty e^{-pt} \cos at\, dt = \lim_{t_0 \to \infty} \int_0^{t_0} e^{-pt} \cos at\, dt$$

---

*Pierre Simon de Laplace (1749–1827) was a French mathematician and theoretical astronomer who was so famous during his own time that he was known as the *Newton of France*.

Introduction to Differential Equations

$$= \lim_{t_0 \to \infty} \left[\frac{e^{-pt} \sin at}{a} \right]_0^{t_0} + \lim_{t_0 \to \infty} \int_0^{t_0} p \frac{e^{-pt} \sin at}{a} dt$$

$$= \lim_{t_0 \to \infty} \left[\frac{e^{-pt} \cdot \sin at_0}{a} \right] + \lim_{t_0 \to \infty} \left[\frac{-pe^{-pt} \cos at}{a^2} \right]_0^{t_0} - \lim_{t_0 \to \infty} \int_0^{t_0} \frac{p^2 e^{-pt} \cos at}{a^2} dt$$

$$\Rightarrow \left(1 + \frac{p^2}{a^2}\right) \int_0^\infty e^{-pt} \cos at \, dt = \lim_{t_0 \to \infty} \frac{e^{-pt_0} \sin at_0}{a} + \lim_{t_0 \to \infty} \frac{-pe^{-pt_0} \cos at_0}{a^2} + \frac{p}{a^2}$$

$$\Rightarrow \int_0^\infty e^{-pt} \cos at \, dt = \frac{p}{p^2 + a^2} \qquad \text{if Re } p > 0 \qquad (8.2)$$

Thus the integral on the left hand side of (8.2) is convergent for Re $p > 0$. In this case we say the Laplace transform of cos at exists for Re $p > 0$ and

$$L(\cos at) = \frac{p}{p^2 + a^2} \qquad \text{for Re } p > 0 \qquad (8.3)$$

and the Laplace transform of cos at does not exist for Re $p \le 0$.

Example 8.2 $\quad L(\sin at) = \dfrac{a}{p^2 + a^2} \qquad \text{for Re } p > 0 \qquad (8.4)$

This can be calculated in the similar manner.

Example 8.3 Find $L(e^{at})$.

$$L(e^{at}) = \int_0^\infty e^{-pt} e^{at} \, dt = \lim_{t_0 \to \infty} \int_0^{t_0} e^{-(p-a)t} \, dt$$

$$= \lim_{t_0 \to \infty} \left[\frac{-e^{-(p-a)t}}{p-a} \right]_0^{t_0} = 0 + \frac{1}{p-a} \qquad \text{Re } p > \text{Re } a$$

Thus we find $L(e^{at}) = \dfrac{1}{p-a} \qquad \text{for Re } p > \text{Re } a \qquad (8.5)$

Note that the special case $a = 0$ yields

$$L(1) = \frac{1}{p} \qquad \text{for Re } p > 0 \qquad (8.6)$$

Example 8.4 Find $L(t^n)$ for $n = 1, 2, 3, \ldots$

$$L(t^n) = \int_0^\infty e^{-pt} t^n \, dt = \lim_{t_0 \to \infty} \int_0^{t_0} e^{-pt} t^n \, dt$$

The Laplace Transform

$$= \lim_{t_0 \to \infty} \left[\frac{-t^n e^{-pt}}{p} \right]_0^{t_0} + \frac{n}{p} \lim_{t_0 \to \infty} \int_0^{t_0} e^{-pt} t^{n-1} dt$$

$$= 0 + \frac{n}{p} \int_0^{\infty} e^{-pt} t^{n-1} dt \qquad \text{for Re } p > 0$$

$$\Rightarrow L(t^n) = \frac{n}{p} L(t^{n-1}) \qquad \text{for Re } p > 0 \qquad (8.7)$$

$$= \frac{n(n-1)}{p^2} L(t^{n-2}) \qquad \text{applying the same formula (8.7)}$$

$$\Rightarrow L(t^n) = \frac{n(n-1)\cdots 1}{p^n} L(t^0) = \frac{n!}{p^{n+1}} \qquad \text{for Re } p > 0 \qquad (8.8)$$

The following Laplace transforms can be calculated easily. The reader should work them out.

Example 8.5 $L(\sinh at) = \dfrac{a}{p^2 - a^2}$ for Re $p >$ |Re a| $\qquad (8.9)$

Example 8.6 $L(\cosh at) = \dfrac{p}{p^2 - a^2}$ for Re $p >$ |Re a| $\qquad (8.10)$

Example 8.7 $L(e^{-at} \cos bt) = \dfrac{p+a}{(p+a)^2 + b^2}$ for Re $p > -$Re a $\qquad (8.11)$

Example 8.8 $L(e^{-at} \sin bt) = \dfrac{b}{(p+a)^2 + b^2}$ for Re $p > -$Re a $\qquad (8.12)$

Example 8.9 $L(e^{-at} t^n) = \dfrac{n!}{(p+a)^{n+1}}$ for Re $p >$ Re a $\qquad (8.13)$

Example 8.10 Find $L(\cos^2 at)$

$$\cos^2 at = \frac{1}{2} + \frac{1}{2} \cos 2at$$

$$\Rightarrow L(\cos^2 at) = \frac{1}{2} \int_0^{\infty} e^{-pt} (1 + \cos 2at) dt$$

$$= \frac{1}{2} \int_0^{\infty} e^{-pt} dt + \frac{1}{2} \int_0^{\infty} e^{-pt} \cos 2at \, dt$$

$$= \frac{1}{2}L(1) + \frac{1}{2}L(\cos 2at)$$

$$= \frac{1}{2}\frac{1}{p} + \frac{1}{2}\frac{p}{p^2 + 4a^2} \quad \text{for Re } p > 0 \text{ from (8.3) and (8.6)}$$

Example 8.11 Does $L(e^{t^2})$ exist?

$$L(e^{t^2}) = \int_0^\infty e^{-pt} e^{t^2} dt = \int_0^\infty e^{t^2 - pt} dt = \int_0^\infty e^{t^2 - 2\frac{p}{2}t + \frac{p^2}{4} - \frac{p^2}{4}} dt$$

$$= \int_0^\infty e^{\left(t - \frac{p}{2}\right)^2 - \frac{p^2}{4}} dt = e^{-\frac{p^2}{4}} \int_0^\infty e^{\left(t - \frac{p}{2}\right)^2} dt$$

$$= e^{-\frac{p^2}{4}} \int_{-\frac{p}{2}}^\infty e^{x^2} dx \qquad \text{putting } t - \frac{p}{2} = x$$

$$= e^{-\frac{p^2}{4}} \int_{-\frac{p}{2}}^0 e^{x^2} dx + e^{-\frac{p^2}{4}} \int_0^\infty e^{x^2} dx = \infty$$

The first integral on the right is finite, whereas the second integral tends to infinity; hence the total integral tends to infinity.

Thus the Laplace transform of e^{t^2} does not exist for any value of p.

Example 8.12 Show that the Laplace transform of t^{-1} does not exist.

The reader is advised to work it out.

Examples 8.11 and 8.12 suggest that not all functions f have the Laplace transform. So under what conditions can f have the Laplace transform? In other words, we should find a large class of functions for which we can prove once for all that the integral (8.1) exists or is convergent for some range of values of p.

Accordingly the convergence of the integral (8.1) requires first of all that the integral $\int_0^b g(t) dt$, where $g(t) = e^{-pt} f(t)$, must exist for each finite $b > 0$. To guarantee this, it is enough to assume that $g(t)$ is continuous or at least piecewise continuous.

A function g of a real variable defined on $a < t < b$ is called a *piecewise continuous* function on $a < t < b$ iff it has a finite number of discontinuities in the interval, and at a point t_0 of discontinuity, $[g(t_0^+) - g(t_0^-)]$ is finite.

For example, $g(t) = [t]$, the step function is piecewise continuous on $(-3, 3)$, whereas $g(t) = \frac{1}{t}$ is not piecewise continuous on $[-1, 1]$.

The following facts concerning piecewise continuous functions are of particular importance, and will be used often. Therefore, we shall state them here without proof.

1. Every continuous function on $a < t < b$ is piecewise continuous function.
2. If g_1 and g_2 are two piecewise continuous functions on (a, b), then so are their sum $g_1 + g_2$, difference $g_1 - g_2$, product $g_1 g_2$ and kg_1, k a nonzero constant on (a, b).
3. If g is piecewise continuous function on (a, b), then the integral $\int_a^b g(t)\,dt$ exists and is independent of whatever values, if any, g assumes at its points of discontinuity. In particular, if g_1 and g_2 are two piecewise continuous functions on (a, b) such that they are identical everywhere in (a, b) except at their points of discontinuity, then

$$\int_a^b g_1(t)\,dt = \int_a^b g_2(t)\,dt$$

A function of a real variable is said to be piecewise continuous on $[0, \infty)$ if it is piecewise continuous on $[0, t)$ for arbitrary $t > 0$.

Now, if f is piecewise continuous in $(0, t_0)$ for every finite $t_0 > 0$, then the integral $\int_0^{t_0} e^{-pt} f(t)\,dt$ certainly exists, but this automatically does not guarantee the existence of

$$\lim_{t_0 \to \infty} \int_0^{t_0} e^{-pt} f(t)\,dt$$

as this depends on the behaviour of the integrand $e^{-pt} f(t)$ for large values of t. In order to make sure that the integrand reduces rapidly enough for convergence — or that $f(t)$ does not grow too rapidly — we shall further assume $f(t)$ to be of exponential order.

A function f of real variable t is said to be of *exponential order* on $[0, \infty)$ if there exist real numbers $M > 0$, and α such that

$$|f(t)| \leq M\, e^{\alpha t} \qquad \text{for all } t > 0 \tag{8.14}$$

Thus $f(t)$ may become infinitely large as $t \to \infty$, but it must grow less rapidly than a multiple of some exponential function $e^{\alpha t}$.

The smallest value of α for which (8.14) holds is called the *order of growth* of $f(t)$.

For example, every bounded function is of exponential order with $\alpha = 0$. Further, the functions t^n ($n > 0$), e^{at}, $\sin bt$, $\cos bt$, $\ln(1 + t)$ are all of exponential order on $[0, \infty)$. All these are of exponential order because of the following:

Theorem 8.1 A piecewise continuous function f on $[0, \infty)$, which satisfies the condition $|f(t)| \leq M\, e^{\alpha t}$ for $t > t_0 > 0$ and some $M > 0$, α is of exponential order on $[0, \infty)$.

Proof: f is piecewise continuous on $[0, \infty) \Rightarrow f$ is bounded on $[0, t_0]$ $\Rightarrow \exists\, M' > 0$ such that $|f(t)| \leq M'$ for $0 \leq t \leq t_0$

Also $\qquad\qquad |f(t)| \leq M\, e^{\alpha t} \qquad$ for $t > t_0$

So $\qquad\qquad |f(t)| \leq M' < (M' + M)e^{0t} \qquad$ for $0 \leq t \leq t_0$

and $\qquad\qquad |f(t)| \leq M\, e^{\alpha t} < (M' + M)e^{\alpha t} \qquad$ for $t > t_0$

These two imply that $|f(t)| \leq (M' + M)e^{\beta t}$ for $t \geq 0$, where $\beta = \max\,(0,\, \alpha)$. Hence f is of exponential order.

It may be noted that $f \pm g$, fg, kf are of exponential order on $[0, \infty)$ if f and g are so and k is a constant. Therefore the functions $t^n e^{\alpha t} \sin bt$ and $t^n\, e^{\alpha t} \cos bt$ are of exponential order for $n > 0$.

On the other hand, e^{t^2} is not of exponential order. For, $\left|e^{-\alpha t}\, e^{t^2}\right| = e^{t^2 - \alpha t}$ can be made as large as we please by increasing t.

Similarly e^{t^n}, $n = 3, 4, 5, \ldots$, $\tan t$, $\ln t$ are not of exponential order on $[0, \infty)$.

We shall now prove our theorem regarding the existence of Laplace transform of a function.

Theorem 8.2 If f is a piecewise continuous function of exponential order on $[0, \infty)$, then there exists a real number α such that $\displaystyle\int_0^\infty e^{-pt} f(t)\, dt$ converges absolutely for Re $p > \alpha$.

Proof: f is of exponential order on $[0, \infty)$, which implies that there exist real numbers $M > 0$ and α such that $|f(t)| \leq M e^{\alpha t}$ for all $t > 0$.

$$\Rightarrow |e^{-pt} f(t)| \leq M\, e^{\alpha t}\, |e^{-pt}| = M e^{\alpha t - xt} \qquad \text{if } p = x + iy$$

$$\Rightarrow \int_0^\infty |e^{-pt} f(t)|\, dt \leq \int_0^\infty M e^{(\alpha - x)t}\, dt$$

$$= \lim_{t_0 \to \infty} \int_0^{t_0} M e^{(\alpha - x)t}\, dt$$

$$= \lim_{t_0 \to \infty} \left[\frac{M e^{(\alpha - x)t_0}}{\alpha - x} - \frac{M}{\alpha - x}\right] = \frac{M}{x - \alpha} \qquad \text{if } x > \alpha$$

The Laplace Transform

$$\Rightarrow \int_0^\infty |e^{-pt} f(t)| \, dt \quad \text{is convergent for Re } p > \alpha$$

$$\Rightarrow \int_0^\infty e^{-pt} f(t) \, dt \quad \text{is absolutely convergent for Re } p > \alpha$$

This Theorem 8.2 suggests that any piecewise continuous function of exponential order has a Laplace transform; so these conditions are sufficient for the existence of $L(f(t)) = F(p)$. However, there are functions which do not satisfy the conditions of the above theorem, yet their Laplace transforms exist. For example, $f(t) = t^{-1/2}$ has infinite discontinuity at $x = 0$; so it is not piecewise continuous in any interval containing the origin. It is also not a function of exponential, order in $[0, \infty)$. But its Laplace transform exists.

$$\int_0^\infty e^{-pt} \frac{1}{\sqrt{t}} \, dt = \int_0^1 e^{-pt} \frac{1}{\sqrt{t}} \, dt + \int_1^\infty e^{-pt} \frac{1}{\sqrt{t}} \, dt \tag{8.15}$$

For $0 < t < 1$ and Re $p > 0$, we have

$$\left| e^{-pt} \frac{1}{\sqrt{t}} \right| = e^{-xt} \frac{1}{\sqrt{t}} \quad p = x + iy$$

$$< \frac{1}{\sqrt{t}}$$

$$\Rightarrow \int_0^1 \left| e^{-pt} \frac{1}{\sqrt{t}} \right| dt < \int_0^1 \frac{1}{\sqrt{t}} \, dt \quad \text{is finite}$$

$$\Rightarrow \int_0^1 e^{-pt} \frac{1}{\sqrt{t}} \, dt \quad \text{exists} \tag{8.16}$$

For $t \geq 1$,

$$\left| e^{-pt} \frac{1}{\sqrt{t}} \right| \leq e^{-xt}$$

$$\Rightarrow \int_1^\infty \left| e^{-pt} \frac{1}{\sqrt{t}} \right| dt \leq \int_1^\infty e^{-xt} \, dt \tag{8.17}$$

which implies that $\int_1^\infty e^{-pt} \frac{1}{\sqrt{t}} \, dt$ converges absolutely for Re $p > 0$, as the integral on the right of (8.17) is convergent for Re $p > 0$. Thus, from (8.15), the integral $\int_0^\infty e^{-pt} \frac{1}{\sqrt{t}} \, dt$ is convergent for Re $p > 0$.

Now $\int_0^\infty e^{-pt} \dfrac{1}{\sqrt{t}} dt = \dfrac{2}{\sqrt{p}} \int_0^\infty e^{-u^2} du$ putting $pt = u^2 \Rightarrow dt = \dfrac{2u}{p} du$

$$\Rightarrow \dfrac{dt}{\sqrt{t}} = \dfrac{2}{\sqrt{p}} du$$

$$= \dfrac{2}{\sqrt{p}} \dfrac{\sqrt{\pi}}{2} = \sqrt{\dfrac{\pi}{p}}$$

Here we may assume p is real and positive. Thus

$$L\left(t^{-\frac{1}{2}}\right) = \sqrt{\dfrac{\pi}{p}} \tag{8.18}$$

This result is also true when p is complex for Re $p > 0$ (see Example 8.15).

Example 8.13 Let us find the Laplace transform of the function $f(t) = e^t \sin e^t$ and the region in which it exists.

$$\int_0^\infty \left| e^{-pt} e^t \sin e^t \right| dt \leq \int_0^\infty e^{-(x-1)t} dt \qquad \text{if } p = x + iy$$

$$= \left[\dfrac{e^{-(x-1)t}}{-(x-1)} \right]_0^\infty = \dfrac{1}{x-1} \qquad \text{if } x > 1$$

Thus, $\int_0^\infty e^{-pt} e^t \sin e^t \, dt$ is absolutely convergent for all p such that Re $p > 1$. On the other hand

$$\int_0^\infty e^{-pt} e^t \sin e^t \, dt = \int_1^\infty u^{-p} \sin u \, du \qquad \text{putting } e^t = u$$

converges for Re $p > 0$.

This example suggests that the region of convergence of the integral (8.1) is in general bigger than that of absolute convergence. Therefore we shall discuss a theorem which will lead us towards the determination of the region of convergence of the integral (8.1).

Theorem 8.3 If $f(t)$ is piecewise continuous on $[0, \infty)$ and $L(f(t))$ exists for some $p = p_0$, then $L(f(t))$ exists for all p such that Re $p >$ Re p_0.

Proof: Let $\phi(x) = \int_0^x e^{-p_0 t} f(t) \, dt$

$\phi(x)$ exists for every finite x, since f and e^{-pt} are piecewise continuous, $\phi'(x) = e^{-p_0 x} f(x)$ exists for all x where f is continuous.
Thus $\phi(x)$ and $\phi'(x)$ are both piecewise continuous on $[0, \infty)$.

$$\lim_{x \to \infty} \phi(x) = \lim_{x \to \infty} \int_0^x e^{-p_0 t} f(t) \, dt = L(f(t)) \quad \text{when } p = p_0$$

So $\phi(x)$ is bounded on $[0, \infty)$ and consequently there exists a number $M > 0$ such that $|\phi(x)| \leq M$ for all x in $[0, \infty)$.

Now
$$\int_0^\infty e^{-pt} f(t) \, dt = \int_0^\infty e^{-(p-p_0)t} f(t) e^{-p_0 t} \, dt = \int_0^\infty e^{-(p-p_0)t} \phi'(t) \, dt$$

$$= \left[e^{-(p-p_0)t} \phi(t) \right]_0^\infty + (p - p_0) \int_0^\infty e^{-(p-p_0)t} \phi(t) \, dt$$

$$= (p - p_0) \int_0^\infty e^{-(p-p_0)t} \phi(t) \, dt \quad \text{for Re } p > \text{Re } p_0$$

$$\therefore \int_0^\infty \left| e^{-(p-p_0)t} \phi(t) \right| dt \leq M \int_0^\infty e^{-(x-x_0)t} \, dt = \frac{1}{x - x_0} \quad \text{for Re } p > \text{Re } p_0$$

$$\Rightarrow \int_0^\infty e^{-pt} f(t) \, dt \text{ is convergent for Re } p > \text{Re } p_0, \text{ and this completes the proof.}$$

This theorem suggests that given a function $f(t)$ of a real variable t if the $L(f(t))$ exists for a particular $p = p_0$, then $L(f(t))$ may exist for all p, or there may be a real number α_0 such that $L(f(t))$ exists for all p for which Re $p > \alpha_0$. If α_0 is the greatest lower bound of all real numbers α such that $L(f(t))$ exists for Re $p > \alpha$, then $L(f(t))$ does not exist for Re $p < \alpha_0$. α_0 is called the *abscissa of convergence* of $L(f(t))$ or of f. It is analogous to the radius of convergence of a power series. Just as for power series, there is no general rule as to the existence of $L(f(t))$ for Re $p = \alpha_0$. Clearly α_0 may be $-\infty$, i.e., $L(f(t))$ exists for all p.

Similarly, if β_0 is the greatest lower bound of all real numbers β such that integral (8.1) converges absolutely for Re $p \geq \beta$, then β_0 is called the *abscissa of absolute convergence* of $L(f(t))$ or of f. The integral (8.1) may or may not converge absolutely on Re $p = \beta_0$, but certainly does not converge absolutely for Re $p < \beta_0$.

Since absolute convergence implies convergence, we must have $\alpha_0 \leq \beta_0$. However, for most of the functions occurring in applications, $\alpha_0 = \beta_0$.

Example 8.14 (a) For $f(t) = e^t \sin e^t$, $\alpha_0 = 0$ and $\beta_0 = 1$.
(b) For $f(t) = e^t\, e^{e^t} \sin e^{e^t}$, $\alpha_0 = 0$ and $\beta_0 = \infty$.
(c) For $f(t) = e^{-t^2}$, $\alpha_0 = -\infty$ and $\beta_0 = -\infty$.
(d) For $f(t) = \cos at$, $\alpha_0 = 0$ and $\beta_0 = 0$.

Before discussing the properties of Laplace transformation let us state, without proof, a very beautiful and useful result. The proof of this theorem is beyond the scope of this book.

Theorem 8.4 Suppose

(i) $L(f)$ exists for Re $p > \alpha$

(ii) $L(f) = \psi(p)$, where p is real and $p > \alpha$

(iii) $F(p)$ is analytic for Re $p > \alpha$, and $F(p) = \psi(p)$ for real p and $p > \alpha$.

Then $L(f) = F(p)$ for Re $p > \alpha$.

In other words, this theorem says if we know $L(f)$ exists for real $p > \alpha$, then we may conclude that $L(f)$ exists for all complex p such that Re $p > \alpha$.

Example 8.15 We found in Example 8.13 that

(i) $L\left(\dfrac{1}{\sqrt{t}}\right)$ exists for Re $p > 0$

(ii) $L\left(\dfrac{1}{\sqrt{t}}\right) = \sqrt{\dfrac{\pi}{p}}$ for real $p > 0$

(iii) $\sqrt{\dfrac{\pi}{p}}$ is analytic for all complex p such that Re $p > 0$.

So by Theorem 8.4, we have

$$L\left(\dfrac{1}{\sqrt{t}}\right) = \sqrt{\dfrac{\pi}{p}} \quad \text{for all } p \text{ such that Re } p > 0$$

Now try some exercises.

Exercises 8.2

1. Use the definition to find the Laplace transform of the following functions. In each case, specify the values for which $F(p)$ exists.

   (a) $t^5 + \cos 2t$

   (b) $4 \sin t \cos t + 2e^{-t}$

   (c) $f(t) = \begin{cases} t & 0 < t < 2 \\ 3 & t > 2 \end{cases}$

   (d) $f(t) = e^{at+b}$

2. Find a function $f(t)$ whose Laplace transform is

   (a) $\dfrac{3}{p^2+4}$ (b) $\dfrac{2}{p+3}$

   (c) $\dfrac{1}{p^2+6p+13}$ (d) $\dfrac{1}{p^4+p^2}$

3. Prove that a piecewise continuous function on $[a, b]$ is bounded on $[a, b]$.
4. Let f be a piecewise continuous function on $[0, \infty)$. Then prove that
   (a) f is of exponential order on $[0, \infty)$ if there exists a real number α such that $\lim\limits_{t \to \infty} \dfrac{f(t)}{e^{\alpha t}} = 0$.

   (b) f is not of exponential order on $[0, \infty)$ if, for all real numbers α,
   $$\lim_{t \to \infty} \dfrac{f(t)}{e^{\alpha t}} \neq 0.$$

5. Prove that the integral (8.1) converges absolutely for $f(t) = e^t \sin e^t$ if $\operatorname{Re} p > 1$. Does the integral converge in a bigger domain? Justify.

8.3 PROPERTIES OF LAPLACE TRANSFORM

Theorem 8.5 If f_1 and f_2 are two functions of real variable $t \in [0, \infty)$ such that

$L(f_1) = F_1(p)$ $\operatorname{Re} p > \alpha_1$

$L(f_2) = F_2(p)$ $\operatorname{Re} p > \alpha_2$

then $L(f_1 + f_2) = F_1(p) + F_2(p)$ for $\operatorname{Re} p > \max(\alpha_1, \alpha_2)$

and $L(\alpha f_1) = \alpha F_1(p)$ for $\operatorname{Re} p > \alpha_1$ for any constant α.

The proof directly follows from the definition (8.1) of Laplace transform. This is the linearity property of Laplace transform. Thus we have

$$L(\alpha f_1 + \beta f_2) = \alpha L(f_1) + \beta L(f_2) \quad \text{for arbitrary } \alpha, \beta \quad (8.19)$$

Example 8.16 $L(\sin^2 t) = L\left(\dfrac{1-\cos 2t}{2}\right) = \dfrac{1}{2} L(1) - \dfrac{1}{2} L(\cos 2t)$

$$= \dfrac{1}{2}\dfrac{1}{p} - \dfrac{1}{2}\dfrac{p}{p^2+4} \quad \text{for } \operatorname{Re} p > 0$$

Theorem 8.2 and the example following it raise the following natural questions: First, given an analytic function $F(p)$, analytic for $\operatorname{Re} p > \alpha$, does there exist a function $f(t)$ such that $L(f(t)) = F(p)$ for $\operatorname{Re} p > \alpha$? Second, if such a function exists, then is it unique? The following two theorems have answers to these questions.

Theorem 8.6 If $f(t)$ is piecewise continuous and of exponential order on $[0, \infty)$, and $L(f(t)) = F(p)$ for Re $p > \alpha$, some real number α, then
$$\lim_{\text{Re } p \to \infty} |F(p)| = 0$$

Proof: We have $|f(t)| \leq Me^{\alpha t}$ for some real $M > 0$ and real α.

$$|F(p)| = \left| \int_0^\infty e^{-pt} f(t) \, dt \right| \leq \int_0^\infty \left| e^{-pt} f(t) \right| dt \leq M \int_0^\infty \left| e^{-pt} e^{\alpha t} \right| dt$$

$$= M \int_0^\infty e^{-(x-\alpha)t} \, dt = \frac{-M}{x - \alpha} \qquad \text{Re } p = x > \alpha$$

Thus
$$\lim_{x \to \infty} |F(p)| = 0 \qquad (8.20)$$

The consequence of this theorem is more useful. Thus if $F(p)$ is an analytic function for all p such that Re $p > \alpha$, some real α, and $\lim_{\text{Re } p \to \infty} |F(p)| \neq 0$, then $F(p)$ cannot be the Laplace transform of any function $f(t)$.

Corollary 8.1 If $f(t)$ is piecewise continuous and of exponential order on $[0, \infty)$ and $L(f(t)) = F(p)$ for Re $p > \alpha$, some real α, then $pF(p)$ is bounded as Re $p \to \infty$.

Proof: $|pF(p)| \leq \dfrac{|p|M}{x - \alpha} = \sqrt{x^2 + y^2} \dfrac{M}{x - \alpha} \qquad p = x + iy$

$$= \sqrt{1 + \frac{y^2}{x^2}} \frac{M}{1 - \frac{\alpha}{x}} \to M \text{ as } x \to \infty$$

Thus there are no functions whose Laplace transforms are p, p^2, $\sin p$, $\cos p$, e^p (Re $p > 0$), $\dfrac{p}{p+1}$.

The next theorem answers our second question in the affirmative.

Theorem 8.7 *(Lerch's theorem)* If $f(t)$ and $g(t)$ are two piecewise continuous functions of exponential order on $[0, \infty)$ such that $L(f(t)) = L(g(t)) = F(p)$ for Re $p > \alpha$, a real number, then $f(t) = g(t)$ for all $t > 0$ except at their points of discontinuities.

The proof is omitted due to the limited scope of the book. Nevertheless, due to this theorem we define two piecewise continuous functions $f(t)$ and $g(t)$ to be *equivalent* iff $f(t) = g(t)$ for all $t > 0$, except at their points of discontinuity in $[0, \infty)$.

The Laplace Transform

Thus if V is a complex vector space of all piecewise continuous functions of exponential order of a real variable $t \in [0, \infty)$ and C is a complex vector space of all complex-valued analytic functions, analytic in some domain, then Theorem 8.5 guarantees that $L : V \to C$ is a linear transformation, and the above discussions suggest the following:

Theorem 8.8 $L : V \to C$ is one to one and therefore $L : V \to L(V) \subset C$ is one to one and onto.

Then $L^{-1} : L(V) \to V$ exists and is a linear transformation.

Hence $L(f(t)) = F(p)$ iff $L^{-1}(F(p)) = f(t)$ and L^{-1} is called the *Inverse Laplace Transform* of $F(p)$. This inverse Laplace transform is equally useful.

Example 8.17 $L(\cos at) = \dfrac{p}{p^2 + a^2}$ $\text{Re } p > 0$

so
$$L^{-1}\left(\dfrac{p}{p^2 + a^2}\right) = \cos at$$

Example 8.18 $L(e^{at}) = \dfrac{1}{p - a}$ $\text{Re } p > \text{Re } a$

so
$$L^{-1}\left(\dfrac{1}{p - a}\right) = e^{at}$$

Example 8.19 $L^{-1}\left(\dfrac{1}{p}\right) = 1$ since $L(1) = \dfrac{1}{p}$ $\text{Re } p > 0$.

Example 8.20 If $L^{-1}(F(p)) = f(t)$ and $L^{-1}(G(p)) = g(t)$, then prove that
$$L^{-1}(F(p) + G(p)) = f(t) + g(t)$$

For $\qquad F(p) = L(f(t)) \qquad G(p) = L(g(t))$

and $\qquad F(p) + G(p) = L(f(t)) + L(g(t)) = L(f(t) + g(t))$

$\Rightarrow L^{-1}(F(p) + G(p)) = f(t) + g(t)$

Theorem 8.9 If $L(f(t)) = F(p)$ for all p such that $\text{Re } p > \alpha$ then

(i) $L(f(at)) = \dfrac{1}{a} F\left(\dfrac{p}{a}\right)$ for $\text{Re } p > a\alpha$ and $a > 0$ (8.21)

(ii) $L(e^{-\lambda t} f(t)) = F(p + \lambda)$ for $\text{Re }(p + \lambda) > \alpha$ (8.22)

Proof: (i) $L(f(at)) = \int_0^\infty e^{-pt} f(at)\, dt = \int_0^\infty e^{-\frac{p}{a}t} f(t) \frac{dt}{a} = \frac{1}{a} \int_0^\infty e^{-\frac{p}{a}t} f(t)\, dt = \frac{1}{a} F\left(\frac{p}{a}\right)$

for Re $p > a\alpha$ and $a > 0$

(ii) $L\left(e^{-\lambda t} f(t)\right) = \int_0^\infty e^{-(p+\lambda)t} f(t)\, dt = F(p+\lambda)$ for Re $(p + \lambda) > \alpha$

This property is known as *first shifting formula*.

Corollary 8.2 $L^{-1}(F(p - \lambda)) = e^{\lambda t} L^{-1}(F(p))$
or
$$L^{-1}(F(p)) = e^{-\lambda t} L^{-1}(F(p - \lambda)) \tag{8.23}$$

Proof: We have due to (8.22)

$$L(e^{\lambda t} f(t)) = F(p - \lambda) \quad \text{for} \quad \text{Re}\,(p - \lambda) > \alpha$$

$$\Rightarrow L^{-1}(F(p - \lambda)) = e^{\lambda t} f(t) = e^{\lambda t} L^{-1}(F(p))$$

$$\Rightarrow e^{-\lambda t} L^{-1}(F(p - \lambda)) = L^{-1}(F(p))$$

Example 8.21 Evaluate $L^{-1}\left(\dfrac{1}{p^2 + 4p + 13}\right)$.

$$L^{-1}\left(\frac{1}{p^2 + 4p + 13}\right) = L^{-1}\left(\frac{1}{(p+2)^2 + 9}\right) = \frac{1}{3} L^{-1}\left(\frac{3}{(p+2)^2 + 3^2}\right) = \frac{1}{3} e^{-2t} \sin 3t$$

Example 8.22 Find $L(e^{4t} \cos 5t)$.

We have from (8.22)

$$L(e^{4t} \cos 5t) = \frac{p - 4}{(p - 4)^2 + 5^2} \quad \text{for Re}\,(p - 4) > 0$$

Example 8.23 Find $L^{-1}\left(\dfrac{4p + 2}{p^2 + 4p + 13}\right)$

$$L^{-1}\left(\frac{4p + 2}{p^2 + 4p + 13}\right) = L^{-1}\left(\frac{4p + 2}{(p+2)^2 + 3^2}\right)$$

$$= L^{-1}\left(\frac{4(p+2) - 6}{(p+2)^2 + 3^2}\right)$$

$$= 4 L^{-1}\left(\frac{p+2}{(p+2)^2 + 3^2}\right) - 2 L^{-1}\left(\frac{3}{(p+2)^2 + 3^2}\right)$$

$$= 4 e^{-2t} \cos 3t - 2 e^{-2t} \sin 3t$$

for Re $(p + 2) > 0$ due to (8.22)

The Laplace Transform

Theorem 8.10 If $L(f) = F(p)$ and $g(t) = \begin{cases} f(t-a) & t > a > 0 \\ 0 & 0 \le t \le a \end{cases}$

then $L(g) = e^{-ap} L(f)$

Proof: $L(g) = \int_0^\infty e^{-pt} g(t) dt = \int_0^a e^{-pt} g(t) dt + \int_a^\infty e^{-pt} g(t) dt$

$$= \int_a^\infty e^{-pt} f(t-a) dt = \int_0^\infty e^{-pt-pa} f(t) dt = e^{-pa} L(f(t)) \quad (8.24)$$

Here the function g is obtained by shifting the function f, 'a' unit to the right and then defining $g(t) = 0$ in $[0, a]$. This property is known as *second shifting formula*.

Corollary 8.3 If $L^{-1}(F(p)) = f(t)$, then

$$L^{-1}(e^{-ap} F(p)) = \begin{cases} f(t-a) & t > a \\ 0 & 0 \le t \le a \end{cases} \quad (8.25)$$

This corollary follows from the above theorem.

Example 8.24 Find $L^{-1}\left(\dfrac{e^{-2p}}{p^2+1}\right)$

We know $L^{-1}\left(\dfrac{1}{p^2+1}\right) = \sin t$

Then, by Corollary 8.3, $L^{-1}\left(\dfrac{e^{-2p}}{p^2+1}\right) = \begin{cases} \sin(t-2) & t > 2 \\ 0 & 0 \le t \le 2 \end{cases}$

We state the following useful theorem without proof.

Theorem 8.11 If f is piecewise continuous function of exponential order on $[0, \infty)$ and $a > 0$, then

$$L\left(\int_a^t f(u) du\right) = \frac{1}{p} L(f) - \frac{1}{p} \int_0^a f(u) du \quad (8.26)$$

for Re $p > \alpha$, for some real number α.

In particular, if $a = 0$, then $L\left(\int_0^t f(u) du\right) = \dfrac{1}{p} L(f)$ $\quad (8.27)$

In general,

$$L\left(\int_0^t \int_0^{t_{n-1}} \cdots \int_0^{t_1} f(u)\,du\,dt_1 \cdots dt_{n-1}\right) = \frac{1}{p^n} L(f) = \frac{1}{p^n} F(p)$$

$$\Rightarrow L^{-1}\left(\frac{1}{p^n} F(p)\right) = \int_0^t \int_0^{t_{n-1}} \cdots \int_0^{t_1} f(u)\,du\,dt_1 \cdots dt_{n-1} \qquad (8.28)$$

Example 8.25 Evaluate $L^{-1}\left(\frac{p+1}{p^{5/2}}\right)$.

$$L^{-1}\left(\frac{p+1}{p^{5/2}}\right) = L^{-1}\left(\frac{1}{p^{3/2}} + \frac{1}{p^{5/2}}\right)$$

$$= L^{-1}\left(\frac{1}{p\sqrt{p}} + \frac{1}{p^2\sqrt{p}}\right) = \frac{1}{\sqrt{\pi}} L^{-1}\left(\frac{1}{p}\sqrt{\frac{\pi}{p}}\right) + \frac{1}{\sqrt{\pi}} L^{-1}\left(\frac{1}{p^2}\sqrt{\frac{\pi}{p}}\right)$$

$$= \frac{1}{\sqrt{\pi}} \int_0^t \frac{1}{\sqrt{u}}\,du + \frac{1}{\sqrt{\pi}} \int_0^t \int_0^{t_1} \frac{1}{\sqrt{u}}\,du\,dt_1 \qquad \text{from (8.18)}$$

$$= \frac{1}{\sqrt{\pi}} 2\sqrt{t} + \frac{1}{\sqrt{\pi}} \frac{4}{3} t^{3/2} = 2\sqrt{\frac{t}{\pi}} + \frac{4t}{3}\sqrt{\frac{t}{\pi}}$$

Theorem 8.12 If f is a function defined on $[0, \infty)$ be such that f' is a piecewise continuous function of exponential order on $[0, \infty)$, then $L(f') = pL(f) - f(0^+)$ for Re $p > \alpha$, a real number.

Proof: $f(t) = \int_0^t f'(u)\,du$ whenever t is a point of continuity for $f'(u)$. Thus f is a piecewise continuous function of exponential order on $[0, \infty)$, and consequently $L(f)$ exists for Re $p > \alpha$, some real number.

$$L(f'(t)) = \int_0^\infty e^{-pt} f'(t)\,dt = \left[e^{-pt} f(t)\right]_0^\infty + p\int_0^\infty e^{-pt} f(t)\,dt$$

$$= pL(f(t)) - f(0^+) \qquad (8.29)$$

$\because \lim_{t \to \infty} \frac{f(t)}{e^{pt}} = 0$ for Re $p > \alpha$, some real number α.

In general, if $f^{(n)}(t)$ is a piecewise continuous function of exponential order on $[0, \infty)$, then by induction we have

$$L\left(f^{(n)}(t)\right) = p^n L(f) - p^{n-1} f(0^+) - p^{n-2} f'(0^+) - \cdots - f^{(n-1)}(0^+) \qquad \text{for Re } p > \alpha$$

(8.30)

Theorem 8.13 If $L(f) = F(p)$ for Re $p > \alpha$, then $L(t^n f) = (-1)^n F^{(n)}(p)$ for Re $p > \alpha$ and positive integer n

Or equivalently

$$L^{-1}\left(F^{(n)}(p)\right) = (-1)^n t^n f(t)$$

Proof: $F(p) = L(f) = \int_0^\infty e^{-pt} f(t)\, dt \qquad$ Re $p > \alpha$

$$F'(p) = \frac{d}{dp} \int_0^\infty e^{-pt} f(t)\, dt = -\int_0^\infty e^{-pt} t f(t)\, dt = -L(t\, f(t))$$

$$\Rightarrow L(t\, f(t)) = (-1) F'(p)$$

This proves the theorem for $n = 1$.

Now suppose the theorem is true for $n - 1$, i.e.,

$$L\left(t^{n-1} f(t)\right) = (-1)^{n-1} F^{(n-1)}(p)$$

$$F^{(n-1)}(p) = (-1)^{n-1} L\left(t^{n-1} f(t)\right) = (-1)^{n-1} \int_0^\infty e^{-pt} t^{n-1} f(t)\, dt$$

$$\Rightarrow F^{(n)}(p) = (-1)^{n-1} \frac{d}{dp} \int_0^\infty e^{-pt} t^{n-1} f(t)\, dt$$

$$= (-1)^n \int_0^\infty e^{-pt} t^n f(t)\, dt = (-1)^n L(t^n f(t)) \qquad (8.31)$$

Hence by induction the theorem follows.

Example 8.26 Evaluate $L\left(\sqrt{t}\right)$.

$$L(\sqrt{t}) = L\left(t/\sqrt{t}\right) = -\frac{d}{dp} L\left(\frac{1}{\sqrt{t}}\right) = -\frac{d}{dp}\sqrt{\frac{\pi}{p}} \qquad \text{by (8.18)}$$

$$= \frac{1}{2p}\sqrt{\frac{\pi}{p}}$$

Introduction to Differential Equations

We now turn to the problem of integrating Laplace transforms, and the main result is the following:

Theorem 8.14 If $L(f(t)) = F(p)$, for Re $p > \alpha$, then

$$L\left(\frac{f(t)}{t}\right) = \int_p^\infty F(z)\,dz$$

provided the integral on the left hand side exists.

Proof: Suppose $\quad \psi(p) = L\left(\dfrac{f(t)}{t}\right) = \displaystyle\int_0^\infty e^{-pt}\dfrac{f(t)}{t}\,dt$

$$\Rightarrow \psi'(p) = -L\left(t\,\frac{f(t)}{t}\right) = -L(f(t)) = -F(p)$$

Now integrating this along the line Im $p = y$ from p to ∞, since this is a line integral, we have

$$\int_p^\infty -F(z)\,dz = \int_p^\infty \psi'(z)\,dz = [\psi(z)]_p^\infty = -\psi(p) + \lim_{\text{Re } z \to \infty} \psi(z)$$

$$= -\psi(p) \quad \text{by Theorem 8.6}$$

Thus

$$\psi(p) = \int_p^\infty F(z)\,dz = L\left(\frac{f(t)}{t}\right) \tag{8.32}$$

Equivalently

$$L^{-1}(F(p)) = f(t) \Rightarrow L^{-1}\left(\int_p^\infty F(z)\,dz\right) = \frac{f(t)}{t} \tag{8.33}$$

Further, we have from (8.32)

$$L\left(\frac{f(t)}{t}\right) = \int_p^\infty F(z)\,dz$$

$$\Rightarrow \int_0^\infty e^{-pt}\frac{f(t)}{t}\,dt = \int_p^\infty F(z)\,dz$$

Making $p \to 0$, we have $\quad \displaystyle\int_0^\infty \frac{f(t)}{t}\,dt = \int_0^\infty F(p)\,dp \tag{8.34}$

The Laplace Transform

which is valid whenever the integral on the left exists. This formula can sometimes be used to evaluate integrals that are difficult to handle by other methods, as shown in the following:

Example 8.27 We know $L(\sin t) = \dfrac{1}{p^2+1}$ by (8.4). Therefore by using (8.34), we have

$$\int_0^\infty \frac{\sin t}{t}\,dt = \int_0^\infty \frac{1}{p^2+1}\,dp = \left[\tan^{-1} p\right]_0^\infty = \frac{\pi}{2}$$

Now we are ready to show, through examples, how the Laplace transform can be used to solve initial value problems for linear differential equations.

Example 8.28 Solve the initial value problem $y'' - 2y' - 8y = 0$, $y(0) = 3$, $y'(0) = 6$.

$y'' - 2y' - 8y = 0$

$\Rightarrow L(y'' - 2y' - 8y) = L(0) = 0$

$\Rightarrow L(y'') - 2L(y') - 8L(y) = 0 \qquad$ by (8.19)

$\Rightarrow p^2 L(y) - py(0) - y'(0) - 2pL(y) + 2y(0) - 8L(y) = 0 \qquad$ by (8.30)

$\Rightarrow L(y)(p^2 - 2p - 8) - 3p - 6 + 6 = 0 \qquad$ using the given conditions

$\Rightarrow L(y) = \dfrac{3p}{p^2 - 2p - 8} = \dfrac{3p}{(p-4)(p+2)}$

$\Rightarrow L(y) = \dfrac{2}{p-4} + \dfrac{1}{p+2}$

$\Rightarrow y = L^{-1}\left(\dfrac{2}{p-4} + \dfrac{1}{p+2}\right) = L^{-1}\left(\dfrac{2}{p-4}\right) + L^{-1}\left(\dfrac{1}{p+2}\right) \qquad$ by Theorem 8.8

$\Rightarrow y = 2e^{4t} + e^{-2t} \qquad$ by Example 8.18

is the required solution.

Example 8.29 Solve the initial value problem

$$y'' + y = e^{-2t}\sin t \qquad y(0) = 0 \quad y'(0) = 0$$

$y'' + y = e^{-2t}\sin t$

$\Rightarrow L(y'' + y) = L(e^{-2t}\sin t)$

$\Rightarrow L(y'') + L(y) = L(e^{-2t}\sin t) \qquad$ by (8.19)

$\Rightarrow p^2 L(y) - py(0) - y'(0) + L(y) = \dfrac{1}{(p+2)^2 + 1} \qquad$ by (8.30) and (8.22)

$$\Rightarrow L(y)(p^2+1) = \frac{1}{(p+2)^2+1} \quad \text{by using the given conditions}$$

$$\Rightarrow L(y) = \frac{1}{(p^2+1)\{(p+2)^2+1\}} = \frac{1}{(p^2+1)(p^2+4p+5)}$$

$$\Rightarrow L(y) = \frac{1}{8}\left\{\frac{-p+1}{p^2+1} + \frac{p+3}{p^2+4p+5}\right\} \quad \text{by using the method of partial fractions}$$

$$\Rightarrow y = \frac{-1}{8}L^{-1}\left(\frac{p}{p^2+1}\right) + \frac{1}{8}L^{-1}\left(\frac{1}{p^2+1}\right) + \frac{1}{8}L^{-1}\left(\frac{p}{p^2+4p+5}\right)$$

$$+ \frac{3}{8}L^{-1}\left(\frac{1}{p^2+4p+5}\right) \quad \text{by Theorem 8.8}$$

$$= -\frac{1}{8}\cos t + \frac{1}{8}\sin t + \frac{1}{8}L^{-1}\left(\frac{p+2}{(p+2)^2+1}\right) + \frac{1}{8}L^{-1}\left(\frac{1}{(p+2)^2+1}\right)$$

from (8.3) and (8.4)

$$\Rightarrow y = \frac{1}{8}\sin t - \frac{1}{8}\cos t + \frac{1}{8}e^{-2t}\cos t + \frac{1}{8}e^{-2t}\sin t \quad \text{by (8.23)}$$

is the required solution.

Example 8.30 Solve $xy'' + (3x-1)y' - (4x+9)y = 0$, $y(0) = 0 = y'(0)$.

We have from the given differential equation

$$L(xy'') + 3L(xy') - L(y') - 4L(xy) - 9L(y) = 0$$

$$\Rightarrow -\frac{d}{dp}L(y'') - 3\frac{d}{dp}L(y') - L(y') + 4\frac{d}{dp}L(y) - 9L(y) = 0 \quad \text{by (8.31)}$$

$$\Rightarrow -\frac{d}{dp}\{p^2 L(y) - py(0) - y'(0)\} - 3\frac{d}{dp}\{pL(y) - y(0)\} - \{pL(y) - y(0)\}$$

$$+ 4\frac{d}{dp}L(y) - 9L(y) = 0 \quad \text{by (8.30)}$$

$$\Rightarrow -p^2\frac{d}{dp}L(y) - 2pL(y) - 3p\frac{d}{dp}L(y) - 3L(y) - pL(y) + 4\frac{d}{dp}L(y) - 9L(y) = 0$$

$$\Rightarrow (-p^2 - 3p + 4)\frac{dL(y)}{dp} + (-2p - 3 - p - 9)L(y) = 0$$

$$\Rightarrow \frac{dL(y)}{L(y)} = -\frac{3p+12}{p^2+3p-4}dp \Rightarrow \ln L(y) = -\int\frac{3p+12}{(p-1)(p+4)}dp + \ln c$$

$\Rightarrow \ln L(y) = -3 \int \dfrac{dp}{p-1} + \ln c$

$\Rightarrow \ln L(y) = \ln \dfrac{c}{(p-1)^3}$

$\Rightarrow L(y) = \dfrac{c}{p^3}\left(1 - \dfrac{1}{p}\right)^{-3} = \dfrac{c}{p^3}\left\{1 + \dfrac{3}{p} + \dfrac{3 \cdot 4}{2!}\dfrac{1}{p^2} + \cdots\right\}$

$\Rightarrow y = cL^{-1}\left\{\dfrac{1}{p^3} + 3 \cdot \dfrac{1}{p^4} + \dfrac{3 \cdot 4}{2!}\dfrac{1}{p^5} + \cdots\right\}$

$= c\left\{\dfrac{1}{2}L^{-1}\left(\dfrac{2!}{p^3}\right) + \dfrac{1}{2}L^{-1}\left(\dfrac{3!}{p^4}\right) + \dfrac{1}{2 \cdot 2!}L^{-1}\left(\dfrac{4!}{p^5}\right) + \dfrac{1}{2 \cdot 3!}L^{-1}\left(\dfrac{5!}{p^6}\right) + \cdots\right\}$

$= \dfrac{c}{2}\left\{x^2 + x^3 + \dfrac{x^4}{2!} + \dfrac{x^5}{3!} + \cdots\right\}$

$\Rightarrow y = \dfrac{cx^2}{2}\left\{1 + x + \dfrac{x^2}{2!} + \dfrac{x^3}{3!} + \cdots\right\} = k\, x^2 e^x$

We list, for easy reference, the elementary Laplace transforms and their main general properties in Table 8.1. The last item in the list is new. We shall discuss this formula and its applications in the next section.

Table 8.1 Elementary Laplace transforms and their main properties

	$f(t) = L^{-1}(F(p))$	$F(p) = L(f(t))$	
1.	1	$\dfrac{1}{p}$	Re $p > 0$
2.	e^{at}	$\dfrac{1}{p-a}$	Re $p >$ Re a
3.	t^n	$\dfrac{n!}{p^{n+1}}$	Re $p > 0$
4.	sin at	$\dfrac{a}{p^2 + a^2}$	Re $p > 0$
5.	cos at	$\dfrac{p}{p^2 + a^2}$	Re $p > 0$
6.	sinh at	$\dfrac{a}{p^2 - a^2}$	Re $p >$ | Re a |
7.	cosh at	$\dfrac{p}{p^2 - a^2}$	Re $p >$ | Re a |

(Table contd.)

Table 8.1 (contd.)

	$f(t) = L^{-1}(F(p))$	$F(p) = L(f(t))$	
8.	$t^n e^{at}$	$\dfrac{n!}{(p-a)^{n+1}}$	Re $p >$ Re a
9.	$t \sin at$	$\dfrac{2ap}{(p^2+a^2)^2}$	Re $p > 0$
10.	$t \cos at$	$\dfrac{p^2-a^2}{(p^2+a^2)^2}$	Re $p > 0$
11.	$e^{-at} \sin bt$	$\dfrac{b}{(p+a)^2+b^2}$	Re $p > -$ Re a
12.	$e^{-at} \cos bt$	$\dfrac{p+a}{(p+a)^2+b^2}$	Re $p > -$ Re a
13.	$\alpha f_1(t) + \beta f_2(t)$	$\alpha F_1(p) + \beta F_2(p)$	Re $p > \max(\alpha_1, \alpha_2)$
14.	$f(at)$	$\dfrac{1}{a} F\left(\dfrac{p}{a}\right)$	Re $p > a\alpha,\ a > 0$
15.	$e^{-\lambda t} f(t)$	$F(p+\lambda)$	Re $(p+\lambda) > \alpha$
16.	$f^{(n)}(t)$	$p^n F(p) - p^{n-1} f(0^+)$ $- p^{n-2} f'(0^+) - \cdots - f^{(n-1)}(0^+)$	Re $p \geq \alpha$
17.	$t^n f(t)$	$(-1)^n F^{(n)}(p)$	Re $p \geq \alpha$
18.	$\displaystyle\int_0^t f(u)\,du$	$\dfrac{1}{p} F(p)$	Re $p > \alpha$
19.	$\dfrac{f(t)}{t}$	$\displaystyle\int_p^\infty F(z)\,dz$	Re $p > \alpha$
20.	$\displaystyle\int_0^t f(t-u)g(u)\,du$	$F(p)\,G(p)$	Re $p > \max(\alpha, \beta)$

Now try some exercises.

Exercises 8.3

1. Find the Laplace transforms of each of the following functions:
   (a) $t^2 e^{-4t}$
   (b) $e^{-t}(3 \sinh 2t - 5 \cosh 2t)$
   (c) $(1 + t e^{-t})^3$
   (d) $\dfrac{\sin^2 t}{t}$
   (e) $\dfrac{1 - e^{-t}}{t}$
   (f) $\displaystyle\int_0^t \dfrac{1}{t} e^{-4t} \sin 3t\, dt$

(g) $t e^t \dfrac{d}{dt}(\sin 2t)$ (h) $t^{3/2}$

(i) $f(t) = \begin{cases} 0 & 0 < t < 5 \\ t - 3 & t > 5 \end{cases}$ (j) $(t+1)^2 e^{-3t} \cos 2t$

2. Find the inverse Laplace transforms of the following functions:

(a) $\dfrac{p+1}{p^3 + p}$ (b) $\dfrac{2p}{4p^2 + 16}$

(c) $\dfrac{e^{-5p}}{(p-2)^4}$ (d) $\dfrac{1}{(p-1)(p+2)(p^2+1)}$

(e) $\dfrac{15}{p^2 + 4p + 13}$ (f) $\dfrac{p}{(p+a)^2 + b^2}$

(g) $\dfrac{p^2}{(p^2 + a^2)(p^2 + b^2)}$, $a^2 \neq b^2$, $ab \neq 0$ (h) $\dfrac{p^2 + p + 1}{p^{5/2}}$

(i) $\dfrac{2p^3}{(p^2 + 1)^3}$ (j) $\dfrac{1}{p^2(p^2 + 4)}$

3. Evaluate the following integrals:

(a) $\displaystyle\int_0^\infty \dfrac{e^{-at} - e^{-bt}}{t} dt$ $a, b > 0$

(b) $\displaystyle\int_0^\infty \dfrac{e^{-at} \sin bt}{t} dt$ $a, b > 0$

(c) $f(x) = \displaystyle\int_0^\infty \dfrac{\sin xt}{t} dt$

(d) $f(x) = \displaystyle\int_0^\infty \dfrac{\cos xt}{1 + t^2} dt$

4. Solve the following initial value problems by using Laplace transforms:

(i) $y'' - 2y' + y = e^t$ $y(0) = 2$ $y'(0) = -1$

(ii) $y'' + k^2 y = \cos kt$ $y(0) = 0$ $y'(0) = \dfrac{1}{2}$

(iii) $y'' + 2y' + 5y = 3e^{-t} \sin t \qquad y(0) = 0 \quad y'(0) = 3$

(iv) $y'' - 6y' + 10y = 1 \qquad y(0) = 0 \quad y'(0) = 0$

(v) $y'' + 2y' + 5y = 8 \sin t + 4 \cos t \qquad y(0) = 1 \quad y'(\pi) = e^{-\pi}$

(vi) $y'' - y = f(t) = \begin{cases} 0 & 0 \le t < 2 \\ -1 & t \ge 2 \end{cases} \qquad y(0) = 0 \quad y'(0) = 0$

(vii) $y''' + y'' + 4y' + 4y = -2 \qquad y(0) = 0 \quad y'(0) = 1 \quad y''(0) = -1$

(viii) $ty'' + (1 - 2t)y' - 2y = 0 \qquad y(0) = 1 \quad y'(0) = 2$

(ix) $ty'' + (2t + 3)y' + (t + 3)y = 3e^{-t} \qquad y(0) = 0 \quad y'(0) = 0$

(x) $ty'' + y' + ty = 0 \qquad y(0) = 1 \quad y'(0) = 0$

[Note: This last exercise is the Bessels equation of order zero (see Eq. 5.42), and its solution is

$$y(t) = \sum_{n=0}^{\infty} \frac{(-1)^n t^{2n}}{2^{2n}(n!)^2} = J_0(t)$$

While working out this solution, observe that $L(J_0(t)) = \dfrac{1}{\sqrt{1+p^2}}$. We obtained this series in Chapter 5 (see Section 5.14) in a totally different way, and here we see how easily it can be derived by the Laplace transform method.]

8.4 CONVOLUTION

Suppose $L(f) = F$ and $L(g) = G$, then is $L(fg) = FG$? Let us consider the following:

Example 8.31 Let $f(t) = t^n$ and $g(t) = e^{-at}$; we know

$$L(t^n) = F(p) = \frac{n!}{p^{n+1}} \qquad \text{Re } p > 0$$

$$L(e^{-at}) = \frac{1}{p+a} = G(p) \qquad \text{Re } p > \text{Re } a$$

Also, we know $\quad L(t^n e^{-at}) = (-1)^n \dfrac{d^n}{dp^n}\left(\dfrac{1}{p+a}\right) \quad$ by (8.31)

$$= \frac{n!}{(p+a)^{n+1}} \qquad \text{Re } p > \text{Re } a$$

This suggests that the Laplace transform of the product of two functions need not be the product of their Laplace transforms. In order to find a formula for the Laplace transform of the product of two functions, we have to introduce a new operation called the *convolution*.

The Laplace Transform

If f and g are two functions which are piecewise continuous and of exponential order on $[0, \infty)$, then the *convolution* of f and g is denoted by $f * g$ and defined by

$$f * g = \int_0^t f(t-u) g(u) \, du \tag{8.35}$$

It can be easily proved that this operation of convolution is associative, distributive and commutative. Further, $f * g$ of two piecewise continuous functions is continuous and of exponential order whenever f and g are so. Thus $L(f * g)$ exists for sufficiently large p. We now prove an important result concerning $L(f * g)$.

Theorem 8.15 (Convolution Theorem) If $L(f) = F(p)$ for Re $p > \alpha$, $L(g) = G(p)$ for Re $p > \beta$, then $L(f * g) = L(f) L(g) = F(p) G(p)$ for Re $p > \max(\alpha, \beta)$.

Proof: $L(f * g) = \displaystyle\int_0^\infty e^{-pt} f * g \, dt = \int_0^\infty e^{-pt} \left(\int_0^t f(t-u) g(u) \, du \right) dt$ (see Fig. 8.1)

$$= \int_0^\infty \int_0^t e^{-pt} f(t-u) g(u) \, du \, dt$$

$$= \int_0^\infty \int_u^\infty e^{-pt} f(t-u) g(u) \, dt \, du$$

changing the order of integration

$$= \int_0^\infty g(u) \int_u^\infty e^{-pt} f(t-u) \, dt \, du$$

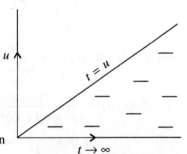

Region of integration: $0 \leq u \leq t$, $0 \leq t < \infty$

Fig. 8.1 Theorem 8.15.

$$= \int_0^\infty g(u) \int_0^\infty e^{-p(u+v)} f(v) \, dv \, du \qquad \text{putting } t - u = v$$

$$= \left(\int_0^\infty e^{-pu} g(u) \, du \right) \left(\int_0^\infty e^{-pv} f(v) \, dv \right) = L(g) L(f)$$

for Re $p > \max(\alpha, \beta)$

Thus $\qquad L(f * g) = L(f) L(g) \qquad$ for Re $p > \max(\alpha, \beta) \qquad (8.36)$

It clearly follows from this that

$$L^{-1}(FG) = f * g = L^{-1}(F) * L^{-1}(G) \tag{8.37}$$

Theorem 8.16 If $L(f) = F(p)$ for Re $p > \alpha$ and

$$u_a(t) = \begin{cases} 0 & 0 \le t < a \\ 1 & t \ge a \end{cases}$$

then $L(u_a(t)f(t-a)) = e^{-ap}F(p)$ for Re $p > \alpha$.

Proof: $L(u_a(t)f(t-a)) = \int_0^\infty e^{-pt} u_a(t) f(t-a) dt$

$$= \int_0^a e^{-pt} u_a(t) f(t-a) dt + \int_a^\infty e^{-pt} u_a(t) f(t-a) dt$$

$$= \int_a^\infty e^{-pt} f(t-a) dt = \int_0^\infty e^{-p(a+s)} f(s) ds \quad \text{putting } t - a = s$$

$$= e^{-pa} \int_0^\infty e^{-ps} f(s) ds = e^{-pa} F(p) \qquad \text{Re } p > \alpha$$

Example 8.32 If $L(f(t)) = \dfrac{1}{(p^2 + 4p + 13)^2}$, then find $f(t)$.

$$L(f(t)) = \frac{1}{(p^2 + 4p + 13)^2} = \frac{1}{\{(p+2)^2 + 3^2\}^2}$$

$$\Rightarrow f(t) = L^{-1}\left(\frac{1}{\{(p+2)^2 + 3^2\}^2}\right) = L^{-1}\left(\frac{1}{(p+2)^2 + 3^2}\right) * L^{-1}\left(\frac{1}{(p+2)^2 + 3^2}\right)$$

$$= \frac{1}{3}(e^{-2t} \sin 3t) * \frac{1}{3}(e^{-2t} \sin 3t) \qquad \text{by (8.23)}$$

$$= \frac{1}{9}\int_0^t e^{-2(t-u)} \sin 3(t-u) e^{-2u} \sin 3u \, du$$

$$= \frac{1}{9}\int_0^t e^{-2t} \sin 3(t-u) \sin 3u \, du$$

$$= \frac{1}{9}\int_0^t e^{-2t} \left\{\frac{\cos 3(t-2u) - \cos 3t}{2}\right\} du$$

$$= \frac{1}{9 \cdot 2} \int_0^t e^{-2t} \cos 3(t - 2u) \, du - \frac{1}{9 \cdot 2} \int_0^t e^{-2t} \cos 3t \, du$$

$$= \frac{1}{18} \left[e^{-2t} \frac{\sin(3t - 6u)}{-6} \right]_0^t - \frac{1}{18} \left[e^{-2t} u \cos 3t \right]_0^t$$

$$\Rightarrow f(t) = \frac{1}{18} e^{-2t} \frac{\sin 3t}{3} - \frac{1}{18} e^{-2t} t \cos 3t$$

$$= \frac{e^{-2t}}{18} \left\{ \frac{\sin 3t}{3} - t \cos 3t \right\}$$

Example 8.33 Find $L^{-1} \left(\frac{1}{(p^2 + 1)\{(p + 2)^2 + 1\}} \right)$.

This can be seen as

$$L^1 \left(\left(\frac{1}{p^2 + 1} \right)\left(\frac{1}{(p+2)^2 + 1} \right) \right) = L^{-1} \left(\frac{1}{p^2 + 1} \right) * L^{-1} \left(\frac{1}{(p+2)^2 + 1} \right) \quad \text{by (8.37)}$$

So
$$F(p) = \frac{1}{p^2 + 1} \Rightarrow f(t) = L^{-1} \left(\frac{1}{p^2 + 1} \right) = \sin t$$

$$G(P) = \frac{1}{(p+2)^2 + 1} \Rightarrow g(t) = L^{-1} \left(\frac{1}{(p+2)^2 + 1} \right) = e^{-2t} \sin t$$

$$f(t) * g(t) = \int_0^t \sin(t - u) e^{-2u} \sin u \, du$$

$$= \frac{1}{2} \int_0^t e^{-2u} \{ \cos(t - 2u) - \cos t \} \, du$$

$$= \frac{1}{2} \int_0^t e^{-2u} \cos(t - 2u) \, du - \frac{1}{2} \int_0^t e^{-2u} \cos t \, du$$

$$= \frac{1}{2} \int_0^t e^{-2u} \cos t \cos 2u \, du + \frac{1}{2} \int_0^t e^{-2u} \sin t \sin 2u \, du - \frac{1}{2} \int_0^t e^{-2u} \cos t \, du$$

$$= \frac{1}{2}\int_0^t e^{-2u}\cos t\,(\cos 2u-1)du + \frac{1}{2}\int_0^t e^{-2u}\sin t\sin 2u\,du$$

$$= -\frac{\cos t}{2}\int_0^t e^{-2u}du + \frac{\cos t}{2}\int_0^t e^{-2u}\cos 2u\,du + \frac{\sin t}{2}\int_0^t e^{-2u}\sin 2u\,du$$

$$= \frac{\cos t}{4}(e^{-2t}-1) + \frac{\cos t}{2}\left[\frac{e^{-2u}}{8}(-2\cos 2u + 2\sin 2u)\right]_0^t$$

$$+ \frac{\sin t}{2}\left[\frac{e^{-2u}}{8}(-2\sin 2u - 2\cos 2u)\right]_0^t$$

$$= \frac{\cos t}{4}(e^{-2t}-1) + \frac{\cos t}{16}e^{-2t}(-2\cos 2t + 2\sin 2t)$$

$$+ \frac{\sin t}{16}e^{-2t}(-2\sin 2t - 2\cos 2t) + \frac{\cos t + \sin t}{8}$$

$$\Rightarrow L^{-1}\frac{1}{(p^2+1)\{(p+2)^2+1\}} = \frac{\cos t}{4}(e^{-2t}-1) + \frac{e^{-2t}}{8}\sin 2t\,(\cos t - \sin t)$$

$$- \frac{e^{-2t}}{8}\cos 2t\,(\cos t + \sin t) + \frac{1}{8}(\cos t + \sin t)$$

$$= \frac{1}{8}(\sin t - \cos t) + \frac{e^{-2t}}{8}(\sin t + \cos t) \qquad \text{on simplification}$$

Compare this example with Example 8.29. The convolution method can be applied to solve initial value problems.

Exercises 8.4

1. Find the convolution of each of the following pairs of functions:
   (a) 1, $\sin at$
   (b) e^{at}, e^{bt}, $a \neq b$
   (c) 1, e^{at}
   (d) $\sin at$, $\sin bt$, $a \neq b$

2. Find the Laplace transforms of the following by using convolution method:

   (a) $\int_0^t (t-u)\sin 3u\,du$
   (b) $\int_0^t e^{-(t-u)}\sin u\,du$

   (c) $\int_0^t (t-u)^3 e^u\,du$
   (d) $\int_0^t \sin(t-u)\cos u\,du$

3. Determine the inverse Laplace transform of each of the following functions using convolution:

(a) $\dfrac{p^2}{(p^2+9)(p^2+16)}$

(b) $\dfrac{2p^2+p-10}{(p-4)(p^2+2p+2)}$

(c) $\ln\left(1+\dfrac{1}{p^2}\right)$

(d) $\dfrac{p}{p^4+64}$

4. Solve the following initial value problems by using convolution method:

(a) $y'' - 5y' + 6y = 0 \qquad y(0) = 1 \quad y'(0) = 2$

(b) $y'' + 5y' + 6y = 5e^{3t} \quad y(0) = 0 \quad y'(0) = 0$

(c) $y'' + y' - 6y = t \qquad y(0) = 0 \quad y'(0) = 0$

(d) $y'' - y' = t^2 \qquad y(0) = 0 \quad y'(0) = 0$

CHAPTER 9

Sturm–Liouville Boundary Value Problems and Fourier Series

9.1 INTRODUCTION

In Chapter 1 the concepts of boundary value problems were introduced. In this chapter we shall consider a special kind of boundary value problem known as *Sturm–Liouville Problem*. The study of this type of problem will lead us to several important concepts like eigenvalues, eigenfunctions, orthogonality and Fourier series, which are often used in the applications of differential equations to physical sciences and engineering. In Chapter 10 we shall use them to obtain solutions of boundary value problems involving partial differential equations.

9.2 STURM–LIOUVILLE PROBLEM

A boundary value problem consisting of second order homogeneous linear differential equation of the form

$$\frac{d}{dx}\left(p(x)\frac{dy}{dx}\right) + [q(x) + \lambda r(x)]y = 0 \qquad (9.1)$$

where p, q and r are continuous real-valued functions defined on $a \leq x \leq b$ such that p has a continuous derivative, $p(x) > 0$ and $r(x) > 0$, and λ is a parameter independent of x and two boundary conditions

$$A_1 y(a) + A_2 y'(a) = 0 \qquad (9.2)$$

$$B_1 y(b) + B_2 y'(b) = 0 \qquad (9.3)$$

where A_1, A_2, B_1 and B_2 are real constants such that A_1 and A_2 are not both zero, and B_1 and B_2 are not both zero, is called a *Sturm–Liouville Problem* or *Sturm–Liouville System*.

Example 9.1 The boundary value problem

$$y'' + \lambda y = 0 \qquad (9.4)$$

with
$$y(0) = 0 \text{ and } y(\pi) = 0$$
is a Sturm–Liouville problem. Here $p(x) = 1$, $q(x) = 0$, $r(x) = 1$, $a = 0$ and $b = \pi$.

Example 9.2 The boundary value problem
$$xy'' + y' + \left(x^2 + 1 + \lambda\right)y = 0$$
with $y(0) = 0$, $y'(L) = 0$, L a constant > 1, is not a Sturm–Liouville problem. Here the equation can be rewritten as
$$(xy')' + (x^2 + 1 + \lambda)y = 0$$
Then $p(x) = x$, $q(x) = x^2 + 1$ and $r(x) = 1$. Since $p(x)$ is zero in $0 \leq x \leq L$, this is not a Sturm–Liouville problem.

Let us now see what is involved in solving a Sturm–Liouville problem. Clearly we have to find a function $y(x)$ which should satisfy (9.1) – (9.3), and this should not be a trivial function, i.e., zero function. The trivial solution will not be useful. Any solution of (9.1) which will satisfy (9.2) and (9.3) will also depend on the parameter λ. Thus our first problem is to find those values of λ for which the Sturm–Liouville problem has nontrivial solutions. It can be proved that λ is real (the proof is consciously omitted). The next problem is to get those nontrivial solutions. We shall explain the method through the following example:

Example 9.3 Solve the Sturm–Liouville problem given in (9.4).

Our objective is to seek for values of λ for which (9.4) has a nontrivial solution. Let us follow the following steps:

Case 1: $\lambda < 0$, say $\lambda = -\alpha^2$, then the equation becomes
$$y'' - \alpha^2 y = 0$$
whose general solution is $y = c_1 e^{\alpha x} + c_2 e^{-\alpha x}$.
Putting the boundary conditions
$$0 = y(0) = c_1 + c_2 \Rightarrow c_2 = -c_1$$
and
$$0 = y(\pi) = c_1 e^{\pi\alpha} + c_2 e^{-\pi\alpha} = c_1(e^{\pi\alpha} - e^{-\pi\alpha})$$
which yields $c_1 = 0$ and hence $c_2 = 0$.

Thus no nontrivial solution is possible for $\lambda < 0$.

Case 2: $\lambda = 0$. In this case the differential equation becomes
$$y'' = 0 \Rightarrow y = c_1 x + c_2$$
Applying the boundary conditions again, we find that no nontrivial solution is possible for $\lambda = 0$.

Case 3: $\lambda > 0$, say $\lambda = \alpha^2$. Then the differential equation becomes $y'' + \alpha^2 y = 0$, whose general solution is given by $y = c_1 \cos \alpha x + c_2 \sin \alpha x$. Applying the first

boundary condition, we have $0 = y(0) = c_1$, which implies that the general solution becomes $y = c_2 \sin \alpha x$. Applying the second boundary condition, we get
$$0 = y(\pi) = c_2 \sin \alpha x$$
which is possible for integral values of α, because $c_2 \neq 0$, since $c_2 = 0$ yields again unwanted trivial solution. Thus for (9.4) to have nontrivial solutions, we must have
$$\lambda = n^2 \quad \text{where } n = 1, 2, 3, 4, \ldots \tag{9.5}$$
In other words, λ must be equal to one of the values 1, 4, 9, 16, ..., n^2, ... These values of λ, for which (9.4) has nontrivial solutions, are called *eigen values* of the problem (9.4) and corresponding solutions
$$\sin x, \sin 2x, \sin 3x, \sin 4x, \ldots, \sin nx,\ldots \tag{9.6}$$
are called *eigen functions*.

It is clear that the eigen values are uniquely determined by the problem, but on the contrary eigen functions are not. For any nonzero constant multiples of (9.6), say
$$k_1 \sin x, k_2 \sin 2x,\ldots, k_n \sin nx,\ldots$$
are also eigen functions.

We notice 3 facts for the eigen values (9.5) and eigen functions (9.6) of the Sturm–Liouville problem (9.4). They are: (i) the eigen values form an increasing infinite sequence of numbers $1 < 4 < 9 < \ldots < n^2 < \ldots$ that approaches ∞, (ii) the eigen function $\sin nx$ corresponding to each eigen value n^2 is unique except for a multiple of an arbitrary constant factor, (iii) the nth eigen function $\sin nx$ vanishes at the end points of the interval $[0, \pi]$ and has exactly $n - 1$ zeros in the interval $(0, \pi)$.

Are these true for all Sturm–Liouville problems or only for Sturm–Liouville problem (9.4)? The following theorem answers this question.

Theorem 9.1 There exists an infinite number of eigen values λ_n of the Sturm–Liouville problem (9.1), (9.2) and (9.3), such that $\lambda_1 < \lambda_2 < \lambda_3 < \ldots$ and $\lambda_n \to \infty$ as $\to \infty$. The eigen function corresponding to each eigen value λ_n is unique except for an arbitrary constant factor and has exactly $n - 1$ zeros in the open interval (a, b).

The proof of this theorem is beyond the scope of this book; therefore it is omitted.

Consider again the Sturm–Liouville problem of Example 9.1. The eigen function corresponding to two different eigen values m^2 and n^2 are $\sin mx$ and $\sin nx$. We know that

$$\int_0^\pi \sin mx \sin nx \, dx = 0 \quad \text{if } m \neq n.$$

Again, this is due to the concept of orthogonality (see Section 5.10) and Theorem 9.2.

In Example 9.1 the eigen functions $\sin mx$ and $\sin nx$ are orthogonal with respect to the weight function 1 on the interval $0 \leq x \leq \pi$. Such orthogonality is due to the following theorem:

Theorem 9.2 Let λ_m and λ_n be any two distinct eigen values of the Sturm–Liouville problem (9.1), (9.2) and (9.3), and y_m and y_n be their corresponding eigen functions. Then y_m and y_n are orthogonal with respect to the weight function $r(x)$ on the interval $a \leq x \leq b$.

Proof: Since y_m and y_n are eigen functions corresponding to the eigen values λ_m and λ_n respectively, the functions satisfy the differential equation (9.1). So we have

$$\frac{d}{dx}\left(p(x)\frac{dy_m}{dx}\right) + [q(x) + \lambda_m r(x)]y_m = 0 \tag{9.7}$$

and

$$\frac{d}{dx}\left(p(x)\frac{dy_n}{dx}\right) + [q(x) + \lambda_n r(x)]y_n = 0 \tag{9.8}$$

Multiplying both sides of (9.7) by y_n and both sides of (9.8) by y_m and subtracting, we get

$$y_n \frac{d}{dx}\left(p(x)\frac{dy_m}{dx}\right) - y_m \frac{d}{dx}\left(p(x)\frac{dy_n}{dx}\right) + (\lambda_m - \lambda_n)r(x)y_m y_n = 0$$

$$\Rightarrow \frac{d}{dx}\left[p(x)\left\{y_n \frac{dy_m}{dx} - y_m \frac{dy_n}{dx}\right\}\right] + (\lambda_m - \lambda_n)r(x)y_m y_n = 0$$

Integrating both the sides from a to b, we have

$$(\lambda_m - \lambda_n)\int_a^b y_m(x)y_n(x)r(x)dx = \int_a^b \frac{d}{dx}\left[p(x)\left\{y_m \frac{dy_n}{dx} - y_n \frac{dy_m}{dx}\right\}\right]dx$$

$$= \left[p(x)\left\{y_m \frac{dy_n}{dx} - y_n \frac{dy_m}{dx}\right\}\right]_a^b$$

$$= p(b)\{y_m(b)y_n'(b) - y_n(b)y_m'(b)\}$$

$$- p(a)\{y_m(a)y_n'(a) - y_n(a)y_m'(a)\} \tag{9.9}$$

Since y_m and y_n are eigen functions of the problem under consideration, they satisfy the conditions (9.2) and (9.3). If $A_2 = B_2 = 0$ in (9.2) and (9.3), then these conditions reduce to $y(a) = 0$ and $y(b) = 0$. Then in this case, $y_m(a) = 0$, $y_m(b) = 0$, $y_n(a) = 0$ and $y_n(b) = 0$; so the right hand side of (9.9) is zero.

If $A_2 = 0$, but $B_2 \neq 0$ in (9.2) and (9.3), these conditions reduce to $y(a) = 0$ and $\beta\, y(b) + y'(b) = 0$, where $\beta = \dfrac{B_1}{B_2}$. Then the second bracket on the right hand side of (9.9) is again zero, and the first bracket may be written as

$$[\beta y_n(b) + y_n'(b)]y_m(b) - [\beta y_m(b) + y_m'(b)]y_n(b)$$

which is again zero and consequently the right hand side of (9.9) is zero.

Introduction to Differential Equations

In the like manner, if either $A_2 \neq 0$, $B_2 = 0$ or $A_2 \neq 0$, $B_2 \neq 0$ in (9.2) and (9.3), then the right hand side of (9.9) is equal to zero. Thus in all cases the right hand side of (9.9) is equal to zero, and so

$$(\lambda_m - \lambda_n)\int_a^b y_m(x)y_n(x)r(x)dx = 0$$

$$\Rightarrow \int_a^b y_m(x)y_n(x)r(x)dx = 0 \quad \text{since } \lambda_m \neq \lambda_n$$

Hence $y_m(x)$ and $y_n(x)$ are orthogonal with respect to the weight function $r(x)$ on $a \leq x \leq b$.

Exercises 9.2

Determine whether the boundary-value problems given in Exercises 1–5 are Sturm–Liouville problems. If not, give reasons.

1. $e^x y'' + e^x y' + \lambda y = 0$ $y(0) = 0$ $y'(1) = 0$
2. $y'' + \lambda(1 + x)y = 0$ $y'(0) = 0$ $y(2) + y'(2) = 0$
3. $\left(\dfrac{1}{x}y'\right)' + (x + \lambda)y = 0$ $y(0) + 3y'(0) = 0$ $y(1) = 0$
4. $(xy')' + (x^2 + 1 + \lambda e^x)y = 0$ $y(1) + 2y'(1) = 0$ $y(2) - 3y'(2) = 0$
5. $(xy')' + (x^2 + 1 - \lambda x^2)y = 0$ $y(0) + 3y'(0) = 0$ $y(1) + y'(1) = 0$
6. Transform the differential equation

    $$a_2(x)y'' + a_1(x)y' + a_0(x)y + \lambda w(x)y = 0$$

    where $a_2(x) > 0$ and $w(x) > 0$ in the interval I, to the differential equation of the Sturm–Liouville problem.

Find the eigen values and eigen functions of each of the Sturm–Liouville problems given in Exercises 7–15.

7. $y'' + \lambda y = 0$ $y(0) = 0$ $y\left(\dfrac{\pi}{2}\right) = 0$
8. $y'' + \lambda y = 0$ $y(0) = 0$ $y(L) = 0$ $L > 0$
9. $y'' + \lambda y = 0$ $y(0) = 0$ $y(\pi) - y'(\pi) = 0$
10. $\dfrac{d}{dx}\left(x\dfrac{dy}{dx}\right) + \dfrac{\lambda}{x}y = 0$ $y(1) = 0$ $y(e^\pi) = 0$
11. $\dfrac{d}{dx}\left[(x^2 + 1)\dfrac{dy}{dx}\right] + \dfrac{\lambda}{x^2 + 1}y = 0$ $y(0) = 0$ $y(1) = 0$

    (Hint: Let $x = \tan t$.)

12. $y'' + \lambda y = 0$ $y'(0) = 0$ $y'(L) = 0$

Sturm–Liouville Boundary Value Problems and Fourier Series

13. $y'' + \lambda y = 0 \quad y'(-\pi) = 0 \quad y'(\pi) = 0$
14. $y'' + 4y' + (4 + 9\lambda)y = 0 \quad y(0) = 0 \quad y(a) = 0 \quad \alpha > 0$
15. $y'' + \lambda y = 0 \quad y(0) = 0 \quad h\,y(L) + y'(L) = 0$

where h and L are positive constants.

9.3 FOURIER SERIES

We now consider a problem which has many applications and especially in partial differential equations. Let $\{\Phi_n\}$ be an orthonormal system with respect to a weight function $r(x) = 1$ on an interval $a \leq x \leq b$, and f be an "arbitrary but nice" function. The problem under consideration is to expand the function f in an infinite series involving the orthonormal set $\{\Phi_n\}$. To explain what exactly this means, let us assume first that such an expansion exists. That is, we assume that

$$f(x) = \sum_{n=1}^{\infty} c_n \Phi_n(x) \tag{9.10}$$

for each x in the interval $a \leq x \leq b$. Without raising any questions like convergence, let us proceed for the moment formally to determine the coefficients c_n.

Multiply both the sides of (9.10) by $\Phi_k(x)$, and then integrate both the sides from a to b to obtain

$$\int_a^b f(x)\Phi_k(x)\,dx = \int_a^b \left(\sum_{n=1}^{\infty} c_n \Phi_n(x)\Phi_k(x) \right) dx = \sum_{n=1}^{\infty} \int_a^b c_n \Phi_n(x)\Phi_k(x)\,dx$$

assuming that the integral of the sum in the middle term is equal to the sum of the integrals on the right hand side. Then we have

$$\int_a^b f(x)\Phi_k(x)\,dx = c_k \int_a^b \Phi_k(x)\Phi_k(x)\,dx \qquad \text{all other integrals are zeros}$$

$$= c_k$$

Thus, under suitable conditions the coefficients in the expansion (9.10) are given by

$$c_n = \int_a^b f(x)\Phi_n(x)\,dx \qquad n = 1, 2, 3, \ldots \tag{9.11}$$

Now given an orthonormal system $\{\Phi_n(x)\}$ and a function f, we can form a series

$$\sum_{n=1}^{\infty} c_n \Phi_n \tag{9.12}$$

where c_n is given by the formula (9.11). The series (9.12) is called *Fourier Series* of f relative to the system $\{\Phi_n\}$, and c_n are called *Fourier Coefficients* relative to the system $\{\Phi_n\}$. In such a situation, we have no assurance that the series (9.12) determined by (9.11) converges pointwise on $a \le x \le b$. Further, if this series does converge at a point $x \in [a, b]$, there is no guarantee that it converges to $f(x)$. For the series (9.12) to converge to $f(x)$ at every point $x \in [a, b]$, the functions f and Φ_n's have to satisfy certain restrictive conditions. It is beyond the scope of this book to attempt to give detailed discussion of the various conditions on f and Φ_n's which are sufficient for the convergence of this expansion. However, we shall state without proof one basic convergence theorem concerned with the case in which the system $\{\Phi_n\}$ is the system of orthonormal eigen functions of a Sturm–Liouville problem.

Theorem 9.3 Consider the Sturm–Liouville problem (9.1), (9.2) and (9.3). Let λ_n, $n = 1, 2, 3, \ldots$ be the eigen values and Φ_n, $n = 1, 2, 3, \ldots$ be the corresponding orthonormal eigen functions. Let f be a function which is continuous on the interval $[a, b]$, has a piecewise continuous derivative f' on $[a, b]$, and is such that $f(a) = 0$ if $\Phi_1(a) = 0$ and $f(b) = 0$ if $\Phi_1(b) = 0$. Then the series

$$\sum_{n=1}^{\infty} c_n \Phi_n$$

where $c_n = \int_a^b f(x)\Phi_n(x)r(x)\,dx \qquad n = 1, 2, 3, \ldots$

converges uniformly and absolutely to f on $[a, b]$.

Example 9.4 Consider the system of functions

$$\left\{1, \cos\frac{\pi x}{L}, \sin\frac{\pi x}{L}, \cos\frac{2\pi x}{L}, \sin\frac{2\pi x}{L}, \ldots\right\}$$

defined on the interval $-L \le x < L$, where L is a positive constant. Let these functions be denoted by

$$\psi_1(x) = 1 \qquad \psi_{2n}(x) = \cos\frac{n\pi x}{L} \qquad \psi_{2n+1}(x) = \sin\frac{n\pi x}{L} \qquad (9.13)$$

$n = 1, 2, 3, \ldots$

It is easy to check that the system (9.13) is orthogonal with respect to the weight function $r(x) = 1$ on $-L \le x \le L$.

Also, we know that

$$\int_{-L}^{L} (1)^2\, dx = 2L$$

$$\int_{-L}^{L} \cos^2\left(\frac{n\pi x}{L}\right) dx = L \qquad n = 1, 2, 3 \ldots$$

Sturm–Liouville Boundary Value Problems and Fourier Series

$$\int_{-L}^{L} \sin^2\left(\frac{n\pi x}{L}\right) dx = L \qquad n = 1, 2, 3 \ldots$$

Thus the corresponding orthonormal system $\{\Phi_n\}$ is given by

$$\Phi_1(x) = \frac{1}{\sqrt{2L}}$$

$$\Phi_{2n}(x) = \frac{1}{\sqrt{L}} \cos\frac{n\pi x}{L} \qquad n = 1, 2, 3, \ldots \qquad (9.14)$$

$$\Phi_{2n+1}(x) = \frac{1}{\sqrt{L}} \sin\frac{n\pi x}{L} \qquad n = 1, 2, 3, \ldots$$

on the interval $-L \leq x \leq L$.

Now the Fourier series of a function f defined on the interval $-L \leq x \leq L$ relative to the orthonormal system (9.14) is given by (9.12), where the Fourier coefficients are given by

$$c_1 = \int_a^b f(x)\Phi_1(x)\,dx = \frac{1}{\sqrt{2L}}\int_{-L}^L f(x)\,dx$$

$$c_{2n} = \int_a^b f(x)\Phi_{2n}(x)\,dx = \frac{1}{\sqrt{L}}\int_{-L}^L f(x)\cos\frac{n\pi x}{L}\,dx \qquad n = 1,2,3,\ldots \qquad (9.15)$$

$$c_{2n+1} = \int_a^b f(x)\Phi_{2n+1}(x) = \frac{1}{\sqrt{L}}\int_{-L}^L f(x)\sin\frac{n\pi x}{L}\,dx \qquad n = 1,2,3,\ldots$$

The Fourier series of f relative to the orthonormal system (9.14) can be rewritten as

$$c_1\Phi_1 + \sum_{n=1}^{\infty}(c_{2n}\Phi_{2n} + c_{2n+1}\Phi_{2n+1})$$

$$= \left(\frac{1}{\sqrt{2L}}\int_{-L}^L f(x)\,dx\right)\left(\frac{1}{\sqrt{2L}}\right) + \sum_{n=1}^{\infty}\left\{\left(\frac{1}{\sqrt{L}}\int_{-L}^L f(x)\cos\frac{n\pi x}{L}\,dx\right)\right.$$

$$\left.\left(\frac{1}{\sqrt{L}}\cos\frac{n\pi x}{L}\right) + \left(\frac{1}{\sqrt{L}}\int_{-L}^L f(x)\sin\frac{n\pi x}{L}\,dx\right)\left(\frac{1}{\sqrt{L}}\sin\frac{n\pi x}{L}\right)\right\}$$

$$= \frac{1}{2}\left(\frac{1}{L}\int_{-L}^L f(x)\,dx\right) + \sum_{n=1}^{\infty}\left\{\left(\frac{1}{L}\int_{-L}^L f(x)\cos\frac{n\pi x}{L}\,dx\right)\cos\frac{n\pi x}{L}\right.$$

$$+\left(\frac{1}{L}\int_{-L}^{L}f(x)\sin\frac{n\pi x}{L}dx\right)\sin\frac{n\pi x}{L}\right\}$$

Thus the Fourier series of f relative to the orthonormal system (9.14) may be written as

$$\frac{1}{2}a_0 + \sum_{n=1}^{\infty}\left(a_n\cos\frac{n\pi x}{L} + b_n\sin\frac{n\pi x}{L}\right) \tag{9.16}$$

where
$$a_n = \frac{1}{L}\int_{-L}^{L}f(x)\cos\frac{n\pi x}{L}dx \qquad n = 0, 1, 2, \ldots \tag{9.17}$$

$$b_n = \frac{1}{L}\int_{-L}^{L}f(x)\sin\frac{n\pi x}{L}dx \qquad n = 1, 2, 3, \ldots \tag{9.18}$$

we denote this by writing

$$f(x) \sim \frac{1}{2}a_0 + \sum_{n=1}^{\infty}\left(a_n\cos\frac{n\pi x}{L} + b_n\sin\frac{n\pi x}{L}\right) \qquad -L \leq x \leq L$$

Due to the importance of Fourier series, we shall use (9.16) for the Fourier series of a function f rather than (9.12).

Thus, to find the Fourier series of a function f defined on the interval $-L \leq x \leq L$, we only have to determine the Fourier coefficients from (9.17) and (9.18) and substitute these coefficients into the series (9.16).

There are two important special cases in which the determination of the Fourier coefficients is considerably simplified. These are cases in which f is either an even function or an odd function.

If f is an even function, then from (9.17)

$$a_n = \frac{2}{L}\int_{0}^{L}f(x)\cos\frac{n\pi x}{L}dx \qquad n = 0, 1, 2, \ldots \tag{9.19}$$

and from (9.18), $b_n = 0$.

Thus the Fourier series of an even function f defined on the interval $-L \leq x \leq L$ is given by

$$\frac{1}{2}a_0 + \sum_{n=1}^{\infty}a_n\cos\frac{n\pi x}{L} \tag{9.20}$$

where the coefficients are given by (9.19).

Similarly, if f is an odd function, then from (9.17), $a_n = 0$, and from (9.18)

$$b_n = \frac{2}{L}\int_{0}^{L}f(x)\sin\frac{n\pi x}{L}dx \qquad n = 1, 2, 3, \ldots \tag{9.21}$$

So the Fourier series of an odd function f defined on the interval $-L \leq x \leq L$ is given by

$$\sum_{n=1}^{\infty} b_n \sin \frac{n\pi x}{L} \qquad (9.22)$$

where the coefficients are given by (9.21).

Example 9.5 Find the Fourier series of the function f defined by

$$f(x) = |x| \text{ in } -\pi \leq x \leq \pi$$

In this problem, $L = \pi$.

Since $f(-x) = |-x| = |x| = f(x)$, the function is an even function.

Therefore, the Fourier series of f on $-\pi \leq x \leq \pi$ is given by (9.20) with the coefficients

$$a_0 = \frac{2}{\pi}\int_0^\pi f(x)\,dx = \frac{2}{\pi}\int_0^\pi x\,dx = \pi$$

and

$$a_n = \frac{2}{\pi}\int_0^\pi f(x)\cos nx\,dx = \frac{2}{\pi}\int_0^\pi x\cos nx\,dx$$

$$= \frac{2}{\pi}\left[\frac{x\sin x}{n} + \frac{\cos nx}{n^2}\right]_0^\pi = \frac{2}{\pi}\left[\frac{\cos n\pi - 1}{n^2}\right] = \frac{2}{\pi}\left[\frac{(-1)^n - 1}{n^2}\right]$$

$$= \begin{cases} -\dfrac{4}{\pi n^2} & \text{if } n \text{ is odd} \\ 0 & \text{if } n \text{ is even} \end{cases}$$

where $n = 1, 2, 3, \ldots$

Thus the required Fourier series of f is given by

$$|x| \sim \frac{\pi}{2} - \frac{4}{\pi}\sum_{n=1}^{\infty}\frac{\cos(2n-1)x}{(2n-1)^2} \qquad -\pi \leq x \leq \pi$$

In the above discussion, we have been only stating that the function f is defined on the interval $[-L, L]$ and we have integrated without any explicit mention of the fact. As a matter of fact, all the functions f under consideration are piecewise continuous on the interval $-L < x < L$ (see Section 8.2).

Keeping this fact in mind, let us discuss the following example.

Example 9.6 Find the Fourier series of the function f defined by

$$f(x) = \begin{cases} \pi & -\pi \leq x < 0 \\ \dfrac{\pi}{2} & x = 0 \\ x & 0 < x \leq \pi \end{cases}$$

in the interval $[-\pi, \pi]$.

Consider the function $g(x) = \begin{cases} \pi & -\pi \leq x < 0 \\ x & 0 \leq x \leq \pi \end{cases}$

on the interval $[-\pi, \pi]$.

Both the functions f and g are piecewise continuous on $[-\pi, \pi]$ and they are identical everywhere in this interval except at $x = 0$. So as per the third fact mentioned for piecewise continuous functions in Section 8.2, we should have

$$a_0 = \frac{1}{\pi}\int_{-\pi}^{\pi} f(x)\,dx = \frac{1}{\pi}\int_{-\pi}^{\pi} g(x)\,dx = \frac{1}{\pi}\left[\int_{-\pi}^{0} \pi\,dx + \int_{0}^{\pi} x\,dx\right] = \frac{3\pi}{2}$$

$$a_n = \frac{1}{\pi}\int_{-\pi}^{\pi} f(x)\cos nx\,dx = \frac{1}{\pi}\int_{-\pi}^{\pi} g(x)\cos nx\,dx = \frac{1}{\pi}\left[\int_{-\pi}^{0} \pi\cos nx\,dx + \int_{0}^{\pi} x\cos nx\,dx\right]$$

$$= \frac{1}{\pi}\left[\frac{\cos n\pi - 1}{n^2}\right] = \frac{(-1)^n - 1}{\pi n^2} = \begin{cases} -\dfrac{2}{\pi n^2} & n\ \text{odd} \\ 0 & n\ \text{even} \end{cases}$$

$$b_n = \frac{1}{\pi}\int_{-\pi}^{\pi} f(x)\sin nx\,dx = \frac{1}{\pi}\int_{-\pi}^{\pi} g(x)\sin nx\,dx = \frac{1}{\pi}\left[\int_{-\pi}^{0} \pi\sin nx\,dx + \int_{0}^{\pi} x\sin nx\,dx\right]$$

$$= \frac{1}{\pi}\left(\frac{-\pi}{n}\right) = -\frac{1}{n}$$

Thus the required Fourier series is

$$\frac{3\pi}{4} + \sum_{n=1}^{\infty}\left[\frac{(-1)^n - 1}{\pi n^2}\cos nx - \frac{1}{n}\sin nx\right]$$

Exercises 9.3

Find the Fourier series of the function f defined on the interval mentioned in Exercises 1–10.

1. $f(x) = x \qquad -\pi \leq x \leq \pi$

Sturm–Liouville Boundary Value Problems and Fourier Series

2. $f(x) = \begin{cases} -2 & -\pi \leq x < 0 \\ 2 & 0 \leq x \leq \pi \end{cases}$

3. $f(x) = \begin{cases} 0 & -\pi \leq x < 0 \\ x^2 & 0 \leq x \leq \pi \end{cases}$

4. $f(x) = x|x| \qquad -\pi \leq x \leq \pi$

5. $f(x) = |x| \qquad -6 \leq x \leq 6$

6. $f(x) = \begin{cases} 0 & -3 \leq x < 0 \\ 1 & 0 \leq x \leq 3 \end{cases}$

7. $f(x) = \begin{cases} 0 & -L \leq x < 0 \\ L & 0 \leq x \leq L \end{cases}$

8. $f(x) = ax + b \qquad -L \leq x \leq L \qquad a, b$ constant

9. $f(x) = e^x \qquad -\pi < x < \pi$

10. $f(x) = \begin{cases} x + \pi & -\pi < x < 0 \\ x - \pi & 0 < x < \pi \end{cases}$

9.4 FOURIER SINE AND COSINE SERIES

In Example 9.3, we have seen that the Sturm–Liouville problem

$$y'' + \lambda y = 0 \qquad y(0) = 0 \quad y(\pi) = 0 \qquad (9.23)$$

has the set of orthogonal eigen functions

$$\{\sin x, \sin 2x, \sin 3x, \ldots\}$$

defined on the interval $0 \leq x \leq \pi$. The corresponding orthonormal set of eigen functions is given by

$$\left\{\sqrt{\frac{2}{\pi}} \sin nx\right\}$$

If we replace the second condition in (9.23) by the condition $y(L) = 0$, where $L > 0$, we obtain the problem $y'' + \lambda y = 0$, $y(0) = 0$, $y(L) = 0$.
This problem has the set of orthonormal eigen functions $\{\Phi_n\}$ defined by

$$\Phi_n(x) = \sqrt{\frac{2}{L}} \sin \frac{n\pi x}{L} \qquad 0 \leq x \leq L \qquad n = 1, 2, 3, \ldots \qquad (9.24)$$

Let us now find the Fourier series of a piecewise continuous function f defined on the interval $0 \leq x \leq L$ relative to the orthonormal system $\{\Phi_n(x)\}$ defined by (9.24).

The Fourier series is of the form (9.12), where the coefficients are given by

$$c_n = \int_a^b f(x)\Phi_n(x)\,dx = \int_0^L f(x)\sqrt{\frac{2}{L}}\sin\frac{n\pi x}{L}\,dx \qquad n = 1, 2, 3, \ldots$$

Thus the series takes the form

$$\sum_{n=1}^{\infty}\left[\sqrt{\frac{2}{L}}\int_0^L f(x)\sin\frac{n\pi x}{L}\,dx\right]\left(\sqrt{\frac{2}{L}}\sin\frac{n\pi x}{L}\right)$$

$$= \sum_{n=1}^{\infty}\left(\frac{2}{L}\int_0^L f(x)\sin\frac{n\pi x}{L}\,dx\right)\left(\sin\frac{n\pi x}{L}\right)$$

$$= \sum_{n=1}^{\infty} b_n \sin\frac{n\pi x}{L} \tag{9.25}$$

where

$$b_n = \frac{2}{L}\int_0^L f(x)\sin\frac{n\pi x}{L}\,dx \qquad n = 1, 2, 3, \ldots$$

This series (9.25) is called the *Fourier Sine Series* of f on the interval $0 \le x \le L$.

This is denoted by

$$f(x) \sim \sum_{n=1}^{\infty} b_n \sin\frac{n\pi x}{L} \qquad 0 \le x \le L$$

We observe that this Fourier sine series is identical with the Fourier series (9.22) of the odd function defined on the interval $-L \le x \le L$ and coinciding with f on $0 \le x \le L$.

Next, if we consider the Sturm–Liouville problem $y'' + \lambda y = 0$, $y'(0) = 0$, $y'(L) = 0$, where $L > 0$, we find that the set of orthonormal eigen functions of this poroblem is the set $\{\Phi_n\}$ defined by

$$\Phi_1(x) = \frac{1}{\sqrt{L}} \tag{9.26}$$

$$\Phi_{n+1}(x) = \sqrt{\frac{2}{L}}\cos\frac{n\pi x}{L} \qquad 0 \le x \le L \quad n = 1, 2, 3, \ldots$$

The Fourier series of a piecewise continuous function f defined on the interval $0 \le x \le L$ relative to the orthonormal set (9.26) is of the form (9.12), where the coefficients are given by

$$c_1 = \int_a^b f(x)\Phi_1(x)\,dx = \frac{1}{\sqrt{L}}\int_0^L f(x)\,dx$$

Sturm–Liouville Boundary Value Problems and Fourier Series

$$c_{n+1} = \int_a^b f(x)\Phi_{n+1}(x)\,dx = \int_0^L f(x)\sqrt{\frac{2}{L}}\cos\frac{n\pi x}{L}\,dx \qquad n = 1, 2, \ldots$$

Thus, the series takes the form

$$\left(\frac{1}{\sqrt{L}}\int_0^L f(x)\,dx\right)\frac{1}{\sqrt{L}} + \sum_{n=1}^{\infty}\left(\sqrt{\frac{2}{L}}\int_0^L f(x)\cos\frac{n\pi x}{L}\,dx\right)\left(\sqrt{\frac{2}{L}}\cos\frac{n\pi x}{L}\right)$$

$$= \frac{1}{L}\int_0^L f(x)\,dx + \sum_{n=1}^{\infty}\left(\frac{2}{L}\int_0^L f(x)\cos\frac{n\pi x}{L}\,dx\right)\left(\cos\frac{n\pi x}{L}\right)$$

$$= \frac{1}{2}\left(\frac{2}{L}\int_0^L f(x)\,dx\right) + \sum_{n=1}^{\infty}\left(\frac{2}{L}\int_0^L f(x)\cos\frac{n\pi x}{L}\,dx\right)\left(\cos\frac{n\pi x}{L}\right)$$

$$= \frac{1}{2}a_0 + \sum_{n=1}^{\infty} a_n \cos\frac{n\pi x}{L} \qquad (9.27)$$

where

$$a_n = \frac{2}{L}\int_0^L f(x)\cos\frac{n\pi x}{L}\,dx \qquad n = 0, 1, 2, \ldots$$

This series (9.27) is called the *Fourier Cosine Series* of f on the interval $0 \le x \le L$.

This is denoted by

$$f(x) \sim \frac{1}{2}a_0 + \sum_{n=1}^{\infty} a_n \cos\frac{n\pi x}{L} \qquad 0 \le x \le L$$

We observe that this Fourier cosine series is identical with the Fourier series (9.20) of the even function defined on the interval $-L \le x \le L$ and coinciding with f on $0 \le x \le L$.

Example 9.7 Find the Fourier sine series expansion of the function $f(x) = \cos x$, $0 < x < \pi$.

Using (9.25)
$$b_n = \frac{2}{\pi}\int_0^{\pi} \cos x \sin x \, dx \qquad n = 1, 2, 3, \ldots$$

$$= \frac{2}{\pi}\int_0^{\pi} \frac{\sin(n+1)x + \sin(n-1)x}{2}\,dx$$

$$= -\frac{1}{\pi}\left[\frac{\cos(n+1)x}{n+1} + \frac{\cos(n-1)x}{n-1}\right]_0^\pi$$

$$= -\frac{1}{\pi}\left\{\frac{(-1)^{n+1}}{n+1} + \frac{(-1)^{n-1}}{n-1} - \frac{1}{n+1} - \frac{1}{n-1}\right\}$$

$$= \frac{2n\left[1-(-1)^{n+1}\right]}{\pi(n^2-1)} = \begin{cases} \dfrac{4n}{\pi(n^2-1)} & n \text{ even} \\ 0 & n \text{ odd} \end{cases}$$

Hence the required Fourier sine series is given by

$$\frac{8}{\pi}\sum_{n=1}^{\infty} \frac{n}{4n^2-1}\sin 2nx$$

Exercises 9.4

1. Find the Fourier sine series and the Fourier cosine series of the function $f(x) = 1$ in $0 \le x \le \pi$.
2. Find the Fourier sine series of $f(x) = e^x$, where $0 < x < \pi$.
3. Find the Fourier sine series expansion of the function $f(x) = \pi - x$, in $0 < x < \pi$.
4. Find the Fourier cosine series of the function $f(x) = x - 1$ in $1 < x < 2$.
5. Find the Fourier sine series of the function $f(x) = x - 1$ in $1 < x < 2$.
6. Find the Fourier cosine series of the function

$$f(x) = \begin{cases} 0 & 0 \le x \le \dfrac{\pi}{2} \\ 2 & \dfrac{\pi}{2} < x \le \pi \end{cases}$$

7. Find the Fourier sine series and Fourier cosine series of the function $f(x) = \sin x$, $0 \le x \le \pi$.
8. Find the Fourier sine series of the function $f(x) = 2$, where $0 \le x \le L$.
9. Find the Fourier cosine series of the function

$$f(x) = \begin{cases} 0 & 0 \le x < \dfrac{L}{2} \\ L^2 & \dfrac{L}{2} \le x \le L \end{cases}$$

10. Find the Fourier cosine series expansion of the function $f(x) = Lx - x^2$, $0 < x < L$.

9.5 CONVERGENCE OF FOURIER SERIES

In this section, we shall discuss the problems of convergence of a Fourier series of a function. We shall answer the following questions:

1. Is the Fourier series of f convergent for some or all values of x in the interval $-L \le x \le L$?
2. If the Fourier series of f converges at a certain point $x \in [-L, L]$, then does it converge to $f(x)$?

As a first step towards starting a theorem to answer these questions, let us consider the series (9.16). We say a function F is *periodic of period p* if there exists a constant $p > 0$ such that $F(x + p) = F(x)$ for every x in the domain of definition of F. In particular, the functions $\sin(n\pi x/L)$ and $\cos(n\pi x/L)$ are periodic of period $2\dfrac{L}{n}$. Now observe that if F is periodic of period p, then F is also periodic of period np, where n is a positive integer. Hence for every positive integer n, the functions $\sin(n\pi x/L)$ and $\cos(n\pi x/L)$ are also periodic of period $2L$. We thus find that every term in the series (9.16) is periodic of period $2L$. Therefore, if the series converges for all x, then the function defined by the series must be periodic of period $2L$.

In particular, the series (9.16) is the Fourier series of a function f defined on $-L \le x \le L$. Assume that this series converges for every x in this interval. Then this series will also converge for every value of x in $-\infty < x < \infty$ and the function so defined will be periodic of period $2L$. This suggests that a Fourier series of the form (9.16) might possibly be used to represent a periodic function F of period $2L$ for all values of x. F is called the *periodic extension of f to* $(-\infty, \infty)$. Such a representation is indeed possible for a periodic function which satisfies some suitable additional conditions. We now state, without proof, the convergence theorem for the Fourier series of a function defined on $[-L, L]$.

Theorem 9.4 The Fourier series expansion of a piecewise smooth function f defined on the interval $-L < x < L$ converges at every point x in $(-\infty, \infty)$. Moreover, if F is the periodic extension of f of period $2L$ to $(-\infty, \infty)$, then the Fourier series at every point x converges to

$$\frac{F(x^+) + F(x^-)}{2}$$

This theorem suggests that the Fourier series converges to the value of the function $F(x)$ if x is a point of continuity for F and converges to the middle point of the jump of $F(x^+)$ and $F(x^-)$ if x is a point of discontinuity for F. Thus, Theorem 9.4 can be used to sketch the graph of the Fourier series of any piecewise smooth function f defined on $-L < x < L$. The procedure is: first, sketch the graph of f in $-L < x < L$; second, sketch the graph of the periodic extension F by repeating the graph of f in an interval of length $2L$, starting from L towards the right hand side and starting from $-L$ towards the left hand side on the whole real line; third, plot the middle points of each jump discontinuity of F. The resulting picture, with these isolated points included, will be the graph of the Fourier series in question.

Example 9.8 Find the Fourier series expansion of the function

$$f(x) = \begin{cases} 0 & -\pi < x \leq 0 \\ x^2 & 0 \leq x < \pi \end{cases}$$

and sketch the graph of the series obtained.

Use the series to show that

$$\frac{\pi^2}{6} = 1 + \frac{1}{2^2} + \frac{1}{3^2} + \cdots$$

We have
$$a_0 = \frac{1}{\pi} \int_{-\pi}^{\pi} f(x)\,dx = \frac{1}{\pi} \int_0^{\pi} x^2\,dx = \frac{\pi^2}{3}$$

$$a_n = \frac{1}{\pi} \int_{-\pi}^{\pi} f(x)\cos nx\,dx = \frac{1}{\pi} \int_0^{\pi} x^2 \cos nx\,dx$$

$$= \frac{1}{\pi} \left[\frac{x^2 \sin nx}{n} + \frac{2x \cos nx}{n^2} - \frac{2 \sin nx}{n^3} \right]_0^{\pi} = \frac{2(-1)^n}{n^2}$$

$$b_n = \frac{1}{\pi} \int_{-\pi}^{\pi} f(x)\sin nx\,dx = \frac{1}{\pi} \int_0^{\pi} x^2 \sin nx\,dx$$

$$= \frac{1}{\pi} \left[-\frac{x^2 \cos nx}{n} + \frac{2x \sin nx}{n^2} + \frac{2\cos nx}{n^3} \right]_0^{\pi} = \frac{1}{\pi}\left[\left(\frac{2}{n^3} - \frac{\pi^2}{n} \right)(-1)^n - \frac{2}{n^3} \right]$$

Thus the Fourier series is given by

$$f(x) \sim \frac{\pi^2}{6} + \sum_{n=1}^{\infty} \left[\frac{2(-1)^n}{n^2} \cos nx + \frac{1}{\pi}\left\{ \left(\frac{2}{n^3} - \frac{\pi^2}{n} \right)(-1)^n - \frac{2}{n^3} \right\} \sin nx \right]$$

To sketch the graph of this Fourier series, we follow the following steps. Since f is piecewise smooth, the different steps are the consequences of Theorem 9.4.

Step 1: Draw the graph of the function f in $-\pi < x < \pi$.

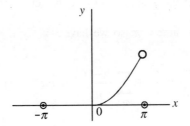

Fig. 9.1 Graph of the function f.

Step 2: Draw the graph of the periodic extension F of f to $(-\infty, \infty)$.

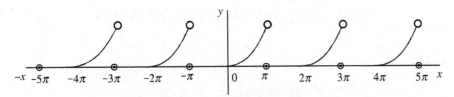

Fig. 9.2 Graph of the periodic extension F.

The above graph is obtained by pasting the graph of f in the intervals $(-5\pi, -3\pi), (-3\pi, -\pi), (-\pi, \pi) (\pi, 3\pi), (3\pi, 5\pi),\ldots$

Step 3: Since F is continuous everywhere except at $\pm\pi, \pm3\pi, \pm5\pi, \ldots, \pm(2n-1)\pi,$ $\ldots$, as per Theorem 9.4, the Fourier series converges to $F(x)$ at every point of continuity x of the function F, whereas the Fourier series converges to

$$\frac{F\left(\pm(2n-1)\pi^+\right) + F\left(\pm(2n-1)\pi^-\right)}{2} = \frac{0 + [\pm(2n-1)\pi]^2}{2} = \frac{[(2n-1)\pi]^2}{2}$$

at the points of discontinuities of F. Hence the graph of F and that of the Fourier series are identical at every point of continuity of F and the Fourier series assumes the value $\dfrac{(2n-1)^2\pi^2}{2}$ at points of discontinuities $\pm(2n-1)\pi, n = 1, 2, 3, \ldots$. Hence the graph of the Fourier series is given by the following figure:

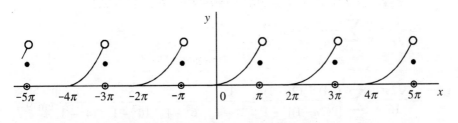

Fig. 9.3 Graph of the Fourier Series.

To prove the last part of the example, let us consider $x = \pi$. We know the Fourier series converges to $\dfrac{\pi^2}{2}$ at $x = \pi$; so we have

$$\frac{\pi^2}{2} = \frac{\pi^2}{6} + \sum_{n=1}^{\infty} \frac{2(-1)^n}{n^2} \cos n\pi \qquad \text{since } \sin n\pi = 0$$

$$\Rightarrow \sum_{n=1}^{\infty} \frac{2}{n^2} = \frac{\pi^2}{2} - \frac{\pi^2}{6} = \frac{2\pi^2}{6}$$

$$\Rightarrow \frac{\pi^2}{6} = \sum_{n=1}^{\infty} \frac{1}{n^2} = 1 + \frac{1}{2^2} + \frac{1}{3^2} + \cdots$$

Example 9.9 Find the Fourier sine series expansion of the function

$$f(x) = \cos x \qquad 0 < x < \pi$$

Use this series to prove that

$$\frac{\sqrt{2}\pi}{16} = \frac{1}{2^2 - 1} - \frac{3}{6^2 - 1} + \frac{5}{10^2 - 1} - \frac{7}{14^2 - 1} + \cdots$$

Sketch the graph of the Fourier sine series.

The Fourier sine series of the given function was found in Example 9.7 to be

$$\frac{8}{\pi} \sum_{n=1}^{\infty} \frac{n}{4n^2 - 1} \sin 2nx.$$

f is piecewise smooth in $(0, \pi)$. The Fourier sine series of f converges to $f(x)$ at a point of continuity $x \in (0, \pi)$. Therefore, we have

$$\cos \frac{\pi}{4} = \frac{8}{\pi} \sum_{n=1}^{\infty} \frac{n}{4n^2 - 1} \sin 2n \frac{\pi}{4}$$

$$\Rightarrow \frac{1}{\sqrt{2}} = \frac{8}{\pi} \sum_{n=1}^{\infty} \frac{n}{4n^2 - 1} \sin \frac{n\pi}{2}$$

$$\Rightarrow \frac{\sqrt{2}\pi}{16} = \sum_{n=1}^{\infty} \frac{(-1)^{n+1}(2n-1)}{[2(2n-1)]^2 - 1} = \frac{1}{2^2 - 1} - \frac{3}{6^2 - 1} + \frac{5}{10^2 - 1} - \frac{7}{14^2 - 1} + \cdots$$

To sketch the graph of the Fourier sine series of f we follow the following steps:

Step 1: Draw the graph of f in $0 < x < \pi$.

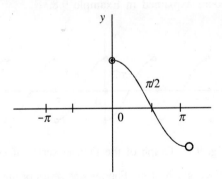

Fig. 9.4 Graph of the function f.

Step 2: Draw the graph of the function

$$O_f(x) = \begin{cases} \cos x & 0 < x < \pi \\ -\cos(-x) & -\pi < x < 0 \end{cases}$$

which is an extension of f to the interval $(-\pi, \pi)$. O_f is an odd function and the Fourier sine series of f and the Fourier series of O_f are identical (check!). The graph of O_f is given below.

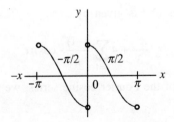

Fig. 9.5 Graph of the function O_f.

Step 3: Draw the graph of the periodic extension F of O_f following the procedure explained in Example 9.8.

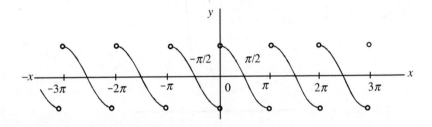

Fig. 9.6 Graph of the periodic extension F of O_f.

Step 4: Draw the graph of the Fourier series of O_f, using the graph of F and following the procedure explained in Example 9.8.

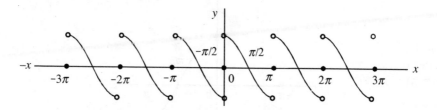

Fig. 9.7 Graph of the Fourier series of O_f.

Figure 9.7 Shows the graph of the Fourier sine series of the function $f(x) = \cos x$.

Example 9.10 Find the Fourier cosine series expansion of the function $f(x) = x^2$, where $0 < x < \pi$, and sketch the graph of the Fourier cosine series.

Here
$$a_0 = \frac{2}{\pi} \int_0^\pi x^2 \, dx = \frac{2\pi^2}{3}$$

$$a_n = \frac{2}{\pi} \int_0^\pi x^2 \cos nx \, dx = \frac{4(-1)^n}{n^2}$$

The Fourier cosine series of f is given by

$$\frac{\pi^2}{3} + 4 \sum_{n=1}^{\infty} \frac{(-1)^n}{n^2} \cos nx$$

To draw the graph of the Fourier cosine series, we follow the same steps as in Example 9.9.

Step 1: Draw the graph of f in $(0, \pi)$.

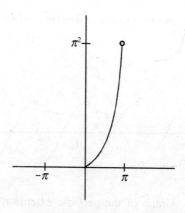

Fig. 9.8 Graph of the function f.

Step 2: Draw the graph of the function

$$E_f(x) = \begin{cases} x^2 & 0 < x < \pi \\ (-x)^2 & -\pi < x < 0 \end{cases}$$

which is an extension of f to $(-\pi, \pi)$. E_f is an even function, and the Fourier cosine series of f and the Fourier series of E_f are identical (check!). The graph of E_f in $(-\pi, \pi)$ is given below.

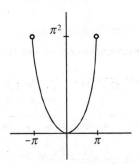

Fig. 9.9 Graph of the function E_f.

Step 3: Draw the graph of the periodic extension F of E_f following the procedure explained in Example 9.8.

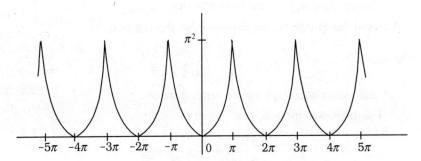

Fig. 9.10 Graph of the periodic extension F of E_f.

Step 4: Draw the graph of the Fourier series of E_f using the graph of F and following the procedure explained in Example 9.8.

Since F does not have any point of discontinuity $(-\infty, +\infty)$, the graph of F and the graph of the Fourier series of E_f, and hence the graph of the Fourier cosine series of f, are identical. Hence the graph of the Fourier cosine series of f is the shown in graph Fig. 9.10.

Now try the following exercises.

Exercises 9.5

1. Find the Fourier series of the function
$$f(x) = \begin{cases} \pi & -\pi \leq x < 0 \\ x & 0 \leq x \leq \pi \end{cases}$$
and hence show that
$$\frac{\pi^2}{8} = 1 + \frac{1}{3^2} + \frac{1}{5^2} + \frac{1}{7^2} + \cdots$$
Sketch the graph of the Fourier series.

2. Find the Fourier series expansion of the function
$$f(x) = \begin{cases} -1 & -\pi < x < 0 \\ 1 & 0 < x < \pi \end{cases}$$
and hence show that
$$\frac{\pi}{4} = 1 - \frac{1}{3} + \frac{1}{5} - \frac{1}{7} + \cdots$$
$$= \frac{1}{\sqrt{2}}\left(1 + \frac{1}{3} - \frac{1}{5} + \frac{1}{7} - \cdots\right)$$
Sketch the graph of the Fourier series.

3. Find the Fourier series expansion of the function
$$f(x) = \begin{cases} 0 & -\pi < x \leq 0 \\ x & 0 \leq x < \pi \end{cases}$$
and sketch the graph of the series obtained.
Use the series to show that
$$\frac{\pi^2}{8} = 1 + \frac{1}{3^2} + \frac{1}{5^2} + \frac{1}{7^2} + \cdots$$

4. Find the Fourier sine series expansion of the function $f(x) = x^2$ in $0 < x < \pi$, and sketch the graph of the series.

5. Sketch the graphs of the series obtained in Exercises 9.3 and 9.4.

6. Find the Fourier series of the function
$$f(x) = \begin{cases} 1 + x & -1 \leq x < 0 \\ 1 - x & 0 \leq x \leq 1 \end{cases}$$
Hence find the Fourier series of the function
$$g(x) = \begin{cases} 1 + x & -1 \leq x < 0 \\ 0 & x = 0 \\ 1 - x & 0 < x \leq 1 \end{cases}$$

7. If $\dfrac{a_0}{2} + \sum_{n=1}^{\infty}\left(a_n \cos\dfrac{n\pi x}{L} + b_n \sin\dfrac{n\pi x}{L}\right)$

and

$\dfrac{A_0}{2} + \sum_{n=1}^{\infty}\left(A_n \cos\dfrac{n\pi x}{L} + B_n \sin\dfrac{n\pi x}{2}\right)$

are the Fourier series of two piecewise continuous functions f and g defined in $-L < x < L$, then prove that the Fourier series of $\alpha f + \beta g$ is given by

$$\dfrac{\alpha a_0 + \beta A_0}{2} + \sum_{n=1}^{\infty}\left\{(\alpha a_1 + \beta A_n)\cos\dfrac{n\pi x}{L} + (\alpha b_n + \beta B_n)\sin\dfrac{n\pi x}{L}\right\}$$

where α and β are any two real numbers.

8. Show that the cosine series for the function

$$f(x) = \begin{cases} x & 0 \le x \le \dfrac{\pi}{2} \\ \pi - x & \dfrac{\pi}{2} < x \le \pi \end{cases}$$

is given by $\dfrac{\pi}{4} - \dfrac{2}{\pi}\sum_{n=1}^{\infty}\dfrac{\cos 2(2n-1)x}{(2n-1)^2}$.

Sketch the graph of the series on the interval $-5\pi \le x \le 5\pi$.

9. For the function $f(x) = \pi - x$, find

   (a) Its Fourier series on the interval $-\pi < x < \pi$
   (b) Its cosine series on the interval $0 \le x \le \pi$
   (c) Its sine series on the interval $0 < x \le \pi$.

   Sketch the graph of each of these series on the interval $-5\pi \le x \le 5\pi$.

10. Find the Fourier series of a piecewise continuous function f defined on the interval $a \le x \le b$, where a and b are any two real numbers.

CHAPTER 10

Partial Differential Equations

10.1 INTRODUCTION

So far we have confined our attention to the discussion of ordinary differential equations. But there exist physical situations for which governing differential equation may be partial differential equations. In this final chapter we shall give a brief introduction to partial differential equations. The subject of partial differential equation is a vast one, and several volumes of books could be devoted to discuss several aspects of the subject. In this short chapter we shall merely introduce certain basic concepts of partial differential equations and discuss a basic method of solution which is very useful in many applied and physical problems.

10.2 BASIC CONCEPTS

We first recall that a partial differential equation is a differential equation which involves partial derivatives of one or more dependent variables with respect to one or more independent variables. A solution of a partial differential equation is an explicit or implicit relation between the variables which does not contain derivatives and which also satisfies the equation.

For example, consider the first-order partial differential equation

$$\frac{\partial u}{\partial x} = x + x^2 y^2 \qquad (10.1)$$

where u is the dependent variable, and x and y are independent variables. We have already solved equations of this type in Chapter 2. The solution is

$$u = \int (x + x^2 y^2)\, dx + \Phi(y)$$

where the integration is with respect to x, treating y as constant, and Φ is an arbitrary function of y only.

Thus the solution of Eq. (10.1) is given by

$$u = \frac{x^2}{2} + \frac{x^3 y^2}{3} + \Phi(y) \qquad (10.2)$$

where Φ is an arbitrary function of y.

Partial Differential Equations

As a second example, consider the second order partial differential equation

$$\frac{\partial^2 u}{\partial x \partial y} = x^2 - y^2 \tag{10.3}$$

We rewrite this equation as

$$\frac{\partial}{\partial x}\left(\frac{\partial u}{\partial y}\right) = x^2 - y^2$$

and integrate with respect to x, treating y as constant, to yield

$$\frac{\partial u}{\partial y} = \frac{x^3}{3} - xy^2 + \Phi(y)$$

where Φ is an arbitrary function of y only.

Then, we integrate this result with respect to y, treating x as constant, to get the solution as

$$u = \frac{x^3 y}{3} - \frac{xy^3}{3} + f(y) + g(x) \tag{10.4}$$

where $f(y) = \int \Phi(y)\, dy$, and $g(x)$ is an arbitrary function of x alone.

We observe from these two simple examples that the solutions of ordinary differential equations involve arbitrary constants, whereas the solutions of partial differential equations involve arbitrary functions. Just like ordinary differential equations, the solution of a first order partial differential equation contains only one arbitrary function, the solution of a second order partial differential equation contains two arbitrary functions, and so on.

We shall discuss the method of obtaining solutions of second order linear partial differential equations with constant coefficients together with boundary conditions and initial conditions. This is the class of partial differential equations, which most frequently occurs in practice.

The general linear partial differential equation of second order in two independent variables x and y is an equation of the form

$$A\frac{\partial^2 u}{\partial x^2} + B\frac{\partial^2 u}{\partial x \partial y} + C\frac{\partial^2 u}{\partial y^2} + D\frac{\partial u}{\partial x} + E\frac{\partial u}{\partial y} + Fu = G \tag{10.5}$$

where A, B, C, D, E, F and G are functions of x and y. If $G(x, y) = 0$ for all (x, y), then Eq. (10.5) reduces to

$$A\frac{\partial^2 u}{\partial x^2} + B\frac{\partial^2 u}{\partial x \partial y} + C\frac{\partial^2 u}{\partial y^2} + D\frac{\partial u}{\partial x} + E\frac{\partial u}{\partial y} + Fu = 0 \tag{10.6}$$

Regarding the solutions of Eq. (10.6), we state the following theorem without proof, which is very basic and analogous to the corresponding theorem for ordinary linear differential equation.

Theorem 10.1 Let $u_1, u_2, u_3, \ldots, u_n, \ldots$ be an infinite set of solutions of Eq. (10.6) in a region R of the xy-plane. Let the infinite series

$$\sum_{n=1}^{\infty} u_n = u_1 + u_2 + u_3 + \cdots + u_n + \cdots$$

be convergent to u in R and be differentiated term by term in R to obtain various derivatives of u which occur in Eq. (10.6). Then the function u $\left(\text{defined by } u = \sum_{n=1}^{\infty} u_n\right)$ is also a solution of Eq. (10.6) in R.

We shall not attempt to justify the existence and uniqueness of the solutions of Eq. (10.6). Theorems which can always be applied to answer questions involving such issues do exist.

We now illustrate the method of *separation of variables* by applying it to obtain the solution of the vibrating string problem or one-dimensional wave equation, one-dimensional heat equation and two-dimensional Laplace equation.

10.3 ONE-DIMENSIONAL WAVE EQUATION

Consider a tightly stretched elastic string, the ends of which are fixed on the x-axis at $x = 0$ and $x = L > 0$. Suppose that for each x in the interval $0 < x < L$, the string is displaced into xy-plane, and that the displacement is given by $f(x)$, where f is a known function of x, which is differentiable.

Suppose at $t = 0$ the string is released from its initial position defined by $f(x)$, with an initial velocity given at each point of the interval $0 < x < L$ by $g(x)$, where $g(x)$ is a known differentiable function of x. The string vibrates, and its displacement in the y-direction at any point x and at any time t is a function of x and t. We want to find this displacement $y(x, t)$.

We assume that the string is perfectly flexible, is of constant linear density ρ and is of constant tension T at all times. We assume that the motion is confined to the xy-plane, each point on the string moves on a straight line perpendicular to the x-axis, the displacement y at each point of the string is small compared to the length L, and the angle between the string and the x-axis at each point is sufficiently small. Finally, we also assume that no external force (like damping forces) act on the string.

We know that all of these assumptions are not valid in any physical problem; nevertheless these assumptions are made in order to make the resulting mathematical problem more tractable, and in many physical problems they are approximately satisfied.

We have seen in Example 1.10 that the displacement $y(x, t)$ satisfied the partial differential equation

$$a^2 \frac{\partial^2 y}{\partial x^2} = \frac{\partial^2 y}{\partial t^2} \qquad (10.7)$$

where $a^2 = T/\rho$, under the above assumptions. This is the *one-dimensional wave equation*. As per the above assumptions, the boundary conditions are

$$y(0, t) = 0$$
$$y(L, t) = 0 \qquad (10.8)$$

At $t = 0$, the string is released from the initial position defined by $f(x)$, $0 \le x \le L$, with initial velocity given by $g(x)$, $0 \le x \le L$. Thus the displacement $y(x, t)$ must also satisfy the initial conditions

$$y(x, 0) = f(x)$$
$$\frac{\partial y(x, 0)}{\partial t} = g(x) \qquad (10.9)$$

Thus our problem is to find a function $y(x, t)$ which satisfies the partial differential equation (10.7), the boundary conditions (10.8) and the initial conditions (10.9). For this, we apply a method called *separation of variables*.

In this method, we assume that

$$y(x, t) = X(x)T(t) \qquad (10.10)$$

where $X(x)$ is a twice-differentiable function of x and $T(t)$ is a twice-differentiable function of t.

Putting this in the differential equation (10.7), we get

$$a^2 T \frac{d^2 X}{dx^2} = \frac{d^2 T}{dt^2} X$$

$$\Rightarrow \frac{d^2 X/dx^2}{X} = \frac{d^2 T/dt^2}{a^2 T} = -\lambda \qquad (10.11)$$

where λ is a constant, since left hand side of (10.11) is a function of x alone and middle term is a function of t alone. The relation (10.11) gives rise to two ordinary differential equations

$$X''(x) + \lambda X(x) = 0 \qquad (10.12)$$

and

$$T''(t) + \lambda a^2 T(t) = 0 \qquad (10.13)$$

The boundary conditions (10.8) reduce to

$$X(0) T(t) = 0 \Rightarrow X(0) = 0 \qquad (10.14)$$
$$X(L) T(t) = 0 \Rightarrow X(L) = 0$$

since $T(t) \ne 0$, otherwise the assumed solution (10.10) will be a trivial solution.

Thus the function X in the assumed solution (10.10) must satisfy both the ordinary differential equation (10.12) and the boundary conditions (10.14) i.e. the function X must be a nontrivial solution of the Sturm–Liouville problem

$$X''(x) + \lambda X(x) = 0$$
$$X(0) = 0 \qquad X(L) = 0$$

A special case of this problem was solved in Example 9.3 (also see

Introduction to Differential Equations

Exercise 8 of Section 9.2). So the eigen values of this Sturm–Liouville problem are given by

$$\lambda_n = \frac{n^2\pi^2}{L^2} \qquad n = 1,2,3,\ldots \tag{10.15}$$

and the eigen functions are given by

$$X_n(x) = c_n \sin\frac{n\pi x}{L} \qquad n = 1,2,3,\ldots \tag{10.16}$$

where c_n's are arbitrary constants.

Now for each integral value of n, the differential equation (10.13) takes the form

$$T''(t) + \frac{n^2\pi^2 a^2}{L^2}T(t) = 0$$

whose solution is given by

$$T_n(t) = a_n \sin\frac{n\pi a}{L}t + b_n \cos\frac{n\pi a}{L}t \tag{10.17}$$

where a_n's and b_n's are arbitrary constants.

Thus for each positive integral value of n, we have a solution

$$y_n(x,t) = X_n(x)T_n(t) = \left(c_n \sin\frac{n\pi x}{L}\right)\left(a_n \sin\frac{n\pi at}{L} + b_n \cos\frac{n\pi at}{L}\right)$$

for Eq. (10.7) which satisfies the boundary conditions (10.8).

Assuming appropriate convergence, we now apply Theorem 10.1 and conclude that

$$y(x,t) = \sum_{n=1}^{\infty} y_n(x,t) = \sum_{n=1}^{\infty} \sin\frac{n\pi x}{L}\left[a_n c_n \sin\frac{n\pi at}{L} + b_n c_n \cos\frac{n\pi at}{L}\right] \tag{10.18}$$

is a solution of Eq. (10.7) which satisfies the boundary conditions (10.8).

Applying the first initial condition of (10.9) to (10.18), we obtain

$$f(x) = y(x,0) = \sum_{n=1}^{\infty} b_n c_n \sin\frac{n\pi x}{L} \qquad 0 \le x \le L \tag{10.19}$$

This will help us to determine the arbitrary constants $(b_n c_n)$. We recognize this as a problem in the Fourier sine series (see Section 9.4). Using (9.25) of Chapter 9, we find that the coefficients are given by

$$b_n c_n = \frac{2}{L}\int_0^L f(x)\sin\frac{n\pi x}{L}dx \qquad n = 1, 2, 3, \ldots \tag{10.20}$$

Next, to apply the second initial condition of (10.9), we differentiate (10.18) partially with respect to t, which yields

$$\frac{\partial y(x,t)}{\partial t} = \sum_{n=1}^{\infty} \frac{n\pi a}{L} \sin\frac{n\pi x}{L}\left[a_n c_n \cos\frac{n\pi a t}{L} - b_n c_n \sin\frac{n\pi a t}{L}\right]$$

and this becomes, due to the second initial condition of (10.9),

$$g(x) = \frac{\partial y(x,0)}{\partial t} = \sum_{n=1}^{\infty} a_n c_n \frac{n\pi a}{L} \sin\frac{n\pi x}{L} \qquad 0 \le x \le L$$

This is another problem in Fourier sine series. Again using (9.25) of Chapter 9 we find that the coefficients are given by

$$\frac{a_n c_n n\pi a}{L} = \frac{2}{L}\int_0^L g(x) \sin\frac{n\pi x}{L} dx \qquad n = 1, 2, 3, \ldots$$

$$\Rightarrow a_n c_n = \frac{2}{n\pi a}\int_0^L g(x) \sin\frac{n\pi x}{L} dx \qquad n = 1, 2, 3, \ldots \qquad (10.21)$$

Hence the solution of Eq. (10.7) which satisfies the boundary conditions (10.8) and initial conditions (10.9) is given by (10.18), where the constants $(a_n c_n)$ and $(b_n c_n)$ are respectively given by (10.21) and (10.20). We may denote these constants as $A_n = a_n c_n$ and $B_n = b_n c_n$ for our future reference.

Summary: We briefly summarize the principal steps in the solution of this problem. First, assume the product solution XT as in (10.10). This leads to the ordinary differential equation (10.12) for the function X and the ordinary equation (10.13) for the function T. Then reduce the boundary conditions (10.8) to (10.14). The next step was to solve the Sturm–Liouville problem consisting of Eqs. (10.12) and (10.14). Then use the eigen values so obtained to solve Eq. (10.13). Thus for each n we obtained $y_n(x, t) = X_n(x)\, T_n(t)$. Then applying Theorem 10.1, a solution of (10.7) in terms of an infinite series $y(x, t) = \sum_{n=1}^{\infty} y_n(n, t)$ was found. The two initial conditions of (10.9) were applied separately to obtain the final solution given by (10.18).

Example 10.1 As a particular case of the vibrating string problem, we now consider a problem of the so-called *plucked string*. Let us suppose that the string is such that the constant $a^2 = 2500$, and that the ends of the string are fixed on the x-axis at $x = 0$ and $x = 1$. Suppose the midpoint of the string is displaced into the xy-plane a distance of 0.01 in the positive direction of the y-axis (see Fig. 10.1).

240 Introduction to Differential Equations

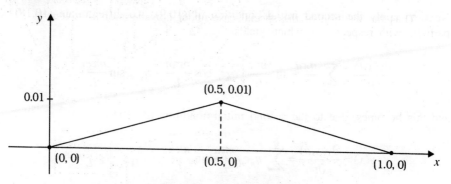

Fig. 10.1 Example 10.1.

Then the displacement from the x-axis on the interval $0 \leq x \leq 1$ is given by the function $f(x)$, where

$$f(x) = \begin{cases} \dfrac{x}{50} & 0 \leq x \leq \dfrac{1}{2} \\ -\dfrac{x}{50} + \dfrac{1}{50} & \dfrac{1}{2} \leq x \leq 1 \end{cases}$$

Suppose at $t = 0$ the string is released from the rest and from the initial position defined by $f(x)$ above. Our problem is to find the solution of Eq. (10.7) with $a = 50$, together with the boundary conditions

$$y(0, t) = 0 \qquad y(1, t) = 0$$

and the initial conditions $y(x, 0) = f(x)$, $y_t(x, 0) = 0$.

The coefficients A_n and B_n in the expression (10.18) are given, for this special case, due to (10.21) and (10.20) respectively by

$$A_n = 0 \qquad n = 1, 2, 3, \ldots$$

$$B_n = 2 \int_0^1 f(x) \sin n\pi x \, dx$$

$$= 2 \int_0^{\frac{1}{2}} \frac{x}{50} \sin n\pi x \, dx + 2 \int_{\frac{1}{2}}^1 \left(-\frac{x}{50} + \frac{1}{50}\right) \sin n\pi x \, dx$$

$$= \frac{2}{25 n^2 \pi^2} \sin \frac{n\pi}{2} \qquad n = 1, 2, 3, \ldots$$

$$= \begin{cases} 0 & \text{for } n = 2, 4, \ldots \\ \dfrac{2(-1)^{n+1}}{25\pi^2 (2n-1)^2} & \text{for } n = 1, 2, 3, \ldots \end{cases}$$

Thus the solution of the special case under consideration is given by

$$y(x,t) = \frac{2}{25\pi^2}\sum_{n=1}^{\infty}\frac{(-1)^{n+1}}{(2n-1)^2}\sin[(2n-1)\pi x]\cos[50(2n-1)\pi t]$$

In the next section, we shall apply the method of separation of variables to solve another important equation of mathematical physics, viz., the heat equation. Before that, try the following exercises.

Exercises 10.3

Use the method of separation of variables to find the solution $y(x, t)$ for each of the following problems.

1. $\dfrac{\partial^2 y}{\partial x^2} = \dfrac{\partial^2 y}{\partial t^2}$ $y(0,t) = 0$ $y(\pi,t) = 0$ $y(x,0) = \sin 2x$ $y_t(x,0) = 0$

2. $4\dfrac{\partial^2 y}{\partial x^2} = \dfrac{\partial^2 y}{\partial t^2}$ $y(0,t) = 0$ $y(3,t) = 0$ $y_t(x,0) = 0$

$$y(x,0) = \begin{cases} x & 0 \leq x \leq 1 \\ 1 & 1 \leq x \leq 2 \\ 3-x & 2 \leq x \leq 3 \end{cases}$$

3. $a^2 \dfrac{\partial^2 y}{\partial x^2} = \dfrac{\partial^2 y}{\partial t^2}$ $y(0,t) = 0$ $y(L, t) = 0$ $y(x,0) = \sin\dfrac{n\pi x}{L}$ $y_t(x,0) = 0$

4. $a^2 \dfrac{\partial^2 y}{\partial x^2} = \dfrac{\partial^2 y}{\partial t^2}$ $y(0,t) = 0$ $y(L, t) = 0$ $y(x,0) = \sum_{n=1}^{N}\alpha_n \sin\dfrac{n\pi x}{L}$

$$y_t(x,0) = \sum_{n=1}^{N}\beta_n \sin\dfrac{n\pi x}{L}$$

5. A string is stretched and fastened to two points apart by distance L. The motion is started by displacing the string into the form $f(x) = k(Lx - x^2)$, from which it is released at time $t = 0$ (from rest). Find the displacement $y(x, t)$ at time t of any point on the string, at a distance x from one end.

10.4 ONE-DIMENSIONAL HEAT EQUATION

Next, we study the flow of heat in thermally conducting bodies. In the interior of a body where heat is flowing from one region to another, the temperature generally varies from point to point at any one time, and from time to time at any one point. Thus, the temperature is a function of the space coordinates x, y and z, and the time t. Therefore, the form of this function depends on the shape of the body, the thermal characteristics of its material, the initial distribution of temperature, and the conditions maintained on the surface of the body. Under these conditions the temperature function must satisfy a three-dimensional partial differential equation.

But without deviating too much from the physical problem, we consider the flow of heat in a thin cylindrical rod of a cross-sectional area, whose lateral surface area is perfectly insulated, so that no heat flows through it. The rod is so thin that the temperature is assumed to be uniform on any cross-section, and is therefore a function of the time and only the position of the cross-section, say $y = y(x, t)$.

If ρ is the density of the rod, m is the mass of the rod, then the mass of the slice of the rod between the positions x and $x + \Delta x$ is given by $\Delta m = \rho A \, \Delta x$, where A is the cross-sectional area of the rod.

Since the quantity of heat gained or lost by a body when its temperature changes, i.e., the change in its thermal energy, is proportional to the mass of the body and to the change of temperature, the change in thermal energy is given by $c \Delta m \Delta y = c \rho A \, \Delta x \Delta y$, where Δy is the temperature change at the point x in a small time interval Δt, and c, the constant of proportionality, is called the *specific heat* of the substance.

So the rate at which heat is being stored is given by

$$c \rho A \, \Delta x \frac{\Delta y}{\Delta t} \qquad (10.22)$$

We assume that no heat is generated inside the slice by any other processes, so that the slice gains heat only by means of the flow of heat through its faces.

We know the rate at which heat flows across an area is proportional to the area, and to the rate of change of temperature with respect to the distance in a direction perpendicular to the area. Further, we know that heat flows from hot regions to cold regions, i.e., in the direction of decreasing temperature. So the rate at which heat flows into the slice through the left face is given by

$$-KA \frac{\partial y}{\partial x}\bigg|_{x}$$

Here K is the constant of proportionality and is called the *thermal conductivity* of the substance. The negative sign is chosen so that the quantity is positive if the change of temperature is negative. Similarly, the rate at which heat flows into the slice through the right face is given by

$$kA \frac{\partial y}{\partial x}\bigg|_{x+\Delta x}$$

so the total rate at which heat flows into the slice is given by

$$kA \frac{\partial y}{\partial x}\bigg|_{x+\Delta x} - kA \frac{\partial y}{\partial x}\bigg|_{x} \qquad (10.23)$$

Equating the expressions in (10.22) and (10.23), we get

$$kA\frac{\partial y}{\partial x}\bigg|_{x+\Delta x} - kA\frac{\partial y}{\partial x}\bigg|_{x} = c\rho A \Delta x \frac{\Delta y}{\Delta t}$$

or

$$\frac{k}{c\rho}\left(\frac{\frac{\partial y}{\partial x}\big|_{x+\Delta x} - \frac{\partial y}{\partial x}\big|_{x}}{\Delta x}\right) = \frac{\Delta y}{\Delta t}$$

Now making $\Delta x \to 0$, $\Delta t \to 0$, we obtain the desired equation

$$a^2 \frac{\partial^2 y}{\partial x^2} = \frac{\partial y}{\partial t} \tag{10.24}$$

where $a^2 = \dfrac{k}{c\rho}$, which is the one-dimensional heat equation.

Let us now try to obtain the solution $y(x, t)$ of Eq. (10.24) by using the method of separation of variables subject to the following conditions.

We assume that the thin rod is L units long and lies along the x-axis with its left hand end at the origin. The initial temperature distribution in the rod is given by a function $f(x)$, so that

$$y(x, 0) = f(x) \tag{10.25}$$

and the ends of the rod have the constant temperature zero for all values of $t \geq 0$; thus

$$y(0, t) = 0 \quad \text{and} \quad y(L, t) = 0 \tag{10.26}$$

We proceed exactly the way we proceeded in Section 10.3 and assume that

$$y(x, t) = X(x)\, T(t) \tag{10.27}$$

where $X(x)$ and $T(t)$ have the same meaning as we had in Section 10.3. Putting this in Eq. (10.24), we get

$$a^2 X''(x)\, T(t) = X(x) T'(t)$$

$$\Rightarrow \frac{X''(x)}{X(x)} = \frac{1}{a^2}\frac{T'(t)}{T(t)} = -\lambda \tag{10.28}$$

where λ is a constant, since the left hand side of (10.28) is a function of x alone and the middle term is a function of t alone. The relation (10.28) gives rise to two ordinary differential equations

$$X''(x) + \lambda X(x) = 0 \tag{10.29}$$

and

$$T'(t) + \lambda a^2 T(t) = 0 \tag{10.30}$$

Just as in Section 10.3, we solve (10.29) and satisfy the boundary conditions (10.26) by setting $\lambda = \dfrac{n^2\pi^2}{L^2}$ for any positive integer n, and the corresponding eigenfunction is

$$X_n(x) = \sin\frac{n\pi x}{L}$$

Then Eq. (10.30) becomes, with this value of λ,

$$T'(t) + \frac{n^2 a^2 \pi^2}{L^2} T(t) = 0$$

whose solution is given by

$$T_n(t) = e^{\frac{-n^2 a^2 \pi^2 t}{L^2}}$$

Thus for each positive integral value of n, we have a solution

$$y_n(x,t) = X_n(x)T_n(t) = e^{\frac{-n^2 a^2 \pi^2 t}{L^2}} \sin\frac{n\pi x}{L}$$

for Eq. (10.24) which satisfies the boundary conditions (10.26).

Assuming appropriate convergence, we now apply Theorem 10.1 to conclude that

$$y(x,t) = \sum_{n=1}^{\infty} b_n y_n(x,t) = \sum_{n=1}^{\infty} b_n e^{\frac{-n^2 a^2 \pi^2 t}{L^2}} \sin\frac{n\pi x}{L} \qquad (10.31)$$

is a solution of Eq. (10.24) which satisfies the boundary conditions (10.26). This will be a solution of Eq. (10.24) provided it satisfies the initial condition (10.25), i.e., at $t = 0$ we should have

$$f(x) = y(x,0) = \sum_{n=1}^{\infty} b_n \sin\frac{n\pi x}{L} \qquad 0 \le x \le L \qquad (10.32)$$

This will help us to determine the arbitrary constants b_n. We recognize this as a problem in Fourier sine series (see Section 9.4). Using (9.25) of Chapter 9, we find that the coefficients are given by

$$b_n = \frac{2}{L}\int_0^L f(x) \sin\frac{n\pi x}{L}\, dx \qquad n = 1, 2, 3, \ldots \qquad (10.33)$$

Hence the solution of Eq. (10.24), which satisfies the boundary conditions (10.26) and initial condition (10.25), is given by (10.32), where the coefficients are given by (10.33).

Partial Differential Equations

We now discuss some examples to illustrate the above concepts of the heat flow equation.

Example 10.2 For $0 < x < 20$ and $t > 0$, solve

$$\frac{\partial y}{\partial t} - 9\frac{\partial^2 y}{\partial x^2} = 0$$

when $y(x, 0) = 10$, $y(0, t) = 0$, $y(20, t) = 0$.

Following the above procedure, we get

$$y(x,t) = \sum_{n=1}^{\infty} b_n e^{\frac{-9n^2\pi^2}{(20)^2}t} \sin\frac{n\pi x}{20} \qquad \text{see Eq. (10.31)}$$

where
$$b_n = \frac{2}{20}\int_0^{20} 10\sin\frac{n\pi x}{20} dx \qquad \text{using Eq. (10.33)}$$

$$= \frac{20}{n\pi}\left[(-1)^{n+1} + 1\right]$$

$$= \begin{cases} 0 & \text{if } n \text{ is even} \\ \frac{40}{(2m-1)\pi} & \text{if } n \text{ is odd } 2m-1, \; m=1, 2, 3, \ldots \end{cases}$$

Hence $y(x,t) = \sum_{m=1}^{\infty} \frac{40}{(2m-1)\pi} e^{-\left(\frac{3(2m-1)\pi}{20}\right)^2 t} \sin\frac{(2m-1)\pi x}{20}$ is the required solution.

Example 10.3 Find the steady state temperature of a thin rod of length L units if the temperatures at the ends $x = 0$ and $x = L$ are y_1 and y_2 respectively.

A solution $y(x, t) = y(x)$, independent of t, for the heat equation is called a steady state solution. Then $\frac{\partial y}{\partial t} = 0$ and the heat equation (10.24) reduces to

$$\frac{d^2 y}{dx^2} = 0.$$

The general solution therefore is $y(x) = Ax + B$. Using the boundary conditions, we obtain the desired solution

$$y(x) = y_1 + \frac{1}{L}(y_2 - y_1)x \tag{10.34}$$

It may be mentioned that the solutions of the type (10.34) are useful for solving differential equation (10.24), together with boundary conditions

$$y(0, t) = y_1 \quad \text{and} \quad y(L, t) = y_2 \tag{10.35}$$

246   *Introduction to Differential Equations*

where at least one of y_1 and y_2 is a nonzero constant, and the initial condition (10.25). The boundary conditions (10.35) are called nonhomogeneous boundary conditions.

Let us assume that the solution $y(x, t)$ of the problem (10.24), (10.25) and (10.35) can be expressed as

$$y(x, t) = u(x) + v(x, t) \tag{10.36}$$

where $u(x)$ is the steady state temperature distribution given by (10.34), and $v(x, t)$ is another (transient) temperature distribution to be determined.

Putting (10.36) in (10.24), we obtain

$$a^2 \frac{\partial^2 (u+v)}{\partial x^2} = \frac{\partial (u+v)}{\partial t}$$

$$\Rightarrow a^2 \frac{\partial^2 v}{\partial x^2} = \frac{\partial v}{\partial t} \text{ since } \frac{\partial^2 u}{\partial x^2} = 0 \quad \frac{\partial u}{\partial t} = 0 \tag{10.37}$$

Similarly, from (10.36), (10.35) and (10.34), we obtain

$$v(0, t) = y(0, t) - u(0) = y_1 - y_1 = 0 \tag{10.38}$$

$$v(L, t) = y(L, t) - u(L) = y_2 - y_2 = 0$$

Finally, from Eqs. (10.36) and (10.25), we obtain

$$v(x, 0) = y(x, 0) - u(x) = f(x) - u(x) = f(x) - y_1 - (y_2 - y_1)\frac{x}{L} \tag{10.39}$$

Thus the transient part of the solution to the original problem is found by solving the problem consisting of Eqs. (10.37), (10.38), and (10.39), which is precisely the problem consisting of (10.24), (10.25), and (10.26), which we have solved already, except that the initial temperature distribution is $\{f(x) - u(x)\}$ instead of $f(x)$. Hence the required solution is given by

$$y(x,t) = y_1 + (y_2 - y_1)\frac{x}{L} + \sum_{n=1}^{\infty} b_n e^{\frac{-n^2 \pi^2 a^2 t}{L^2}} \sin \frac{n\pi x}{L} \tag{10.40}$$

where

$$b_n = \frac{2}{L} \int_0^L \left\{ f(x) - y_1 - (y_2 - y_1)\frac{x}{L} \right\} \sin \frac{n\pi x}{L} \, dx \tag{10.41}$$

We illustrate this through the following example.

Example 10.4 The ends A and B of a rod 20 cm long have the temperatures at 40°C and 90°C respectively until steady state prevails. After the steady state has prevailed, the temperature of the ends A and B are changed to 45°C and 95°C respectively. Determine the temperature distribution in the rod at time t.

The initial temperature distribution in the rod is given by $y(x, 0) = Cx + D$, where C and D are constants.

Given that $y(0, 0) = 40 = D$ and
$$y(20, 0) = 20C + 40 \Rightarrow 90 = 20C + 40 \Rightarrow C = \frac{5}{2}$$

Thus
$$y(x, 0) = \frac{5x}{2} + 40 \qquad (10.42)$$

is the initial temperature distribution and is the value of $f(x)$ in Eq. (10.25).

The steady state temperature distribution $u(x)$ is given from Eq. (10.34) by
$$u(x) = 45 + (95 - 45)\frac{x}{20} = 45 + \frac{5}{2}x \qquad (10.43)$$

The transient temperature distribution $v(x, t)$ is the solution of the problem
$$a^2 \frac{\partial^2 v}{\partial x^2} = \frac{\partial v}{\partial t}$$
$$v(0, t) = 0 = v(20, t)$$
$$v(x, 0) = y(x, 0) - u(x) = \frac{5x}{2} + 40 - 45 - \frac{5x}{2} \qquad \text{from (10.42) and (10.43)}$$
$$= -5$$

Thus we obtain the required solution from Eq. (10.40) as
$$y(x, t) = 45 + \frac{5}{2}x + \sum_{n=1}^{\infty} b_n e^{\frac{-a^2 n^2 \pi^2 t}{20^2}} \sin\frac{n\pi x}{20}$$

where
$$b_n = \frac{2}{20}\int_0^{20}(-5)\sin\frac{n\pi x}{L}dx = \frac{1}{2}\left[\cos\frac{n\pi x}{20}\right]_0^{20}\frac{20}{n\pi}$$
$$= \frac{10}{n\pi}\left\{(-1)^n - 1\right\} = \begin{cases} 0 & \text{if } n \text{ is even} \\ \dfrac{-20}{n\pi} & \text{if } n \text{ is odd} \end{cases}$$

Hence the required solution is
$$y(x, t) = 45 + \frac{5x}{2} - \sum_{m=1}^{\infty} \frac{20}{(2m-1)\pi} e^{-\left(\frac{a\pi(2m-1)}{20}\right)^2 t} \sin\frac{(2m-1)\pi x}{20}$$

Example 10.5 Find the temperature $y(x, t)$ in a metal rod of length 25 cm, that is insulated on the ends as well as on the sides, and whose initial temperature distribution is $y(x, 0) = x$ for $0 < x < 25$.

Since the ends of the rod are insulated, there is no flow of heat through them. Further, we know the rate of flow of heat across a cross-section is proportional to the area and to the rate of change of temperature with respect to the distance in a direction perpendicular to the area. Therefore, the boundary conditions are

$$y_x(0, t) = 0 \qquad y_x(25, t) = 0 \qquad t > 0 \qquad (10.44)$$

Following the usual procedure, we shall arrive at Eqs. (10.29) and (10.30).

We solve Eq. (10.29) together with the boundary conditions (10.44) to obtain the eigen values $\lambda_0 = 0$ and $\lambda_n = \dfrac{n^2\pi^2}{25^2}$, with the corresponding eigen functions $X_0(x) = 1$ and $X_n(x) = \cos\dfrac{n\pi x}{25}$, where $n = 1, 2, 3, \ldots$ (see Exercise 12 of Exercises 9.2).

For these values of λ, the solutions $T_n(t)$ of Eq. (10.30) are given by

$$T_n(t) = e^{-\left(\frac{an\pi}{25}\right)^2 t} \qquad n = 0, 1, 2, 3, \ldots$$

Thus for each value of n, we have a solution

$$y_0(x, t) = 1$$

$$y_n(x, t) = e^{-\left(\frac{an\pi}{25}\right)^2 t} \cos\frac{n\pi x}{25} \qquad n = 1, 2, 3, \ldots$$

for Eq. (10.24) which satisfies the boundary conditions (10.44). Then applying Theorem 10.1, we obtain the solution

$$y(x, t) = \frac{b_0}{2} y_0(x, t) + \sum_{n=1}^{\infty} b_n y_n(x, t)$$

$$= \frac{b_0}{2} + \sum_{n=1}^{\infty} b_n e^{-\left(\frac{an\pi}{25}\right)^2 t} \cos\frac{n\pi x}{25} \qquad (10.45)$$

To determine the coefficients b_n, we use the initial condition, i.e. at $t = 0$ we have

$$x = y(x, 0) = \frac{b_0}{2} + \sum_{n=1}^{\infty} b_n \cos\frac{n\pi x}{25} \qquad 0 < x < 25$$

We recognize this as a problem in the Fourier cosine series (see Section 9.4). Using (9.27), we find that the coefficients are given by

$$b_0 = \frac{2}{25} \int_0^{25} x \, dx = 25$$

$$b_n = \frac{2}{25} \int_0^{25} x \cos\frac{n\pi x}{25} dx = 50 \frac{(\cos n\pi - 1)}{(n\pi)^2} = \begin{cases} 0 & \text{if } n \text{ is even} \\ -\dfrac{100}{(n\pi)^2} & \text{if } n \text{ is odd} \end{cases}$$

Hence the required solution is

$$y(x,t) = \frac{25}{2} - \frac{100}{\pi^2}\sum_{m=1}^{\infty}\frac{1}{(2m-1)^2}e^{-\left\{\frac{(2m-1)a\pi}{25}\right\}^2 t}\cos\frac{(2m-1)\pi x}{25}$$

You may now try to solve the following exercises.

Exercises 10.4

1. Solve the heat equation $y_{xx} = y_t$ for a rod, 40 cm in length, whose ends are maintained at 0°C for all $t > 0$, and whose initial temperature distribution is given by

$$y(x,0) = \begin{cases} x & 0 \le x < 20 \\ 40-x & 20 \le x \le 40 \end{cases}$$

2. Let a metallic rod 20 cm long be heated to a uniform temperature of 100°C. Suppose that at $t = 0$ the ends of the rod are plunged into an ice bath at 0°C and thereafter maintained at this temperature, but that no heat is allowed to escape through the lateral surface. Find an expression for the temperature at any point in the rod at any later time. Determine the temperature at the centre of the rod at time $t = 30$ s.

3. A rod of length 20 cm is initially at the uniform temperature of 25°C. Suppose that at time $t = 0$ the end $x = 0$ is cooled to 0°C, while the end $x = 20$ is heated to 60°C and both are thereafter maintained at those temperatures. Find the temperature distribution in the rod at any time t.

4. In the following problems, find the steady-state solution of the heat equation (10.24) that satisfies the given set of boundary conditions
   (i) $y_x(0, t) = 0$ $y(L, t) = 0$
   (ii) $y_x(0, t) = 0$ $y(L, t) = T$

5. Consider a rod of length 40 cm whose initial temperature is given by $y(x, 0) = x(60 - x)/30$. Suppose that $a^2 = \frac{1}{4}$ cm²/s and that both ends of the rod are insulated. Find the temperature distribution in the rod at any time t.

10.5 LAPLACE EQUATION

The one-dimensional heat equation (10.24) discussed in the previous section takes the following form in a two-dimensional space.

$$a^2\left(\frac{\partial^2 \omega}{\partial x^2} + \frac{\partial^2 \omega}{\partial y^2}\right) = \frac{\partial \omega}{\partial t} \qquad (10.46)$$

If a steady state exists, then the temperature function ω is a function of x and y and independent of t. So Eq. (10.46) reduces to

$$\frac{\partial^2 \omega}{\partial x^2} + \frac{\partial^2 \omega}{\partial y^2} = 0 \qquad (10.47)$$

This is called the *two-dimensional Laplace equation,* and it occurs in the study of electrostatics, gravitation and theory of elasticity. We shall solve this equation subject to some boundary conditions by applying the method of separation of variables.

We assume that w is the temperature distribution function at any point (x, y) in a rectangular plate of length L and width M units, i.e., $0 \leq x \leq L$, $0 \leq y \leq M$. Further, we assume that no heat is lost or gained through the boundaries of the plate, the three sides of the plate are maintained at $0°C$, and the temperature distribution on the fourth side is given by a function f. Thus the four boundary conditions are

$$w(0, y) = 0 \qquad (10.48)$$

$$w(L, y) = 0 \qquad (10.49)$$

$$w(x, 0) = 0 \qquad (10.50)$$

$$w(x, M) = f(x) \qquad (10.51)$$

Let us assume that the solution of Eq (10.47) has the form

$$w(x, y) = X(x)\, Y(y) \qquad (10.52)$$

where X is a twice-differentiable function of x only and y is a twice-differentiable function of y only.

Putting (10.52) in (10.47), we obtain

$$X''(x)\, Y(y) + X(x) Y''(y) = 0$$

$$\Rightarrow \frac{X''(x)}{X(x)} = -\frac{Y''(y)}{Y(y)} = -\lambda \quad \text{(say)} \qquad (10.53)$$

Here λ is a constant, being equal to a function of x as well as y.

Equation (10.53) gives rise to two ordinary differential equations

$$X''(x) + \lambda X(x) = 0 \qquad (10.54)$$

$$Y''(y) - \lambda Y(y) = 0 \qquad (10.55)$$

The two boundary conditions (10.48) and (10.49) reduce (10.52) to

$$0 = w(0, y) = X(0)Y(y) \Rightarrow X(0) = 0$$
$$0 = w(L, y) = X(L)Y(y) \Rightarrow X(L) = 0 \qquad (10.56)$$

Equations (10.54) and (10.56) essentially constitute the Sturm–Liouville problem we discussed in Example 9.3. Therefore, λ in (10.54) is given by

$$\lambda_n = \frac{n^2 \pi^2}{L^2} \quad n = 1, 2, 3, \ldots \qquad (10.57)$$

Partial Differential Equations

and the eigen functions are given by

$$X_n(x) = \sin\frac{n\pi x}{L} \qquad n = 1, 2, 3, \ldots \qquad (10.58)$$

Now, for each integral value of n, the differential equation (10.55) takes the form

$$Y''(y) - \frac{n^2\pi^2}{L^2}Y(y) = 0$$

It is convenient to write its general solution for each integral value of n as

$$Y_n(y) = A_n \cosh\frac{n\pi y}{L} + B_n \sinh\frac{n\pi y}{L} \qquad (10.59)$$

where A_n and B_n are arbitrary constants.
Due to the boundary condition (10.50), Eq. (10.52) yields

$$0 = w(x, 0) = X(x)Y(0) \Rightarrow Y(0) = 0$$

Putting this condition in (10.59), we obtain $A_n = 0$, and therefore (10.59) becomes

$$Y_n(y) = B_n \sinh\frac{n\pi y}{L} \qquad (10.60)$$

Thus for each integral value of n, we have a solution

$$w_n(x, y) = X_n(x)Y_n(y) = B_n \sin\frac{n\pi x}{L} \sinh\frac{n\pi y}{L} \qquad \text{from (10.58) and (10.60)}$$

which satisfy Eq. (10.47) and the boundary conditions (10.48)–(10.50).

Assuming appropriate convergence, we now apply Theorem 10.1 to conclude that

$$w(x, y) = \sum_{n=1}^{\infty} w_n(x, y) = \sum_{n=1}^{\infty} B_n \sin\frac{n\pi x}{L} \sinh\frac{n\pi y}{L} \qquad (10.61)$$

is a solution of Eq. (10.47) which satisfies the boundary conditions (10.48)–(10.50).

We now apply the nonhomogeneous boundary condition (10.51) to the series solution (10.61) to obtain

$$f(x) = w(x, M) = \sum_{n=1}^{\infty} B_n \sin\frac{n\pi x}{L} \sinh\frac{n\pi M}{L} \qquad 0 \le x \le L$$

$$\Rightarrow f(x) = \sum_{n=1}^{\infty} C_n \sin\frac{n\pi x}{L} \qquad \text{letting } C_n = B_n \sinh\frac{n\pi M}{L} \qquad (10.62)$$

To determine coefficients C_n, we recognize that (10.62) is a Fourier sine series; using (9.25) of Chapter 9, we find that the coefficients C_n are given by

$$C_n = \frac{2}{L}\int_0^L f(x) \sin\frac{n\pi x}{L} dx \qquad n = 1, 2, 3, \ldots$$

$$\Rightarrow B_n \sinh \frac{n\pi M}{L} = \frac{2}{L} \int_0^L f(x) \sin \frac{n\pi x}{L} dx$$

$$\Rightarrow B_n = \frac{2}{L \sinh \frac{n\pi M}{L}} \int_0^L f(x) \sin \frac{n\pi x}{L} dx \qquad n = 1, 2, 3, \ldots \qquad (10.63)$$

Therefore, the solution of Eq. (10.47) satisfying the boundary conditions (10.48) –(10.51) is given by (10.61), where the coefficients B_n are given by (10.63).

Example 10.6 Find the temperature distribution $w(x, y)$ at any point of a rectangular plate of length 20 cm and width 15 cm, subject to the boundary conditions $w(0, y) = w(20, y) = w(x, 15) = 0$ and $w(x, 0) = \sin \frac{n\pi x}{20}$.

Solution: Proceeding exactly as above, we obtain the values of λ_n from (10.57), taking $L = 20$, as

$$\lambda_n = \frac{n^2 \pi^2}{20^2} \qquad n = 1, 2, 3, \ldots$$

and the corresponding eigen functions as

$$X_n(x) = \sin \frac{n\pi x}{20} \qquad n = 1, 2, 3, \ldots$$

With this, Eq. (10.59) becomes

$$Y_n(y) = A_n \cosh \frac{n\pi y}{20} + B_n \sinh \frac{n\pi y}{20} \qquad (10.64)$$

where A_n and B_n are arbitrary constants.

Due to the boundary condition $w(x, 15) = 0$, Eq. (10.52) yields
$$0 = w(x, 15) = X(x) Y(15) \Rightarrow Y(15) = 0$$
Putting this condition in (10.64), we have

$$0 = Y_n(15) = A_n \cosh \frac{n\pi 15}{20} + B_n \sinh \frac{n\pi 15}{20}$$

$$\Rightarrow A_n = \sinh \frac{n\pi 15}{20} \qquad B_n = -\cosh \frac{n\pi 15}{20}$$

Putting these values in (10.64), we obtain

$$Y_n(y) = \sinh \frac{3n\pi}{4} \cosh \frac{n\pi y}{20} - \cosh \frac{3n\pi}{4} \sinh \frac{n\pi y}{20} = \sinh \frac{n\pi(15-y)}{20}$$

So
$$w_n(x, y) = X_n(x) Y_n(y) = \sin \frac{n\pi x}{20} \sinh \frac{n\pi(15-y)}{20}$$

and Eq. (10.61) becomes

$$w(x, y) = \sum_{n=1}^{\infty} c_n \sin \frac{n\pi x}{20} \sinh \frac{n\pi(15-y)}{20} \qquad (10.65)$$

which is a solution of Eq. (10.47) satisfying all the three given homogeneous boundary conditions.

We now apply the given nonhomogeneous boundary condition

$$w(x, 0) = \sin\frac{n\pi x}{20}$$

to the series solution (10.65) to obtain

$$\sin\frac{n\pi x}{20} = w(x, 0) = \sum_{n=1}^{\infty} c_n \sin\frac{n\pi x}{20} \sinh\frac{n\pi 15}{20} \qquad 0 \le x \le 20$$

$$\Rightarrow \sin\frac{n\pi x}{20} = \sum_{n=1}^{\infty} \left(c_n \sinh\frac{15 n\pi}{20} \right) \sin\frac{n\pi x}{20}$$

This is a Fourier sine series. Using (9.25) of Chapter 9, we find that

$$c_n \sinh\frac{15 n\pi}{20} = \frac{2}{20} \int_0^{20} \sin\frac{n\pi x}{20} \sin\frac{n\pi x}{20} dx \qquad n = 1, 2, 3, \ldots$$

$$= \frac{1}{20} \int_0^{20} \left(1 - \cos\frac{2 n\pi x}{20} \right) dx = 1$$

$$\Rightarrow c_n = \frac{1}{\sinh\frac{15 n\pi}{20}} \qquad n = 1, 2, 3, \ldots$$

Therefore the required solution is given by (putting the above value of c_n in (10.65))

$$w(x, y) = \sum_{n=1}^{\infty} \frac{1}{\sinh\frac{15 n\pi}{20}} \sin\frac{n\pi x}{20} \sinh\frac{n\pi(15 - y)}{20}$$

You may now try the following exercises.

Exercises 10.5

1. Solve the Laplace equation (10.47) with the boundary conditions
   $w(0, y) = 0 \qquad w(L, y) = g(y) \qquad 0 \le y \le M$
   $w(x, 0) = 0 \qquad w(x, M) = 0 \qquad 0 \le x \le L$

2. Solve the Laplace equation (10.47) with the boundary conditions
   $w(0, y) = k(y) \qquad w(L, y) = 0 \qquad 0 \le y \le M$
   $w(x, 0) = 0 \qquad w(x, M) = 0 \qquad 0 \le x \le L$

3. Solve the Laplace equation (10.47) with the boundary conditions

$$w(0, y) = k(y) \qquad w(L, y) = g(y) \qquad 0 \le y \le M$$
$$w(x, 0) = h(x) \qquad w(x, M) = f(x) \qquad 0 \le x \le L$$

(**Hint**: Consider the possibility of adding the solutions of four problems:

(i) With homogeneous boundary conditions except for $w(0, y) = k(y)$
(ii) With homogeneous boundary conditions except for $w(L, y) = g(y)$
(iii) With homogeneous boundary conditions except for $w(x, 0) = h(x)$
(iv) With homogeneous boundary conditions except for $w(x, M) = f(x)$.)

4. Solve the Laplace equation (10.47) with the boundary conditions

$$w_x(0, y) = 0 \qquad w_x(40, y) = 0 \qquad 0 < y < 20$$
$$w(x, 0) = 0 \qquad w(x, 20) = 1 + x^2(x - 40)^2 \qquad 0 \le x \le 40$$

5. Use the method of separation of variables to obtain the solution $w(r, \theta)$ of the Laplace equation (10.47) in a circular plate of radius a subject to the boundary condition $w(a, \theta) = f(\theta)$, where f in a given function on $0 \le \theta \le 2\pi$. The Laplace equation in polar form is given by

$$\omega_{rr} + \frac{1}{r}\omega_r + \frac{1}{r^2}\omega_{\theta\theta} = 0$$

6. A plate with concentric circular boundaries with radii $r = a$ and $r = b$, with $0 < a < b$, has its inner boundary held at temperature t_1 and its outer one at temperature t_2. Find the temperature distribution throughout the annular region of the plate.

Suggested Readings

Bell, W.W., *Special Functions for Scientists and Engineers*, Van Nostrand, 1968.

Boyce, William E. and Richard C. Diprima, *Elementary Differential Equations and Boundary Value Problems*, 7th edition, John Wiley & Sons, 2001.

Bronson, Richard, *Differential Equations*, Schaum's Outline Series, 2nd edition, McGraw-Hill International edition, 1994.

Jackson, Dunham, *Fourier Series and Orthogonal Polynomials*, Mathematical Association of America, 1963.

Kreider, D.L., R.G. Kuller, D.R. Ostberg, and F.W. Perkins, *An Introduction to Linear Analysis*, Addison-Wesley, 1966.

Murray, Daniel A., *Introductory Course to Differential Equations*, Longmans, Green & Company, 1953.

Rainville, Earl D. and Phillip E. Bedient, *Elementary Differential Equations*, 7th edition, Macmillan Publishing Company, 1989.

Rao, S. Balachandra and H.R. Anuradha, *Differential Equations with Applications and Programs*, Universities Press, 1996.

Ross, Shepley L., *Differential Equations*, 3rd edition, John Wiley & Sons, 1984.

Simmons, George F., *Differential Equations with Applications and Historical Notes*, 2nd edition, McGraw-Hill Inc., 1991.

Answers to Exercises

Exercises 1.5

1. Ordinary, nonlinear, nonhomogeneous, 1st order, fourth degree.
2. Ordinary, nonlinear, nonhomogeneous, 3rd order, first degree.
3. Partial, linear, nonhomogeneous, 2nd order, first degree.
4. Ordinary, linear, homogeneous, 2nd order, first degree.
5. Partial, linear, homogeneous, 2nd order, first degree.
6. Ordinary, nonlinear, nonhomogeneous, first order, 7th degree.
7. Partial, nonlinear, nonhomogeneous, 1st order, 2nd degree.
8. Ordinary, nonlinear, nonhomogeneous, 2nd order, first degree.
9. Partial, nonlinear, homogeneous, 1st order, 2nd degree.
10. Ordinary, linear, nonhomogeneous, 2nd order, first degree.
11. If $R(t)$ is the amount of radium present at any time t, then

    $$\frac{dR}{dt} = kR \quad k \text{ is the proportionality constant.}$$

12. $\frac{dT}{dt} = k(T - 24)$, where T is the temperature of the rod at any time t, and k is the constant of proportionality.

13. $m\dfrac{d^2x}{dt^2} = k_1 x - k_2 \dfrac{dx}{dt}$

14. $y^2 \left(\dfrac{dy}{dx}\right)^2 + y^2 = r^2$

15. $2x(y')^2 - 2y\,y' + 1 = 0$

16. $\dfrac{dq}{dt} = k(100 - q)$

17. $m\dfrac{d^2x}{dt^2} = k\dfrac{dx}{dt}$

18. $\dfrac{d^2x}{dt^2} + 16x = 2\sin 4t \qquad x(0) = -\dfrac{1}{2} ft \qquad x'(0) = 0$

19. $\dfrac{d^2q}{dt^2} + 9\dfrac{dq}{dt} + 14q = \dfrac{1}{2}\sin t$

20. $\dfrac{d^2x}{dt^2} + 16x = 0 \qquad x(0) = -\dfrac{1}{2} ft \qquad x'(0) = 0$

21. (a), (c) and (e) are boundary conditions.
    (b) and (d) are initial conditions.

22. $y = \tan(x + k)$
23. $y = \dfrac{k}{t}$
24. $y^2 = ke^{-2x} - 1$
25. $\sin y = k - \ln|\cos x|$
26. $y = \dfrac{-4}{(4x - 1)}$
27. $xy = k$
28. $x = ky$
29. $y = \dfrac{1}{1 + x^2}$
30. $y = 2\left(e^{3x} - 1\right)$

Exercises 2.2

1. Yes
2. No
3. Yes
4. $xy = ce^{y-x}$
5. $x - y = c(1 + xy)$
6. $\tan x \tan y = c$
7. $y \sin y = x^2 \ln x + c$
8. $e^y = e^x + \dfrac{1}{3}x^3 + c$
9. $r = a\sin(\theta + c)$ where $x = r\cos\theta,\ y = r\sin\theta$
10. $y = \dfrac{1}{(C - e^{\cos x})}$
11. $x + y = \tan(x + c)$
12. $\tan(x - y + 1) = x + c$
13. $(b - a)\ln\{(x + y)^2 - ab\} = 2(x - y) + c$
14. $2y + c = a\ln\dfrac{x - y - a}{x - y + a}$
15. $ax + by = z$

Exercises 2.3

1. $\dfrac{2xy}{(x+y)^2} = \ln \dfrac{c}{x+y}$

2. $y = ce^{x^3/3y^3}$

3. $cy = e^{y/x}$

4. $cx = e^{x/y}$

5. $cy = e^{x^2/2y^2}$

6. $cy = e^{-x^2/2y}$

7. $y = x \sinh(x+c)$

8. $x^2 - y^2 = c(x^2 + y^2)^2$

9. $\tan^{-1}\left\{\dfrac{(2y+1)}{(2x+1)}\right\} = \ln\left\{c\sqrt{x^2 + y^2 + x + y + \dfrac{1}{2}}\right\}$

10. $x + y - 2 = c(y - x)^3$

11. $\tan^{-1}\left\{\dfrac{(y-2)}{(x-3)}\right\} + \ln\left[c\sqrt{(y-2)^2 + (x-3)^2}\right] = 0$

12. $(2x - y)^2 = c(x + 2y - 5)$

13. $3(x^2 + y^2) + 4xy - 10(x+y) = c$

14. $\ln(x-1) + c = -\dfrac{1}{2}\ln\left\{2\left(\dfrac{y+1}{x-1}\right)^2 - 1\right\} + \dfrac{\sqrt{2}}{4}\ln\dfrac{\dfrac{y+1}{x+1} - \dfrac{1}{\sqrt{2}}}{\dfrac{y+1}{x+1} + \dfrac{1}{\sqrt{2}}}$

15. $x^2 + xy - y^2 = c$

Exercises 2.4

1. $xy(ax + by) = c$

2. $x^2 + y^2 + 2a^2 \tan^{-1}\left(\dfrac{x}{y}\right) = c$

3. $x^3 - 3axy + y^3 = c$

4. $(e^y + 1)\sin x = c$

5. $2(x+y) + \sin 2x + \sin 2y - 4 \sin \alpha \sin x \sin y = c$

6. $y(x + \ln x) + x \cos y = c$

7. $x + ye^{x/y} = c$

8. $\dfrac{1}{2}e^{2x} + \ln\sec y - \cos x \cos y = c$

9. $x^3 - y^3 = 3xy(x+y) + c$

10. $x^a y^b = c$

11. $e^x + \dfrac{x^2}{y} = c$

12. $x^2 - y - 1 - x \cos y = cx$

13. $\dfrac{ax^2}{2} + \dfrac{e^x}{y} = c$

14. $x^2 e^x + my^2 = cx^2$

15. $x^2 + y^2 = cy$

16. $\ln \dfrac{x^2}{y} - \dfrac{1}{xy} = c$

17. $xy - \dfrac{1}{xy} + \ln \dfrac{x}{y} = c$

18. $x = cy \cos xy$

19. $xy - \dfrac{1}{xy} - \ln y^2 = c$

20. $y = ce^{-(3xy+1)/(3x^3 y^3)}$

21. $3x^4 + 4x^3 + 6x^2 y^2 = c$

22. $(2x^2 y^2 + 4xy + y^4) e^{x^2} = c$

23. $x^4 y(3 + y^2) + x^6 = c$

24. $\left(4x^5 + 2x^4 y + \dfrac{4}{3}x^3 y^2 + x^3 y^3\right) y = c$

25. $x^3 y^3 + x^2 = cy$

26. $xy^3 + y^4 + 2x = cy^2$

27. $e^{6y}\left(\dfrac{x^2 y^2}{2} - \dfrac{x^3}{3} + \dfrac{y^2}{6} - \dfrac{y}{18} + \dfrac{1}{108}\right) = c$

28. $2y = ce^{x^2} + 1$

29. $xy = c$

30. $x = ce^{y^2}$

31. $x^2 e^y + \dfrac{x^2}{y} + \dfrac{x}{y^3} = c$

32. $x^9 y^6 e^{y^3} = (c/3)^9$

33. $5x^{-36/13} y^{24/13} - 12 x^{-10/13} y^{-15/13} = c$

34. $x^2 y^3 + 2x^3 y^4 = c$

35. $4\sqrt{xy} - \dfrac{2}{3}(\sqrt{x})^{-3}(\sqrt{y})^3 = c$

Exercises 2.5

1. $x = \tan^{-1} y - 1 + ce^{-\tan^{-1} y}$

2. $xy \sec x = \tan x + c$

3. $y = \tan x + c\sqrt{\tan x}$

4. $y\dfrac{x^2 + 1}{x} = \dfrac{x^2}{2} \ln x - \dfrac{x^2}{4} + c$

5. $y\dfrac{\sqrt{(a^2 + x^2)} + x}{a} = a \sinh^{-1}\left(\dfrac{x}{a}\right) + c$

6. $y = t + ce^{-t}$, where $t = \dfrac{x}{\sqrt{1 - x^2}}$

7. $\left(y + \dfrac{1}{3}\right) \tan \dfrac{3x}{2} = 2 \tan \dfrac{x}{2} - x + c$ 　　8. $y \sec x = \tan x + c$

9. $\dfrac{xy}{x-1} = \dfrac{x^3}{3} + c$ 　　10. $xy = c - \tan^{-1} x$

11. $x^2 y = c + (2 - x^2) \cos x + 2x \sin x$ 　　12. $y = ce^{\sin x} - (1 + \sin x)$

13. $y = \cos x + c \sec x$ 　　14. $(1 + x^2)y = c + \sin x$

15. $y = cx + x \ln \tan x$ 　　16. $x = y - a^2 + ce^{-y/a^2}$

17. $(x - 2y^3)y^2 = c$ 　　18. $3(1 + x^2)y = 4x^3$

19. $y = cx\sqrt{x^2 - 1} + ax$ 　　20. $4xy = x^4 + 3$

21. $y = (c + x^2)e^{x^2}$ 　　22. $y = (c + x^3) \ln x$
23. $xy = c + y^2 \ln y$ 　　24. $y = (c + x^2)e^{e^x}$

25. $y = (c + x)e^{(1-x)e^x}$

Exercises 2.6

1. $xy \ln\left(\dfrac{c}{x}\right) = 1$ 　　2. $x = y\left(1 + c\sqrt{x}\right)$

3. $cy = (1 - y)\sqrt{1 - x^2}$ 　　4. $(1 + \ln x + cx)y = 1$

5. $1 = cxy - \int\left\{\dfrac{(\sin x)}{x}\right\} dx$ 　　6. $\sqrt{(1 + x^2)} = (c + \sinh x^{-1})y$

7. $1 = y^2 \left\{(c + x) \cot\left(\dfrac{1}{2}x + \dfrac{1}{4}\pi\right) - 1\right\}$ 　　8. $x^3 = y^3(3 \sin x + c)$

9. $e^x = y(c - x^2)$ 　　10. $1 = y(c \cos x + \sin x)$

11. $y = c_1 e^{-x} + c_2$ 　　12. $y = c_1 \int e^{-x^2/2} dx + c_2$

13. $y = c_1 \sinh(x + c_2)$ 　　14. $y = c_1 \sin(ax + c_2)$

15. $y = c_2 e^{c_1 x}$ 　　16. $y^2 = c_2 e^{2x} + c_1$

17. $c_1 \tan^{-1} c_1 x = y + c_2$ 　　18. $2y = (\ln x)^2 + c_1 \ln x + c_2$

19. $y = c_1 \ln\left\{x + \sqrt{(1 + x^2)}\right\} + c_2$ 　　20. $2y = 2x^3 + c_1 x^2 + c_2$

21. $1 = y(cx + \ln x + 1)$ 　　22. $\tan y = x^2 - 2 + ce^{-x^2/2}$

23. $2x = e^y(2cx^2 + 1)$

24. $\dfrac{1}{x \ln y} = \dfrac{1}{2x^2} - c$

25. $\sec y = x + 1 + ce^x$

Exercises 3.3

1. $y = c_1 e^{-x} + c_2 e^{2x}$
2. $y = c_1 e^{2x} + c_2 e^{-3x}$
3. $y = c_1 e^{x/2} + c_2^{2x}$
4. $y = c_1 e^x + c_2 e^{2x} + c_3 e^{3x}$
5. $y = c_1 e^{2x} + c_2 e^{-2x} + c_3 e^{-3x} + c_4 e^{3x}$
6. $y = e^{3x}(c_1 \cos 4x + c_2 \sin 4x)$
7. $y = c_1 \cos 3x + c_2 \sin 3x$
8. $y = e^{-4x}(c_1 \cos 3x + c_2 \sin 3x)$
9. $y = e^{-2x}(c_1 \cos 2x + c_2 \sin 2x)$
10. $y = e^{(3/2)x}\left(c_1 \cos \dfrac{\sqrt{7}}{2}x + c_2 \sin \dfrac{\sqrt{7}}{2}x\right)$
11. $y = c_1 + c_2 \cos x + c_3 \sin x$
12. $y = c_1 + c_2 \cos 2x + c_3 \sin 2x$
13. $y = c_1 \cos 2x + c_2 \sin 2x + c_3 e^{2x} + c_4 e^{-2x}$
14. $y = (c_1 x + c_2)e^{2x}$
15. $y = (c_1 x + c_2)e^{-15x}$
16. $y = c_1 e^{2x} + (c_2 x + c_3)e^{-3x}$
17. $y = c_1 e^{2x} + c_2 e^{-2x} + (c_3 x + c_4)e^{-3x}$
18. $y = c_1 + c_2 x + c_3 \cos 2x + c_4 \sin 2x$
19. $y = c_1 e^{2x/3} + c_2 e^{-3x/8}$
20. $y = e^3 + 3e^{-1}$

Exercises 3.4

1. $y = c_1 x + c_2 x^{-2}$
2. $y = c_1 x + c_2 x e^x$
3. $y = c_1 x + c_2 e^x$
4. $y = c_1 x + c_2 \left\{\dfrac{x}{2} \ln\left(\dfrac{x+1}{1-x}\right) - 1\right\}$
5. $y = c_1 x^2 + c_2 x^{-2}$
6. $y = c_1 x + c_2 x^2$
7. $y = c_1 x + c_2 x^{-1}$
8. $y = c_1 e^x + c_2(x + 1)$
9. $y = c_1 + c_2 x^{-2}$
10. $y = c_1 x + c_2 x \int x^{-2} e^{\int x f(x) dx} dx$
11. $y = c_1 e^x + c_2 e^x \int e^{-2x + \int f(x) dx} dx$
12. $y = c_1(3x + 2)^2 + c_2(3x + 2)^{-2}$
13. $y = c_1(x + a)^2 + c_2(x + a)^3$
14. $y = c_1 \cos \ln(1 + x) + c_2 \sin \ln(1 + x)$
15. $y = c_1 x^3 + c_2 x^{-2}$

Exercises 3.5

1. $y = c_1 x + (c_2 - 1)xe^x + xe^x \ln|x|$
2. $y = (c_1 - 1)e^x + c_2 xe^x - e^x \ln|x|$
3. $y = c_1 e^x + c_2 xe^x + \dfrac{1}{2x}e^x$
4. $y = c_1 e^{-x} + c_2 e^{2x} - 2x^2 + 2x - 3$
5. $y = c_1 \cos 5x + c_2 \sin 5x + \dfrac{2}{21}\sin 2x$
6. $y = c_1 + c_2 e^{-x} + \dfrac{x^2}{2} - x$
7. $y = c_1 \cos x + c_2 \sin x - \cos x \sinh^{-1}(\tan x)$
8. $y = c_1 \cos x + c_2 \sin x - \dfrac{1}{2}x \cos x$
9. $y = (c_1 + c_2 x)e^{-x} + \dfrac{1}{4}e^x$
10. $y = \left(c_1 + c_2 x + \dfrac{1}{2}x^2\right)e^{-x}$
11. $y = c_1 e^x + c_2 e^{-x} - x$
12. $y = c_1 e^x + c_2 e^{-x} + \dfrac{xe^x}{2}$
13. $y = c_1 e^x + c_2 e^{-x} - \dfrac{\sin x}{2}$
14. $y = e^{-2x}(c_1 \cos x + c_2 \sin x) + 2$
15. $y = e^{-2x}(c_1 \cos x + c_2 \sin x) + \dfrac{1}{5}x + \dfrac{6}{26}$
16. $y = c_1 + c_2 \cos x + c_3 \sin x + \ln|\sec x + \tan x| - x \cos x + \sin x \ln|\cos x|$
17. $y = c_1 + c_2 \cos x + c_3 \sin x - \ln|\csc x + \cot x| - \cos x \ln|\sin x| - x \sin x$
18. $y = c_1 - \dfrac{4}{27} + c_2 x + c_3 e^{3x} + c_4 e^{-3x} - \dfrac{x^4}{2} - \dfrac{2}{3}x^2$
19. $y = c_1 + c_2 x + c_3 \cos x + c_4 \sin x + \dfrac{1}{2}e^x$
20. $y = c_1 + c_2 x + c_3 x^2 + \left(c_4 - \dfrac{7}{4}\right)e^{2x} + c_5 e^{-2x} + xe^{2x}$
21. $y = c_1 x^3 + c_2 x^{-4} + \dfrac{x^4}{8}$
22. $y = c_1 x + c_2\left(x \tanh^{-1} x - 1\right) + x \tanh^{-1} x - \dfrac{x^6}{6} + \dfrac{x^4}{2} - \dfrac{x^2}{2}$
23. $y = c_1 x + c_2 x^{-1} + c_3 x^2 - \dfrac{x(\ln x)^2}{4} - \dfrac{x \ln x}{4} - \dfrac{9x}{24}$
24. $y = c_1 e^x - c_2 x - x^2 - x - 1$
25. $y = c_1 e^x - c_2(x+1) + \left(\dfrac{x^2}{2} + x\right)e^x - e^x(x+1)$

Exercises 3.6

1. $y = c_1 x \ln x + c_2 x + 4\sqrt{x}$
2. $y = c_1 e^x \ln x + c_2 e^x + xe^x$
3. $y = x^{3/2}(c_1 \ln x + c_2)$
4. $y = c_1 x + c_2 x^2 + 4x^2 \ln x$
5. $y = \dfrac{c_1}{6} x^{-5} + c_2 x - \dfrac{4}{27} x^{-1/2}$
6. $y = c_1 e^x + c_2 \dfrac{(x-1)^2 e^x}{2} + \dfrac{(x-1)^2 \ln(x-1) e^x}{2} - \dfrac{(x-1)^2 e^x}{4} - xe^x$
7. $y = c_1 e^x - x^3 - (3+c_2)x^2 - 2(3+c_2)x - 2(3+c_2)$
8. $y = c_1 x^{-1} + c_2 x^{-1} \ln x - x^{-1} \int x^{-1} \cos x \, dx$
9. $y = \dfrac{1}{3x} + c_1 x^2 \int x^{-4} e^{-x^2/2} dx + c_2 x^2$
10. $y = c_1 x^{-1} - c_2 x^{-1} e^{-x} - e^{-x} - x^{-1} e^{-x}$

Exercises 3.7

1. $y = c_0 \left[1 + \sum\limits_{n=1}^{\infty} \dfrac{(-1)^n x^{2n}}{2 \cdot 4 \cdot 6 \ldots 2n} \right] + c_1 \left[x + \sum\limits_{n=1}^{\infty} \dfrac{(-1)^n x^{2n+1}}{3 \cdot 5 \cdot 7 \ldots (2n+1)} \right]$

2. $y = c_0 \left(1 - \dfrac{x^2}{2} - \dfrac{x^4}{24} + \cdots \right) + c_1 \left(x - \dfrac{x^3}{3} - \dfrac{x^5}{30} + \cdots \right)$

3. $y = c_0 \left(1 - x^2 - \dfrac{x^3}{2} + \dfrac{x^4}{3} + \dfrac{11x^5}{40} + \cdots \right) + c_1 \left(x - \dfrac{x^3}{2} - \dfrac{x^4}{4} + \dfrac{x^5}{8} + \cdots \right)$

4. $y = c_0 \left(1 - \dfrac{x^3}{6} + \dfrac{3x^5}{40} + \cdots \right) + c_1 \left(x - \dfrac{x^3}{6} - \dfrac{x^4}{12} + \dfrac{3x^5}{40} + \cdots \right)$

5. $y = c_0 \left(1 + \dfrac{x^3}{6} + \dfrac{x^6}{18} + \cdots \right) + c_1 \left(x + \dfrac{x^4}{6} + \dfrac{17x^7}{252} + \cdots \right)$

6. $y = 1 + \sum\limits_{n=1}^{\infty} \dfrac{x^{2n}}{2 \cdot 4 \cdot 6 \ldots 2n} = \sum\limits_{n=0}^{\infty} \dfrac{x^{2n}}{2^n n!}$

Answers to Exercises 265

7. $y = 2 + 3x - \dfrac{7x^3}{6} - \dfrac{x^4}{2} + \dfrac{21x^5}{40} + \cdots$

8. $y = c_0 \left[1 - \dfrac{(x-1)^2}{2} + \dfrac{(x-1)^3}{2} - \dfrac{5(x-1)^4}{12} + \dfrac{(x-1)^5}{3} + \cdots \right]$

$\quad + c_1 \left[(x-1) - \dfrac{(x-1)^2}{2} + \dfrac{(x-1)^3}{6} - \dfrac{(x-1)^5}{3} + \cdots \right]$

9. $y = 2 + 4(x-1) - 4(x-2)^2 + \dfrac{4(x-1)^3}{3} - \dfrac{(x-1)^4}{3} + \dfrac{2(x-1)^5}{15} + \cdots$

10. (a) $y = c_0 \left[1 + \displaystyle\sum_{n=1}^{\infty} \dfrac{(-1)^n x^{3n}}{2.5.8\ldots(3n-1)3^n \, n!} \right] + c_1 \left[x + \displaystyle\sum_{n=1}^{\infty} \dfrac{(-1)^n x^{3n+1}}{4.7.10\ldots(3n+1)3^n \, n!} \right]$

(b) $y = c_0 \left[1 + \displaystyle\sum_{n=1}^{\infty} \dfrac{x^{3n}}{2.5.8\ldots(3n-1)3^n \, n!} \right] + c_1 \left[-x - \displaystyle\sum_{n=1}^{\infty} \dfrac{x^{3n+1}}{4.7.10\ldots(3n+1)3^n \, n!} \right]$

11. $y_1(x) = 1 - \dfrac{p \cdot p}{2!} x^2 + \dfrac{p(p-2)p(p+2)}{4!} x^4 - \cdots$

$y_2(x) = x - \dfrac{(p-1)(p+1)}{3!} x^3 + \dfrac{(p-1)(p-3)(p+1)(p+3)}{5!} x^5 - \cdots$

12. $y_1(x) = 1 - \dfrac{2p}{2!} x^2 + \dfrac{2^2 p(p-2)}{4!} x^4 - \dfrac{2^3 p(p-2)(p-4)}{6!} x^6 + \cdots$

$y_2(x) = x - \dfrac{2(p-1)}{3!} x^3 + \dfrac{2^2 p(p-1)(p-3)}{5!} x^5 - \dfrac{2^3 (p-1)(p-3)(p-5)}{7!} x^7 + \cdots$

13. $x = 0$ and $x = 3$ are regular singular points.
14. $x = 1$ is a regular singular point; $x = 0$ is an irregular singular point.
15. $x = 0$ and $x = -\dfrac{1}{3}$ are regular points.
16. $x = 0$ is irregular; $x = 1$ is regular.
17. $x = 0$ and $x = 1$ are regular and $x = -1$ is irregular.

18. $y = c_1 x \left(1 - \dfrac{x^2}{14} + \dfrac{x^4}{616} - \cdots \right) + c_2 x^{-\frac{1}{2}} \left(1 - \dfrac{x^2}{2} + \dfrac{x^4}{40} - \cdots \right)$

19. $y = c_1 x^{\frac{4}{3}}\left(1 - \frac{3x^2}{16} + \frac{9x^4}{896} - \cdots\right) + c_2 x^{\frac{2}{3}}\left(1 - \frac{3x^2}{8} + \frac{9x^4}{320} - \cdots\right)$

20. $y = c_1 x^{\frac{1}{3}}\left(1 - \frac{3x^2}{16} + \frac{9x^4}{896} - \cdots\right) + c_2 x^{-\frac{1}{3}}\left(1 - \frac{3x^2}{8} + \frac{9x^4}{320} - \cdots\right)$

21. $y = c_1\left(1 + x + \frac{3x^2}{10} + \cdots\right) + c_2 x^{-\frac{1}{3}}\left(1 + \frac{7x}{12} + \frac{5x^2}{36} + \cdots\right)$

22. $y = c_1 x^{\frac{1}{2}}\left(1 - \frac{x^2}{6} + \frac{x^4}{120} - \cdots\right) + c_2 x^{-\frac{1}{2}}\left(1 - \frac{x^2}{2} + \frac{x^4}{24} - \cdots\right)$

23. $y = c_1\left(1 + \frac{x^2}{2} + \frac{x^4}{8} + \cdots\right) + c_2 x^3\left(1 + \frac{x^2}{5} + \frac{x^4}{35} + \cdots\right)$

24. $y = (c_1 - c_2) e^{-x} + c_2 x^{-1} e^x$

25. $y = c_1 x\left[1 + 2\sum_{n=1}^{\infty} \frac{(-1)^n x^n}{n!(n+2)!}\right] + c_2\left[x^{-1}\left(-\frac{1}{2} - \frac{x}{2} + \frac{29x^2}{144} + \cdots\right) + \frac{1}{4} y_1(x) \ln|x|\right]$

where $y_1(x)$ denotes the solution of which c_1 is the coefficient.

26. $y = c_1 y_1 + c_2 y_2$ $y_1 = x^4\left(1 - \frac{x^2}{2} + \frac{x^4}{10} - \cdots\right)$

$y_2 = x^{-2}\left(-\frac{1}{6} - \frac{x^2}{6} - \frac{x^4}{6} + \cdots\right) + \frac{2}{9} y_1(x) \ln|x|$

27. $y = c_1 y_1 + c_2 y_2$ $y_1 = 1 + \sum_{n=1}^{\infty} \frac{(-1)^n 2^n x^n}{(n!)^2}$

$y_2 = 4x - 3x^2 + \frac{22x^3}{7} + \cdots + y_1(x) \ln|x|$

28. $y = c_1 y_1 + c_2 y_2$ $y_1 = x\left[1 + \sum_{n=1}^{\infty} \frac{(-1)^n x^{2n}}{\{2.4.6...(2n)\}^2}\right]$

$y_2 = \frac{x^3}{4} - \frac{3x^5}{128} + \frac{11x^7}{13824} + \cdots + y_1(x) \ln|x|$

29. $y = x^2 (1 - 4x + 4x^2 - \cdots)$

30. $y = c_1 \sqrt{x} e^x + c_2 \sqrt{x} e^x \ln x$

Answers to Exercises

Exercises 4.3

1. $u'' + u = 0$ Yes.
2. (i) $(\cos^2 x)\, u'' + (\sin x - \sin^2 x - 1)\, u = 0$
   (ii) $u'' + x^2 u = 0$
   (iii) $2x^2\, u'' + u = 0$
   (iv) $2x^2\, u'' + u = 0$
10. Yes.

Exercises 5.7

1. $-2\sqrt{\pi}$ and $\dfrac{16\sqrt{\pi}}{105}$

5. (a) $\Gamma\left(\dfrac{5}{4}\right)$ (b) $\sqrt{\pi}$ (c) 2

13. $\dfrac{\sqrt{\pi}}{k}$

14. $\dfrac{1}{n}\beta\left(\dfrac{m+1}{n}, p+1\right)$, $\dfrac{1}{396}$

Exercises 5.9

1. $c_1 F\left(1, 2; \dfrac{1}{2}; x\right) + c_2 x^{\frac{1}{2}} F\left(\dfrac{3}{2}, \dfrac{5}{2}; \dfrac{3}{2}; x\right)$

2. $c_1 F\left(1, 1; \dfrac{1}{2}; x\right) + c_2 x^{\frac{1}{2}} F\left(\dfrac{3}{2}, \dfrac{3}{2}; \dfrac{3}{2}; x\right)$

3. $c_1 F\left(3, 5; \dfrac{1}{3}; x\right) + c_2 x^{\frac{2}{3}} F\left(\dfrac{11}{3}, \dfrac{17}{3}; \dfrac{5}{3}; x\right)$

4. $c_1 F\left(1, 3; \dfrac{3}{2}; x\right) + c_2 x^{-\frac{1}{2}} F\left(\dfrac{1}{2}, \dfrac{5}{2}; \dfrac{1}{2}; x\right)$

5. $c_1 F\left(2, -1; \dfrac{3}{2}; x\right) + c_2 x^{-\frac{1}{2}} F\left(\dfrac{3}{2}, \dfrac{-3}{2}; \dfrac{1}{2}; x\right)$

6. $c_1 F\left(\dfrac{1}{2}, 1; \dfrac{1}{2}; x+1\right) + c_2 (x+1)^{-\frac{1}{2}} F\left(1, \dfrac{3}{2}; \dfrac{3}{2}; x+1\right)$

7. $c_1 F\left(2, 2; \dfrac{9}{2}; x-1\right) + c_2 (x-1)^{-\frac{7}{2}} F\left(-\dfrac{3}{2}, \dfrac{-3}{2}; \dfrac{-5}{2}; x-1\right)$

8. $c_1 F\left(1, 1; \dfrac{14}{5}; \dfrac{3-x}{5}\right) + c_2 \left(\dfrac{3-x}{5}\right)^{-\frac{9}{5}} F\left(-\dfrac{4}{5}, \dfrac{-4}{5}; \dfrac{-4}{5}; \dfrac{3-x}{5}\right)$

9. $c_1 F\left(1, -1; \dfrac{1}{2}; \dfrac{1-x}{2}\right) + c_2 \left(\dfrac{1-x}{2}\right)^{\frac{1}{2}} F\left(\dfrac{3}{2}, -\dfrac{1}{2}; \dfrac{3}{2}; \dfrac{1-x}{2}\right)$

10. $c_1 F\left(2, -2; \dfrac{1}{2}; \dfrac{1-x}{2}\right) + c_2 \left(\dfrac{1-x}{2}\right)^{\frac{1}{2}} F\left(\dfrac{5}{2}, \dfrac{-3}{2}; \dfrac{3}{2}; \dfrac{1-x}{2}\right)$

11. $c_1 (1-x)^{-1} + c_2 (\ln x)(1-x)^{-1}$

12. $y = c_1 y_1 + c_2 y_2,\ y_1 = F\left(\dfrac{1}{2}, \dfrac{1}{2}; 2; x\right)$

$$y_2 = y_1 \ln x + \dfrac{1}{x} + \sum_{n=1}^{\infty} \dfrac{(-3)1.5.9\ldots(4n-7)}{4^n\, n!} x^n$$

15. (a) $y_1(x) = 1 + \displaystyle\sum_{n=1}^{\infty} \dfrac{a(a+1)\cdots(a+n-1)}{n!\, c(c+1)\cdots(c+n-1)} x^n$

Exercises 5.16

3. $J_2'(x) = \left(\dfrac{x^2 - 4}{x^2}\right) J_1(x) + \dfrac{2}{x} J_0(x)$

$J_4(x) = 8\left(\dfrac{6 - x^3}{x^3}\right) J_1(x) - \left(\dfrac{24 - x^2}{x^2}\right) J_0(x)$

4. $J_5(x) = 12\left(\dfrac{x^2 - 16}{x^2}\right) J_0(x) - \left(\dfrac{384 - 72x^2 + x^4}{x^4}\right) J_0'(x)$

7. (b) $J_{\frac{5}{2}}(x) = \left(\dfrac{3 - x^2}{x^2}\right)\sqrt{\dfrac{2}{\pi x}} \sin x - \dfrac{3}{x}\sqrt{\dfrac{2}{\pi x}} \cos x$

$J_{-\frac{5}{2}}(x) = \left(\dfrac{3 - x^2}{x^2}\right)\sqrt{\dfrac{2}{\pi x}} \cos x + \dfrac{3}{x}\sqrt{\dfrac{2}{\pi x}} \sin x$

Exercises 5.21

15. $f(x) = \frac{1}{2}P_0(x) + \frac{3}{4}P_1(x) - \frac{7}{16}P_3(x) + \cdots$

16. $f(x) = \sum_{n=0}^{\infty} (-1)^{n+1} \frac{(4n+1)(2n-2)!}{2^{2n}(n+1)!(n-1)!} P_{2n}(x)$

17. $f(x) = \frac{1}{2}(e - e^{-1})P_0(x) + 3e^{-1}P_1(x) + \frac{1}{2}(5e - 35e^{-1})P_2(x) + \cdots$

Exercises 5.23

5. $f(x) = \frac{1}{\pi}T_0(x) + \frac{1}{2}T_1(x) + \frac{2}{3\pi}T_2(x) + \cdots$

Exercises 5.25

5. $f(x) = \frac{1}{4}H_0(x) + \frac{1}{2\sqrt{\pi}}H_1(x) + \frac{1}{8}H_2(x) + \cdots$

Exercises 5.27

1. The two independent solutions are

$$y_1 = 1 + \sum_{n=1}^{\infty} \frac{(-1)^n p(p-1)\cdots(p-n+1)}{n!(\alpha+1)(\alpha+2)\cdots(\alpha+n)} x^n$$

$$y_2 = x^{-\alpha}\left(1 + \sum_{n=1}^{\infty} \frac{(-1)^n p(p+\alpha)(p+\alpha+1)\cdots(p+\alpha+n-1)}{n!(1-\alpha)(2-\alpha)\cdots(n-\alpha)} x^n\right)$$

α is not an integer

Exercises 6.8

1. $x = 3c_1 + c_2 e^{-2t}$
   $y = 4c_1 + 2c_2 e^{-2t}$

2. $x = c_1 e^{2t}$
   $y = c_2 e^{3t}$

3. $x = c_1 e^{-3t} + c_2(t+1)e^{-3t}$
   $y = -c_1 e^{-3t} - c_2 t e^{-3t}$

4. $x = e^{3t}(b_1 \cos 2t + b_2 \sin 2t)$
   $y = e^{3t}\{b_1(\sin 2t - \cos 2t) - b_2(\cos 2t + \sin 2t)\}$

5. $x = c_1 e^t + c_2 e^{2t} + c_3 e^{4t}$
   $y = c_1 e^t + 2c_2 e^{2t} + 4c_3 e^{4t}$
   $z = c_1 e^t + 4c_2 e^{2t} + 16 c_3 e^{4t}$

6. $x = c_1 e^{2t} + c_2(t-1)e^{2t} + c_3 e^{3t}$
   $y = 2c_1 e^{2t} + c_2(2t-1)e^{2t} + 3c_3 e^{3t}$
   $z = 4c_1 e^{2t} + c_2 4t e^{2t} + 9 c_3 e^{3t}$

7. $x = 5b_1 \cos 2t + 5b_2 \sin 2t$
   $y = -b_1(4\cos 2t + 2\sin 2t) + b_2(2\cos 2t - 4\sin 2t)$

8. $x = c_1 e^t + c_2 e^{3t} + c_3 e^{-2t}$
   $y = -c_1 e^t + c_2 e^{3t} - c_3 e^{-2t}$
   $z = -2c_1 e^t + c_3 e^{-2t}$

9. $x = -5b_1 \cos t - 5b_2 \sin t$
   $y = b_1(2\cos t - \sin t) + b_2(\cos t + 2\sin t)$

10. $x = c_1 e^{2t} + c_2 e^t + c_3 t e^t$
    $y = c_1 e^{2t} + c_3 e^t$
    $z = c_1 e^{2t}$

11. $\begin{pmatrix} x \\ y \\ z \end{pmatrix} = c_1 \begin{pmatrix} 2 \\ -3 \\ 2 \end{pmatrix} e^t + c_2 e^t \left\{ \begin{pmatrix} 0 \\ 1 \\ 0 \end{pmatrix} \cos 2t - \begin{pmatrix} 0 \\ 0 \\ -1 \end{pmatrix} \sin 2t \right\}$

    $+ c_3 e^t \left\{ \begin{pmatrix} 0 \\ 0 \\ -1 \end{pmatrix} \cos 2t + \begin{pmatrix} 0 \\ 1 \\ 0 \end{pmatrix} \cos 2t \right\}$

12. $\begin{pmatrix} x \\ y \end{pmatrix} = c_1 \begin{pmatrix} 1 \\ 1 \end{pmatrix} e^{-7t} + c_2 \begin{pmatrix} 3 \\ -2 \end{pmatrix} e^{-2t} + \begin{pmatrix} \dfrac{5t}{14} - \dfrac{31}{196} - \dfrac{e^t}{8} \\ -\dfrac{t}{7} + \dfrac{9}{98} + \dfrac{5}{24} e^t \end{pmatrix}$

13. $x = e^{2t}(b_1 \cos 2t + b_2 \sin 2t) + \dfrac{t}{4} e^{2t}$

    $y = e^{2t}(-2b_1 \sin t + 2b_2 \cos 2t) - \dfrac{11}{4} e^{2t}$

14. $x = c_1 e^{6t} + c_2 t e^{6t} + \dfrac{t^2}{2} e^{6t}$

    $y = 2c_1 e^{6t} + 2c_2 t e^{6t} + c_2 e^{6t} + t^2 e^{6t}$

15. $\begin{pmatrix} x \\ y \end{pmatrix} = c_1 \begin{pmatrix} 1 \\ 1 \end{pmatrix} e^t + c_2 \begin{pmatrix} 1 \\ 2 \end{pmatrix} e^{2t} - 3t e^t \begin{pmatrix} 1 \\ 1 \end{pmatrix} - 3 e^t \begin{pmatrix} 1 \\ 2 \end{pmatrix}$

16. $\begin{pmatrix}x\\y\end{pmatrix} = c_1 \begin{pmatrix}2\\0\end{pmatrix} e^{2t} + c_2 \left\{ \begin{pmatrix}2\\0\end{pmatrix} te^{2t} + \begin{pmatrix}0\\2\end{pmatrix} e^{2t} \right\}$

$+ \left\{ \dfrac{t-1}{4} \cos t - \dfrac{t+2}{4} \sin t \right\} \begin{pmatrix}2\\0\end{pmatrix} e^t + \left\{ \dfrac{\sin t - 2\cos t}{10} \right\} \left\{ \begin{pmatrix}2\\0\end{pmatrix} t + \begin{pmatrix}0\\2\end{pmatrix} \right\}$

Exercises 7.5

1. $\dfrac{x}{y} - \ln z = c$ 2. $x\sqrt{y^2 + z^2} = c$

3. $x + \ln(y + z) = c$ 4. $x + \dfrac{z}{y} = c$

5. $x^2 + y^2 + z(x + y) = cxy$

Exercises 7.6

1. $x^2 y = cze^z$ 2. $e^x y + e^y z + e^z x = ce^z$
3. $y + z = ce^{-x}$ 4. $(x + y + z^2) e^{x^2} = c$
5. $zx + zy - y^2 = cz^2$

Exercises 7.7

1. $xyz + x^2 + y^2 + z^2 = c^2$ 2. $1 + xy = c(z - x)$
3. $x^2 + y^2 - xy = cz$ 4. $xy + yz + zx = c(x + y + z)$
5. $x^2 + y^2 + z(x + y) = cxy$

Exercises 7.8

1. $x^2 + y^2 + z(x + y) = cxy$ 2. $x^2 + y^2 + z^2 = c(x + y + z)$
3. $x^2 + y^2 + z^2 = cx$ 4. $x^2 y + y^2 z + z^2 x = c$
5. $x - cy - y - y \ln z = 0$

Exercises 7.9

1. $z = x + y$ $(x - 1)^2 (2y - 1) = c$ 2. $x^2 + y^2 + z^2 = k^2$

3. $x = c_1 y$ and $zx = c_2$ $\dfrac{x^2}{a^2} + \dfrac{y^2}{b^2} + \dfrac{z^2}{c^2} = 1$

4. $x^{-1} + y^{-1} = c$ 5. $2xy + y^2 = c$

Exercises 7.12

1. $x^4 - y^4 = c_1$ $\quad x^2 - z^2 = c_2$
2. $x + y = c_1$ $\quad z^2 + (x + y)^2 = c_2 e^{-2y}$
3. $lx + my + nz = c_1$ $\quad x^2 + y^2 + z^2 = c_2^2$
4. $ax^2 + by^2 + cz^2 = c_1$ $\quad a^2x^2 + b^2y^2 + c^2z^2 = c_2$
5. $x^2 + y^2 + z^2 = c_1$ $\quad xyz = c_2$
6. $xy = c_1$ $\quad x^4 - (z^2 + xy)^2 = c_2$
7. $l^2x + m^2y + n^2z = c_1$ $\quad l^2x^2 + m^2y^2 + n^2z^2 = c_2$
8. $x + y = c_1 z$ $\quad 2y = c_2(x^2 - y^2)$
9. $x - y - z = a$ $\quad x^2 - y^2 = bz^2$
10. $xy = a$ $\quad x^2 + y^2 + z(x + y) = b$

Exercises 8.2

1. (a) $\dfrac{5!}{p^6} + \dfrac{p}{p^2 + 4}$ $\quad$ Re $p > 0$ $\qquad$ (b) $\dfrac{4}{p^2 + 4} + \dfrac{2}{p + 1}$ $\quad$ Re $p > 0$

   (c) $\dfrac{1}{p^2} + \dfrac{e^{-2p}}{p} - \dfrac{e^{-2p}}{p^2}$ $\quad$ Re $p > 0$ $\qquad$ (d) $\dfrac{e^b}{p - a}$ $\quad$ Re $p >$ Re a

2. (a) $f(t) = \dfrac{3}{2} \sin 2t$ $\qquad$ (b) $f(t) = 2e^{-3t}$

   (c) $f(t) = \dfrac{1}{2} e^{-3t} \sin 2t$ $\qquad$ (d) $f(t) = t - \sin t$

Exercises 8.3

1. (a) $\dfrac{2}{(p + 4)^3}$ $\qquad$ (b) $\dfrac{1 - 5p}{p^2 + 2p - 3}$

   (c) $\dfrac{1}{p} + \dfrac{3}{(p+1)^2} + \dfrac{6}{(p+2)^3} + \dfrac{6}{(p+3)^4}$ $\qquad$ (d) $\dfrac{1}{4} \ln\left(\dfrac{p^2 + 4}{p^2}\right)$

   (e) $\ln\left(1 + \dfrac{1}{p}\right)$ $\qquad$ (f) $\dfrac{\pi}{2p} - \dfrac{1}{p} \tan^{-1}\left(\dfrac{p+4}{3}\right)$

(g) $2\dfrac{p^2-2p-3}{(p^2-2p+5)^2}$

(h) $\dfrac{3}{4p^2}\sqrt{\dfrac{\pi}{p}}$

(i) $e^{-5p}\left(\dfrac{1}{p^2}+\dfrac{2}{p}\right)$

(j) $\dfrac{q^5+q^4+10q^3-10q^2-8q-56}{(q^2+4)^3}$ where $q=p+3$

2. (a) $1-\cos t+\sin t$

(b) $\dfrac{1}{2}\cos 2t$

(c) $f(t)=\begin{cases}\dfrac{1}{6}(t-5)^3 e^{2(t-5)} & \text{if } t>5 \\ 0 & \text{if } t<5\end{cases}$

(d) $\dfrac{1}{6}e^t-\dfrac{1}{15}e^{-2t}-\dfrac{1}{10}\cos t-\dfrac{3}{10}\sin t$

(e) $5\,e^{-2t}\sin 3t$

(f) $\dfrac{1}{b}e^{-at}(b\cos bt-a\sin bt)$

(g) $\dfrac{a\sin at-b\sin bt}{a^2-b^2}$

(h) $\dfrac{1}{\sqrt{\pi t}}+2\sqrt{\dfrac{t}{\pi}}+\dfrac{4t}{3}\sqrt{\dfrac{t}{\pi}}$

(i) $\dfrac{3}{4}t\sin t+\dfrac{1}{4}t^2\cos t$

(j) $\dfrac{1}{4}t-\dfrac{1}{8}\sin 2t$

3. (a) $\ln\dfrac{b}{a}$

(b) $\tan^{-1}\dfrac{b}{a}$

(c) $\dfrac{\pi}{2}$

(d) $\dfrac{\pi}{2}e^{-x}$

4. (i) $y=\dfrac{1}{2}(t^2-3t+2)e^t$

(ii) $y=\dfrac{k}{2}\sin kt$

(iii) $y=e^{-t}\sin 2t+e^{-t}\sin t$

(iv) $y=\dfrac{1}{10}\left(1-e^{3t}\cos t+3e^{3t}\sin t\right)$

(v) $y=e^{-t}(\sin 2t+\cos 2t)+2\sin t$

(vi) $y=4(t-2)\left\{1-\dfrac{e^{t-2}}{2}-\dfrac{e^{-t+2}}{2}\right\}$

(vii) $y=-\dfrac{1}{2}+\dfrac{1}{5}e^{-t}+\dfrac{3}{10}\cos 2t+\dfrac{3}{5}\sin 2t$

(viii) $y=e^{2t}$

(ix) $y=te^{-t}$

(x) $y=J_0(t)$

Exercises 8.4

1. (a) $\dfrac{1}{a}(1-\cos at)$ (b) $\dfrac{1}{a-b}(e^{at}-e^{bt})$

   (c) $\dfrac{1}{a^2}(e^{at}-1-at)$ (d) $\dfrac{1}{a^2-b^2}(a\sin bt - b\sin at)$

2. (a) $\dfrac{3}{p^2(p^2+9)}$ (b) $\dfrac{1}{(p+1)(p^2+1)}$

   (c) $\dfrac{6}{p^4(p-1)}$ (d) $\dfrac{p}{(p^2+1)^2}$

3. (a) $\dfrac{1}{7}(4\sin 4t - 3\sin 3t)$ (b) $e^{4t}+e^{-t}(\cos 2t + 2\sin t)$

   (c) $\dfrac{2}{t}(1-\cos t)$ (d) $\dfrac{1}{8}\sin 2t \sinh 2t$

4. (a) $y=e^{2t}$ (b) $y=\dfrac{1}{6}e^{3t}+\dfrac{5}{6}e^{-3t}-e^{-2t}$

   (c) $y=-\dfrac{1}{6}t-\dfrac{1}{36}-\dfrac{1}{45}e^{-3t}+\dfrac{1}{20}e^{2t}$ (d) $y=2e^t-\dfrac{1}{3}t^3-t^2-2t-2$

Exercises 9.2

1. Yes. 2. Yes.
3. No, since $p(x)$ is not continuous in $[0, 1]$ 4. Yes.
5. No, since $r(x) < 0$ in $[0, 1]$.
6. Multiply both the sides by $e^{\int(a_1(x)/a_2(x))dx}$ to reduce the equation to the required form with $p(x) = e^{\int(a_1(x)/a_2(x))dx}$, $q(x) = p(x)\, a_0(x)/a_2(x)$, and $r(x) = p(x)w(x)/a_2(x)$.
7. $\lambda_n = 4n^2$ $y_n = c_n \sin 2nx$ $n = 1, 2, 3, ...$
8. $\lambda_n = \left(\dfrac{n\pi}{L}\right)^2$ $y_n = c_n \sin\dfrac{n\pi x}{L}$ $n = 1, 2, 3, ...$
9. $\lambda_n = \alpha_n^2$, where α_n are the positive roots of the equation $\alpha = \tan \pi\alpha$.

   $y_n = c_n \sin \alpha_n x$ $n = 1,2,3,...$
10. $\lambda_n = n^2$ $y_n = c_n \sin(n \ln|x|)$ $n = 1, 2, 3, ...$

11. $\lambda_n = 16n^2$ $\quad y_n = c_n \sin(4n \tan^{-1} x)$ $\quad n = 1, 2, 3, ...$

12. $\lambda_n = \dfrac{n^2 \pi^2}{L^2}$ $\quad y_n = c_n \cos\dfrac{n\pi x}{L}$ $\quad n = 0, 1, 2, 3, 4, ...$

13. $\lambda_n = n^2$ $\quad y_n = c_n \cos nx$ $\quad n = 1, 2, 3, 4, ...$

14. $\lambda_n = \dfrac{\pi^2 n^2}{9a^2}$ $\quad y_n = c_n e^{-2x} \sin\dfrac{n\pi x}{L}$ $\quad n = 1, 2, 3, ...$

15. $\lambda_n = \dfrac{\alpha_n^2}{L^2}$ $\quad y_n = c_n \sin\dfrac{\alpha_n x}{L}$ $\quad n = 1, 2, 3, ...$

where α_n is a root of the equation $\tan\alpha = -\dfrac{1}{hL}\alpha$.

Exercises 9.3

1. $2\displaystyle\sum_{n=1}^{\infty} \dfrac{(-1)^{n+1} \sin nx}{n}$

2. $\dfrac{4}{\pi}\displaystyle\sum_{n=1}^{\infty} \left(\dfrac{1-\cos n\pi}{n}\right) \sin nx$

3. $\dfrac{\pi^2}{6} + \displaystyle\sum_{n=1}^{\infty} \left\{ \left(\dfrac{2(-1)^n}{n^2}\right) \cos nx + \dfrac{1}{\pi}\left[\left(\dfrac{3}{n^3} - \dfrac{\pi^2}{n}\right)(-1)^n - \dfrac{2}{n^3}\right] \sin nx \right\}$

4. $\dfrac{2}{\pi}\displaystyle\sum_{n=1}^{\infty} \left[\left(\dfrac{2}{n^3} - \dfrac{\pi^2}{n}\right)(-1)^n - \dfrac{2}{n^3}\right] \sin nx$

5. $3 + \dfrac{12}{\pi^2}\displaystyle\sum_{n=1}^{\infty} \left[\dfrac{(-1)^n - 1}{n^2}\right] \cos\dfrac{n\pi x}{6}$

6. $\dfrac{1}{2} + \dfrac{2L}{\pi}\displaystyle\sum_{n=1}^{\infty} \dfrac{1}{2n-1} \sin\dfrac{(2n-1)\pi x}{3}$

7. $\dfrac{L}{2} + \dfrac{2L}{\pi}\displaystyle\sum_{n=1}^{\infty} \dfrac{1}{2n-1} \sin\dfrac{(2n-1)\pi x}{L}$

8. $b + \dfrac{2aL}{\pi}\displaystyle\sum_{n=1}^{\infty} \dfrac{(-1)^{n+1}}{n} \sin\dfrac{n\pi x}{L}$

9. $\dfrac{e^{\pi} - e^{-\pi}}{\pi}\left[\dfrac{1}{2} + \sum_{n=1}^{\infty}\dfrac{(-1)^n}{1+n^2}(\cos nx - n\sin nx)\right]$

10. $-2\sum_{n=1}^{\infty}\dfrac{1}{n}\sin nx$

Exercises 9.4

1. $\dfrac{4}{\pi}\sum_{n=1}^{\infty}\dfrac{\sin(2n-1)}{2n-1}, 1$

2. $\dfrac{2}{\pi}\sum_{n=1}^{\infty}\dfrac{n}{n^2+1}\left[1+(-1)^{n+1}e^{\pi}\right]\sin nx$

3. $2\sum_{n=1}^{\infty}\dfrac{1}{n}\sin nx$

4. $1+2\sum_{n=1}^{\infty}\left(\dfrac{(-1)^n-1}{n^2\pi^2}\right)(-1)^n\cos n\pi x$

5. $-\dfrac{2}{\pi}\sum_{n=1}^{\infty}\dfrac{1}{n}\sin n\pi x \qquad 1<x<2$

6. $1+\dfrac{4}{\pi}\sum_{n=1}^{\infty}\dfrac{(-1)^n}{2n-1}\cos(2n-1)x$

7. $\sin x,\quad \dfrac{2}{\pi}-\dfrac{4}{\pi}\sum_{n=1}^{\infty}\dfrac{\cos 2n\pi}{4n^2-1}$

8. $\dfrac{8L}{\pi}\sum_{n=1}^{\infty}\dfrac{1}{2n-1}\sin\dfrac{(2n-1)\pi x}{L}$

9. $\dfrac{L^2}{2}+\dfrac{2L^2}{\pi}\sum_{n=1}^{\infty}\dfrac{(-1)^n}{2n-1}\cos\dfrac{(2n-1)\pi x}{L}$

10. $\dfrac{L^2}{6}-\dfrac{L^2}{\pi^2}\sum_{n=1}^{\infty}\dfrac{1}{n^2}\cos\dfrac{2n\pi x}{L}$

Answers to Exercises

Exercises 9.5

1. $\dfrac{3\pi}{4} + \sum\limits_{n=1}^{\infty}\left[\dfrac{(-1)^n - 1}{\pi n^2}\cos nx - \dfrac{1}{n}\sin nx\right]$; the series is at $x = \pi$

2. $\dfrac{4}{\pi}\sum\limits_{n=1}^{\infty}\dfrac{\sin(2n-1)x}{2n-1}$; the series are at $x = \dfrac{\pi}{2}$ and $x = \dfrac{\pi}{4}$

3. $\dfrac{\pi}{4} + \sum\limits_{n=1}^{\infty}\left(\dfrac{(-1)^n - 1}{n^2\pi}\cos nx + \dfrac{(-1)^{n+1}}{n}\sin nx\right)$; the series is at $x = \pi$.

4. $2\pi\left(\sin x - \dfrac{\sin 2x}{2} + \dfrac{\sin 3x}{3} - \cdots\right) - \dfrac{8}{\pi}\left(\sin x + \dfrac{\sin 3x}{3^3} + \dfrac{\sin 5x}{5^3} + \cdots\right)$

6. $\dfrac{1}{2} + \dfrac{4}{\pi^2}\sum\limits_{n=1}^{\infty}\dfrac{\cos(2n-1)\pi x}{(2n-1)^2}$ is the Fourier series of $f(x)$ and $g(x)$

9. (a) $\pi + 2\sum\limits_{n=1}^{\infty}(-1)^n\dfrac{\sin nx}{n}$ 　 (b) $\dfrac{\pi}{2} + \dfrac{4}{\pi}\sum\limits_{n=1}^{\infty}\dfrac{\cos(2n-1)x}{(2n-1)^2}$

(c) $2\sum\limits_{n=1}^{\infty}\dfrac{\sin nx}{n}$

10. $\dfrac{1}{2}a_0 + \sum\limits_{n=1}^{\infty}\left(a_n\cos\dfrac{2n\pi x}{b-a} + b_n\sin\dfrac{2n\pi x}{b-a}\right)$

$a_n = \dfrac{2}{b-a}\int_a^b f(x)\cos\dfrac{2n\pi x}{b-a}dx \quad n = 0, 1, 2, \ldots;$

$b_n = \dfrac{2}{b-a}\int_a^b f(x)\sin\dfrac{2n\pi x}{b-a}dx \quad n = 1, 2, \ldots$

Exercises 10.3

1. $y(x, t) = \sin 2x \cos 2t$

2. $y(x, t) = \sum\limits_{n=1}^{\infty} B_n \sin\dfrac{n\pi x}{3}\cos\dfrac{2n\pi t}{3} \qquad B_n = \dfrac{12}{n^2\pi^2}\sin\dfrac{n\pi}{2}\cos\dfrac{n\pi}{6}$

3. $y(x,t) = \sin\dfrac{n\pi x}{L}\cos\dfrac{n\pi at}{L}$

4. $y(x,t) = \displaystyle\sum_{n=1}^{\infty}\left(\alpha_n \cos\dfrac{n\pi at}{L} + \dfrac{L}{n\pi a}\beta_n \dfrac{n\pi at}{L}\right)\sin\dfrac{n\pi x}{L}$

5. $y(x,t) = \displaystyle\sum_{n=1}^{\infty}\dfrac{1}{2}A_n\left\{\sin\dfrac{n\pi}{L}(x+at) + \sin\dfrac{n\pi}{L}(x-at)\right\}$

$A_n = \dfrac{4kL^2}{n^3\pi^3}\{1-(-1)^n\}$

Exercises 10.4

1. $y(x,t) = \dfrac{160}{\pi^2}\displaystyle\sum_{n=1}^{\infty}\dfrac{\sin(n\pi/2)}{n^2}e^{-\left(\frac{n\pi}{40}\right)^2 t}\sin\dfrac{n\pi x}{40}$

2. $y(x,t) = \dfrac{200}{\pi}\displaystyle\sum_{n=1}^{\infty}\dfrac{1-\cos n\pi}{n}e^{-\left(\frac{n\pi a}{20}\right)^2 t}\sin\dfrac{n\pi x}{20}$

$y(10,30) = \dfrac{400}{\pi}\displaystyle\sum_{m=1}^{\infty}\dfrac{(-1)^{m+1}e^{\frac{-3\pi^2 a^2}{40}(2m-1)^2}}{2m-1}$

3. $y(x,t) = 3x + \displaystyle\sum_{n=1}^{\infty}\left(\dfrac{50+70(-1)^n}{n\pi}\right)e^{-\left(\frac{n\pi a}{20}\right)^2 t}\sin\dfrac{n\pi x}{20}$

4. (i) $y(x) = 0$ (ii) $y(x) = T$

5. $y(x,t) = \dfrac{200}{9} - \displaystyle\sum_{n=1}^{\infty}\left\{\dfrac{160}{3n^2\pi^2}(3+(-1)^n)\right\}e^{-\left(\frac{n\pi}{80}\right)^2 t}\cos\dfrac{n\pi x}{40}$

Exercises 10.5

1. $w(x,y) = \displaystyle\sum_{n=1}^{\infty} c_n \sinh\dfrac{n\pi x}{M}\sin\dfrac{n\pi y}{M}$ $c_n = \dfrac{2}{M\sinh\dfrac{n\pi L}{M}}\displaystyle\int_0^M g(y)\sin\dfrac{n\pi y}{M}dy$

2. $w(x, y) = \sum_{n=1}^{\infty} c_n \sinh \frac{n\pi(L-x)}{M} \sin \frac{n\pi y}{M}$

$c_n = \frac{2}{M \sinh \frac{n\pi L}{M}} \int_0^M k(y) \sin \frac{n\pi y}{M} dy$

3. $w(x, y) = \sum_{n=1}^{\infty} c_n^{(1)} \sinh \frac{n\pi(L-x)}{M} \sin \frac{n\pi y}{M} + \sum_{n=1}^{\infty} c_n^{(2)} \sinh \frac{n\pi x}{M} \sin \frac{n\pi y}{M}$

$+ \sum_{n=1}^{\infty} c_n^{(3)} \sin \frac{n\pi x}{L} \sinh \frac{n\pi(M-y)}{L} + \sum_{n=1}^{\infty} c_n^{(4)} \sin \frac{n\pi x}{L} \sinh \frac{n\pi y}{L}$

$c_n^{(1)} = \frac{2}{M \sinh \frac{n\pi L}{M}} \int_0^M k(y) \sin \frac{n\pi y}{M} dy \quad c_n^{(2)} = \frac{2}{M \sinh \frac{n\pi L}{M}} \int_0^M g(y) \sin \frac{n\pi y}{M} dy$

$c_n^{(3)} = \frac{2}{L \sinh \frac{n\pi M}{L}} \int_0^L h(x) \sin \frac{n\pi x}{L} dx \quad c_n^{(4)} = \frac{2}{L \sinh \frac{n\pi M}{L}} \int_0^L f(x) \sin \frac{n\pi x}{L} dx$

4. $w(x, y) = \frac{c_o y}{2} + \sum_{n=1}^{\infty} c_n \cos \frac{n\pi x}{40} \sinh \frac{n\pi y}{40}$

$c_o = \frac{1}{10}\left(1 + \frac{40^4}{30}\right) \quad c_n = -\frac{24.40^4(1 + \cos n\pi)}{n^4 \pi^4 \sinh \frac{n\pi}{2}}$

5. $w(r, \theta) = \frac{c_o}{2} + \sum_{n=1}^{\infty} r^n (c_n \cos n\theta + k_n \sin n\theta)$

$c_n = \frac{1}{\pi a^n} \int_0^{2\pi} f(\theta) \cos n\theta \, d\theta \qquad n = 0, 1, 2, 3, \ldots$

$k_n = \frac{1}{\pi a^n} \int_0^{2\pi} f(\theta) \sin n\theta \, d\theta \qquad n = 1, 2, 3, \ldots$

6. $w(r, \theta) = \dfrac{t_2 \ln\left(\dfrac{r}{a}\right) - t_1 \ln\left(\dfrac{r}{b}\right)}{\ln\left(\dfrac{b}{a}\right)}$

Index

Abscissa of absolute convergence of $L(f(t))$, 189
Abscissa of convergence of $L(f(t))$, 189
Airy function, 82
Airy's equation, 82
Analytic function, 60
Associated homogeneous equation, 38
Auxiliary equation, 42

Bernoulli's equation, 33
Bessel function of first kind and of order (or index) p, 108
Bessel function of second kind and of order (or index) p, 110
Beta function, 88
Boundary conditions, 4
Boundary value problem, 4

Cauchy's equations (see Euler's equations)
Chebyshev equation, 126
Chebyshev polynomial, 127
Chebyshev series, 130
Coefficient matrix, 145
Complementary function, 39
Condition for integrability, 160
Confluent hypergeometric equation, 101
Convergence theorem for Fourier series, 225
Convolution of f and g, 205
Convolution theorem, 205

Degree of differential equation, 3
Digamma function, 90
Duplication formula, 95

Eigenfunctions, 212
Eigenvalues, 212

Elementary Laplace transforms, 201
Equidimensional equations (see Euler's equations)
Equivalence iff $f(t) = g(t)$, 192
Euler's constant, 91
Euler's equations, 48
Euler's method, 151
Exact differential equation, 21, 160
Exponential order, 185

Factorial function, 87
First shifting formula, 194
Fourier coefficients, 215
Fourier cosine series, 223, 230
Fourier series, 215, 226
Fourier sine series, 222, 228
Frobenius method, 64

Gamma function, 87
General solution of differential equation, 6
Generating function, 115, 123, 129, 134
Geometrical interpretation of total differential equation, 164

Hermite equation, 83, 131
Hermite polynomial, 132
Hermite series, 136
Homogeneous equation, 4, 19, 171
Homogeneous system, 145
Hypergeometric equation, 95
Hypergeometric function, 96
Hypergeometric series, 96

Indicial equation, 65
Initial conditions, 4
Initial value problem, 4

Index

Integrating factor, 25
Inverse Laplace transform of $F(p)$, 193
Irregular singular point, 62

Laplace equation in polar form, 254
Laplace transform, 181
Laguerre equation, 138
Laguerre polynomial, 139
Legendre polynomial of degree n, 119
Legendre series, 124
Legendre's equation of degree p, 118
Lerch's theorem, 192
Linear differential equation, 3
Linear system, 144
Linearity property of Laplace transform, 191

Non-integrable equation, 173
Nonhomogeneous system, 145
Nonlinear differential equation, 3
Normal equation, 4
Normal form, 78, 144

One-dimensional heat equation, 243
One-dimensional wave equation, 236
Order of differential equation, 3
Order of growth of $f(t)$, 185
Ordinary differential equation, 3
Ordinary point, 62
Orthogonal polynomials, 121
Orthogonal set, 102
Orthogonality relation, 115, 128, 140
Orthonormal set, 106
Oscillating function, 75

Partial differential equation, 3
Particular solution, 6

Periodic extension of f, 225
Periodic function, 225
Piecewise continuous function, 184

Quasi-power series or Frobenius series, 64

Recurrence relation, 120, 128, 132, 139
Reduction of order, 58
Reflection formula, 92
Regular singular point, 62
Rodrigues' formula, 119, 134, 139

Second shifting formula, 195
Separation of variables, 237
Series solution method, 60
Simple set of polynomials, 102
Simple zero, 74
Simultaneous equations, 175
Singular point, 62
Solution of system of first order linear equations, 149
Solution of differential equation, 5
Standard form, 78
Standard set of polynomials, 102
Steady state solution, 245
Sturm comparison theorem, 81
Sturm separation theorem, 77
Sturm–Liouville problem or Sturm–Liouville system, 210

Total differential equation, 160
Two-dimensional Laplace equation, 250

Variables separable, 15
Variation of parameters, 51, 155

Zero, 74